Case Studies in Infection Control

Case Studies in Infection Control

Meera Chand and John Holton

Garland Science
Taylor & Francis Group

NEW YORK AND LONDON

Vice President: Denise Schanck
Assistant Editor: Jordan Wearing
Production Editor: Daniela Amodeo
Illustrator: Louise Dawney
Layout: Thomson Digital
Cover Designer: Georgina Lucas
Copyeditor: Marjorie Singer Anderson
Proofreader: Susan Wood

Meera Chand, Public Health England, Guy's and St Thomas' NHS Foundation Trust, Imperial College London, UK.

John Holton, Department of Natural Sciences, School of Science & Technology, University of Middlesex, National Mycobacterial Reference Service - South, Public Health England, UK.

ISBN paperback: 978-0-8153-4517-6; eBook: 978-0-2037-3331-8

Library of Congress Cataloging-in-Publication Data

Names: Chand, Meera, author. | Holton, John, MRCPath, author.
Title: Case studies in infection control / Meera Chand and John Holton.
Description: London; New York : Taylor & Francis Group, [2018] | Includes index.
Identifiers: LCCN 2017056613 (print) | LCCN 2017058395 (ebook) | ISBN 9780203733318 () | ISBN 9780815345176
Subjects: | MESH: Communicable Disease Control | Infection Control | Case Reports
Classification: LCC RA643 (ebook) | LCC RA643 (print) | NLM WA 110 | DDC 616.9/045--dc23
LC record available at https://lccn.loc.gov/2017056613

Published by Garland Science, Taylor & Francis Group, LLC, an Informa business, 711 Third Avenue, New York, NY, 10017, USA, and 3 Park Square, Milton Park, Abingdon, OX14 4RN, UK.

15 14 13 12 11 10 9 8 7 6 5 4 3 2 1

Visit our web site at http://www.garlandscience.com

Preface

Case Studies in Infection Control follows from the well-received *Case Studies in Infectious Disease*. This new book has 25 cases, each focusing on an infectious disease, which illustrate the critical aspects of infection control and prevention. Scenarios in the cases are real events from both community and hospital situations and are written by experts. Although brief comments are included in relation to the organism, diagnosis, and treatment, the main emphasis is on the case, its epidemiology, and how the situation should be managed from the perspective of infection control and prevention. As in the previous book, there is a series of multiple choice questions with answers and a short list of appropriate guidelines and publications.

Online resources:

For instructors, all of the figures from the book are available to download in both PowerPoint® and JPEG format. Contact science@garland.com to gain access to the figures.

Contents

Acronyms

aat	aggregative adherence transporter
ABPA	allergic bronchopulmonary aspergillosis
ACDP	Advisory Committee on Dangerous Pathogens
Ach	acetylcholine
ACH	air changes per hour
ACT	artemisinin combination therapy
aggR	aggregative adherence regulator
AGP	aerosol-generating procedures
aHUS	atypical haemolytic uraemic syndrome
AHVLA	Animal Health Veterinary Laboratories Agency
AIDS	acquired immune deficiency syndrome
ALT	alanine transaminase
AML	acute myeloid leukaemia
aOR	adjusted odds ratio
AP	auramine phenol
APHA	Animal and Plant Health Agency
ARDS	acute respiratory distress syndrome
ART	anti-retroviral therapy
AST	aspartate transaminase
BASHH	British Association of Sexual Health and HIV
Bcc	*Burkholderia cepacia* complex
BCG	Bacillus Calmette-Guérin
BCYE	buffered charcoal containing yeast extract
BSE	bovine spongiform encephalopathy
BSL	biosecurity level
CA-MRSA	community-associated methicillin resistant Staphylococcus aureus
cART	combined anti-retroviral therapy
CCDA	charcoal cefoperazone deoxycholate agar
CCDC	Consultant in Communicable Disease Control
CCHF	Crimean-Congo haemorrhagic fever
CDC	Centers for Disease Control and Prevention
CDI	Clostridium difficile infection
CF	cystic fibrosis
CFR	case fatality rate
CHP	Consultant in Health Protection
CI	confidence interval
CIN	cervical intraepithelial neoplasia
CJD	Creutzfeldt-Jakob disease
CMV	cytomegalovirus
CNS	central nervous system
COSHH	Control of Substances Hazardous to Health
CRP	C-reactive protein
CSF	cerebrospinal fluid

CT	computerised tomography
CTL	cytotoxic T-lymphocyte
CT-SMAC	cefixime tellurite sorbitol MacConkey
CVC	central venous catheters
CWD	chronic wasting disease
D+ HUS	haemolytic uraemic syndrome associated with diarrhoea
XDR-TB	N,N-diethyl-meta-toluamide
DEFRA	Department of Environment, Farming, and Rural Affairs
D-HUS	haemolytic uraemic syndrome in the absence of diarrhoea
DIPC	Director of Infection Prevention and Control
DOT	Department of Transport
DOT	directly observed therapy
DPH	Director of Public Health
DPP4	dipeptidyl peptidase 4
DST	diagnostic sensitivity testing
EBLV	European Bat lyssavirus
EBLV-2	European Bat lyssavirus-2
ECDC	European Centre for Disease Prevention and Control
ECMO	extracorporeal membrane oxygenation
EF	edema factor (anthrax)
EHO	Environmental Health Officer
EIA	enzyme-linked immunoassay
EIR	entomological inoculation rate
ELISA	enzyme-linked immunosorbent assay
Env	envelope glycoprotein (HIV)
EPA	Environmental Protection Agency
ETEC	enterotoxigenic Escherichia coli
EUE	exotic ungulate encephalopathy
Euro-GASP	European Gonococcal Antimicrobial Surveillance Programme
EVD	Ebola virus disease
FAT	fluorescent antibody test
FBC	full blood count
fCJD	familial Creutzfeldt-Jakob disease
FES	Field Epidemiology Services
FFI	fatal familial insomnia
FFP3	fit-tested high-filtration face masks
FPU	first pass urine
FSA	Food Standards Agency
FSE	feline spongiform encephalopathy
FSI	fatal sporadic insomnia
GALT	gut-associated lymphoid tissue
GBRU	gastrointestinal bacteria reference unit
GBS	Guillain-Barré syndrome
GCS	Glasgow coma scale
GDH	glutamate dehydrogenase
GFN	Global Foodborne Infections Network
GMHPU	Greater Manchester Health Protection Unit
GP	general practitioner

GPI	glycosylphosphatidylinositol
GRASP	Gonococcal Resistance to Antimicrobials Surveillance Programme
GSS	Gerstmann-Sträussler-Scheinker
GUM	genitourinary medicine
HAART	highly active antiretroviral therapy
HACCP	Hazard Analysis Critical Control Point
HA-MRSA	healthcare-associated methicillin resistant *Staphylococcus aureus*
HAND	HIV-associated neurocognitive disorder
HEPA	high-efficiency particulate air
HG	Hazard Group
HIV	human immunodeficiency virus
HIV-1	human immunodeficiency virus-1
HIV-2	human immunodeficiency virus-2
HIVAN	HIV-associated nephropathy
HLA	human leukocyte antigen
HLIU	high level isolation unit
HNIG	human normal immunoglobulin
HPA	Health Protection Agency
HPT	Health Protection Team
HPU	Health Protection Unit
HPV	human papillomavirus
HR-HPV	high-risk human papillomavirus
HRIG	human rabies immunoglobulin
HSCT	haematopoietic stem cell transplant
HSE	Health and Safety Executive
HSWA	Health and Safety at Work Act
HTM	Health Technical Memorandum
IA	invasive aspergillosis
iCJD	iatrogenic Creutzfeldt-Jakob disease
ICP	intracranal pressure probe
ICT	incident control team
ICU	intensive care unit
IFA	immunofluorescent antibody
IFN	interferon
IFS	Imported Fever Service
IHR	International Health Regulations
IID	infective intestinal disease
IL	interleukin
INT	integrase (HIV)
IPC	Infection prevention and control
IPSID	immunoproliferative small intestinal disease
IPT	intermittent preventive therapy/treatment
IPTc	intermittent preventive treatment for children
IPTi	intermittent preventive treatment for infants
IPTp	intermittent preventive treatment for pregnant women
IPTsc	intermittent preventive treatment for schoolchildren

IRS	indoor residual spraying
ISAGA	IgM-immunosorbent agglutination assay
ITN	insecticide-treated bednets
ITU	intensive therapy unit
IV	intravenous
IVDU	intravenous drug user
IWA	Inland Waterways Association
JCVI	Joint Committee on Vaccination and Immunisation
LA	local authorities
LBC	liquid-based cytology
LDH	lactate dehydrogenase
LES	Liverpool endemic strain
LF	lethal factor (anthrax)
LFT	liver function tests
LLIN	long-lasting insecticidal nets
LP	lumbar puncture
LPA	line probe assay
LPS	lipopolysaccharide
LR-HPV	low-risk human papillomavirus
LTBI	latent TB infection
MALT	mucosa-associated lymphoid tissue
MAO	monoamine oxidase
MCA	Maritime Coastguard Agency
MDR-TB	multidrug-resistant tuberculosis
MERS-CoV	Middle East respiratory syndrome coronavirus
MIBE	measles inclusion body encephalitis
MIC	minimum inhibitory concentration
MIRU-VNTR	mycobacterial interspersed repetitive units-variable number tandem repeats
MLST	multilocus sequence typing
MLVA	multilocus variable number tandem repeat analysis
MMR	measles, mumps, and rubella
MOMP	major outer membrane protein
MR	measles-rubella
MRI	magnetic resonance imaging
MRSA	methicillin resistant *Staphylococcus aureus*
MSM	men who have sex with men
MSSA	methicillin sensitive *Staphylococcus aureus*
MTBC	*Mycobacterium tuberculosis complex*
MTCT	mother-to-child transmission
NAAT	nucleic acid amplification test
NAP1	North American pulsed-field type 1
NCB	Nuclear Chemical Biological
NG-MAST	*Neisseria gonorrhoeae* multi-antigen sequence typing
NHS	National Health Service
NICE	National Institute for Health and Care Excellence
NK (cell)	natural killer (cell)
NSF	non-sorbitol fermenting

NTS	non-typhoidal salmonella
OCT	outbreak control team
OECD	Organisation for Economic Co-ordination and Development
OSHA	Occupational Safety and Health Administration
PA	protective antigen
PCR	polymerase chain reaction
PEP	post-exposure prophylaxis
PEPSE	post exposure prophylaxis following sexual exposure
PET	post-exposure treatment
PFGE	pulsed-field gel electrophoresis
PHE	Public Health England
PHN	post-herpetic neuralgia
PIR	post infection review
PMCA	protein misfolding cyclic amplification
PPE	personal protective equipment
PPI	proton pump inhibitor
PPV	positive predictive value
PrEP	pre-exposure prohylaxis
PT	phage typing
PVL	Panton Valentine Leukocidin
PWTAG	Pool Water Treatment Advisory Group
Qnr	quinolone resistance proteins
RABV	Rabies virus
RAPD	random amplified polymorphic DNA
RBC	red blood cell count
RCOG	Royal College of Obstetricians and Gynaecologists
RCT	randomised controlled trials
RDT	rapid diagnostic dipstick tests
REA	restriction endonuclease analysis
RFLP	restriction fragment length polymorphism
RH	relative humidity
RIG	rabies immunoglobulin
RIPL	Rare and Imported Pathogens Laboratory
RO	reverse osmosis
RSV	respiratory syncytial virus
RT	reverse transcriptase
RT-PCR	reverse transcription polymerase chain reaction
RT-QUIC	real-time quaking-induced conversion
SARS	severe acute respiratory syndrome
SARS-CoV	severe acute respiratory coronavirus
sCJD	sporadic Creutzfeldt-Jakob disease
SCV	Salmonella-containing vacuole
SF	spray factor
SIV	simian immunodeficiency virus
SOP	standard operating procedure
SP	sulfadoxine–pyrimethamine
SPI1	Salmonella Pathogenicity Island 1
SPI2	Salmonella Pathogenicity Island 2

SSPE	subacute sclerosing panencephalitis
SSTI	skin and soft tissue infections
STEC	shiga toxin-producing *Escherichia coli*
STI	sexually transmitted infection
TB	tuberculosis
Td/IPV	tetanus, diphtheria, and polio vaccine
TDS	ter die sumendum (three times a day)
TESSy	The European Surveillance System
TLR	toll-like (pattern) receptor
TLR (2, 4)	toll-like receptor (2, 4)
TME	transmissible mink encephalopathy
TNF	tumor necrosis factor
TOC	test of cure
TSE	transmissible spongiform encephalopathy
TSST-1	toxic shock syndrome toxin 1
U&E	urea and electrolytes
UVGI	ultraviolet germicidal irradiation
VAMP	vesical-associated membrane proteins
vCJD	variant Creutzfeldt-Jakob disease
VFR	visiting friends and relatives
VHF	viral haemorrhagic fever
VLP	virus like particle
VNTR	variable number tandem repeats
VP	ventriculoperitoneal
VPSPr	variable protease-sensitive prionopathy
VSP	Vessel Sanitation Program
VT	verocytotoxin
VTEC	verocytotoxin-producing *Escherichia coli*
VZ	varicella-zoster
VZIG	varicella-zoster immunoglobulin
VZV	varicella-zoster virus
WBC	white blood cell count
WCC	white cell count
WHO	World Health Organization
WSG	Water Safety Group
WSP	Water Safety Plan
XLD	xylose lysine deoxycholate
XDR-TB	extensively drug-resistant tuberculosis

ANTHRAX

Alastair McGregor[1] and Tim Brooks[2]

[1]SpR, Rare and Imported Pathogens Laboratory, Public Health England and Hospital for Tropical Diseases, University College London Hospitals NHS Foundation Trust, London, UK.
[2]Head and Clinical Services Director, Rare and Imported Pathogens Laboratory, Public Health England, UK.

This case occurred during an outbreak of anthrax in the UK, which caused a series of serious soft tissue infections among injecting drug users. Skin and soft tissue infections are common in injecting drug users, about 35% of whom are estimated annually of developing a significant infection. There have been significant outbreaks in UK drug users resulting from contamination of street drugs with spore-forming organisms, the most notable being an outbreak of *Clostridium novii* in 2000, but also *C. tetani* (tetanus) and a variety of *Bacillus* species over the years.

In this case, a 45-year-old woman had a severe soft tissue infection surrounding a groin injection site, which involved the perineum and left buttock. When she arrived at the hospital, she was grossly septic with tissue necrosis and massive oedema surrounding the injection site. Anthrax was suspected early because there were concurrent cases of anthrax in injecting drug users in other UK centres.

INVESTIGATION OF THE CASE

Tissue samples were tested in a specialist Containment Level 3 laboratory. DNA of *Bacillus anthracis* was detected by polymerase chain reaction (PCR) of debrided tissue, and this organism was also subsequently cultured on agar. Toxin testing was performed on serum samples. An enzyme immunoassay for *B. anthracis* Lethal Factor (LF) toxin, a 3-protein exotoxin complex produced by *B. anthracis* that contributes to its virulence, was positive, and another enzyme-linked immunoassay (EIA) for Protective Antigen (PA) was equivocal.

CLINICAL MANAGEMENT

The principles of clinical management of cases of cutaneous anthrax are the same as those for any severe necrotizing soft tissue infection:

- to prevent toxin formation and organism multiplication (antibiotics)
- to remove the infectious nidus where possible (surgical debridement, usually only in cutaneous disease)
- to neutralize existing toxin (anthrax immune globulin, anti-anthrax toxin monoclonal antibodies, for example, Raxibacumab)

Bacillus anthracis is generally susceptible to penicillin (although resistance has been reported on several occasions), chloramphenicol, tetracycline, erythromycin, strepto-mycin, fluoroquinolones, and linezolid, but not to cephalosporins or trimethoprim-sul-famethoxazole.

In this case, extensive soft tissue debridement was performed in the operating theatre and high-dose antibiotics (benzylpenicillin 1.2 g every four hours and clindamycin 600 mg every six hours) were commenced. The patient made a good recovery following debride-ment. Other potential therapies, such as human anthrax immunoglobulin (used to elimi-nate circulating toxins) and Raxibacumab (a human IgG1 monoclonal antibody directed at the protective antigen of *B. anthracis*), were considered but not used.

PREVENTION OF FURTHER CASES

Control of anthrax in the hospital setting and waste management

From a public health perspective, the key concern with anthrax is that it is a severe disease with a high mortality, which can contaminate the environment with infectious spores of extreme longevity.

Infection control in the healthcare setting

In general, universal infection control precautions apply. Person-to-person spread of inhalational anthrax has never been documented, and cutaneous anthrax from contact with a human case is extremely rare. This limited transmission is because anthrax bacteria are not themselves capable of establishing infection and spores, by which infection can be transmitted, are only formed in suitable conditions that are usually found outside the human body. Respiratory isolation of cases is therefore not required, although standard personal protective equipment (PPE), including gloves, aprons, and visors (if there is a risk of splash), is advised.

Particular precautions are required in the management of waste, human tissues, and items contaminated with blood and body fluid from anthrax cases because these may be contaminated with bacteria that, once outside the body, form spores and can lead to onward transmission. PPE is necessary when handling such samples. If clothing or sheets are grossly contaminated with blood or body fluids, washing may not remove the remote risk posed by anthrax spores, and accepted advice is to have the clothing incinerated or autoclaved (in standard autoclave conditions - 121°C for 15 minutes under 1.05 kg/cm² pressure). Minimally contaminated or noncontaminated clothing should be laundered separately from other people's items in a washing machine at the hottest cycle possible.

Waste management

Segregation of waste into separate management streams at the point of production is vital to good waste management. In general, waste is divided into two streams (hazardous and nonhazardous) based on the threat it poses to the environment and the public. Hazardous waste is further divided into clinical waste and nonclinical waste. Clinical waste may be potentially infectious (that is, contaminated with human blood or tissue) or non-infectious (medicinal waste, cytotoxic waste, amalgam, chemical waste including laboratory, X-ray, and photochemicals, radioactive waste, and gypsum). Waste containers and bags are colour coded according to the type of waste they contain and the ultimate treatment designated for that waste, as described in the UK in Health Technical Memorandum (HTM) 07-01: Safe Management of Healthcare Waste. In the US, each state issues regulations that mandate standards for the management of medical waste, and there is further federal-level legislation implemented by agencies such as the Department of Transport (DOT), the Environmental Protection Agency (EPA), and the Occupational Safety and Health Administration (OSHA).

As an example, practicalities of medical waste disposal in the UK are summarized in Figure 1.1. For ease of use, disposal pathways are colour coded according to the type of waste and the disposal procedures necessary to make it safe. The yellow waste stream is used for waste that is infectious but that has an additional characteristic so that it must be

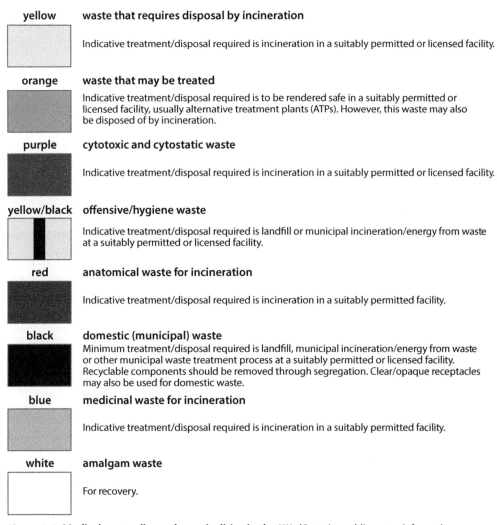

yellow	**waste that requires disposal by incineration**
	Indicative treatment/disposal required is incineration in a suitably permitted or licensed facility.
orange	**waste that may be treated**
	Indicative treatment/disposal required is to be rendered safe in a suitably permitted or licensed facility, usually alternative treatment plants (ATPs). However, this waste may also be disposed of by incineration.
purple	**cytotoxic and cytostatic waste**
	Indicative treatment/disposal required is incineration in a suitably permitted or licensed facility.
yellow/black	**offensive/hygiene waste**
	Indicative treatment/disposal required is landfill or municipal incineration/energy from waste at a suitably permitted or licensed facility.
red	**anatomical waste for incineration**
	Indicative treatment/disposal required is incineration in a suitably permitted facility.
black	**domestic (municipal) waste**
	Minimum treatment/disposal required is landfill, municipal incineration/energy from waste or other municipal waste treatment process at a suitably permitted or licensed facility. Recyclable components should be removed through segregation. Clear/opaque receptacles may also be used for domestic waste.
blue	**medicinal waste for incineration**
	Indicative treatment/disposal required is incineration in a suitably permitted facility.
white	**amalgam waste**
	For recovery.

Figure 1.1. **Medical waste disposal practicalities in the UK.** (Contains public sector information licensed under the Open Government Licence v3.0.)

incinerated in a suitably licensed or permitted facility, rather than simply being treated. Recognized examples are waste containing chemicals from human or animal healthcare, or waste contaminated with Advisory Committee of Dangerous Pathogens (ACDP) Category 3 pathogens, such as anthrax. Red stream waste is also incinerated but is limited to anatomical samples.

Orange stream infectious waste may be treated to render it safe prior to final disposal, but is not necessarily incinerated. This waste stream must not contain chemicals, amalgam, medicines, or anatomical waste. In addition, it should not contain waste that is non-infectious (for example, domestic waste) or that has additional characteristics that require incineration (medicinal, chemical, anatomical). Treatment of this waste usually involves maceration, followed by sterilization in a continuous-flow autoclave, known as a hydroclave. The hardiness of anthrax spores means that contaminated waste and tissue pose a unique threat in this waste stream because they can contaminate the macerator prior to sterilization, with attendant exposure risks for operators and maintenance staff. All materials known or suspected of being contaminated with anthrax must therefore be disposed of in the yellow stream, except for debrided human tissues, which are disposed of in the red stream.

Where practicable, medical equipment and mattresses should be decontaminated prior to disposal. Once decontaminated, any infectious risk should be eliminated, although the equipment may still retain hazardous properties that will be subject to statutory waste management controls. If no hazardous properties remain (for example, decontaminated mattresses with the impervious cover intact), the item may be disposed of as domestic waste. Heavily soiled or infectious mattresses should be disposed of as potentially infectious clinical waste.

Unfortunately, some potentially serious lapses of infection control occurred in the management of this case. During surgery, debrided tissue was disposed of incorrectly by being designated as clinical waste to be treated (orange stream), rather than hazardous anatomical waste to be incinerated (red stream). In order to manage the risk from the contaminated waste, a biocontainment team was sent to the waste processing facility. It was fortunate that the waste processing facility was not operating and that a 10-day backlog of waste had accumulated, meaning that no environmental contamination had occurred. Infectious tissue had not leaked into the environment, but the exact location was unclear as orange clinical waste is only traceable to the source hospital. Expert consensus was that both macerating untreated tissue containing anthrax spores and examining the contents of sealed bags were unacceptably hazardous. Eventually, all bags received at the waste disposal unit from the same location as the debrided tissue (approximately 50 tons of waste) were identified, rebagged, and incinerated.

In the days following the surgery, the bed sheets became blood stained and were sent for incineration. The mattress was treated with sporicidal wipes. After several days, it was noted that the mattress cover was leaking and that mattress foam had been extensively contaminated with blood. The mattress was removed, covered, and stored in a side room, and the decision was made to divide the foam into pieces small enough to be individually incinerated. The following day, the mattress was accidentally identified as general waste and sent for disposal as untreated landfill. The disposal of contaminated waste as landfill represents an ongoing risk to animals and people who may have contact with it. Domestic hospital waste is untraceable and the mattress was not found. The chance of a significant exposure was assessed as being negligible.

Infection with anthrax is extremely uncommon in individuals without specific risks, and rigorous waste management procedures, along with standard isolation precautions, minimize the small risk of onward transmission.

EPIDEMIOLOGY

Anthrax is primarily an infection of herbivorous mammals that eat contaminated pasture and, less commonly, may affect carnivores that feed on these animals. Anthrax was a major public health and veterinary problem until the beginning of the last century, causing the deaths of hundreds of thousands of animals each year.

Anthrax in humans is acquired from direct contact with infected animals or their hides, or inhalation, ingestion, or injection of contaminated substances and, therefore, there is usually an identifiable occupational or behavioural risk factor for infection. Sporadic cases of anthrax continue to occur in developing countries, and outbreaks associated with eating infected meat have occurred recently in the Philippines (2010) and India (2016). There have also been several anthrax outbreaks associated with the accidental or deliberate dissemination of anthrax spores, such as the Sverdlovsk incident (USSR, 1979) and the anthrax attacks on the United States Postal Service (2001).

Anthrax is extremely uncommon in the UK and has previously occurred sporadically in individuals with specific exposures. Infection in intravenous drug users (IVDUs) was a new finding during the outbreak described, although a case had been reported in a heroin user in Norway in 2000, and there are anecdotal reports of anthrax in IVDUs in Iran and elsewhere.

Transmission characteristics

As with other *Bacillus* spp., when environmental conditions are unfavourable to bacterial growth, *B. anthracis* can form endospores. In this state, the organism can remain inactive (dormant) for decades. In the body of an infected animal, the vegetative bacteria are not infectious to other animals, but when they come into contact with air or soil, sporulation commences within a few minutes and is complete within 24 hours or so, depending on temperature and environmental conditions. These spores are infectious through ingestion, inhalation, or inoculation.

BIOLOGY

B. anthracis is a rod-shaped, Gram-positive, aerobic bacterium that readily grows on standard microbiological media. The typical colony at 24 hours of growth in air or 5% CO_2 is grey to white, nonhemolytic on blood agar, with a dry, ground-glass appearance. The colony may be distinguishable by the presence of swirling projections (commonly known as Medusa head appearance). Encapsulated colonies may be mucoid. *In vivo*, bacterial cells often grow in long chains that resemble bamboo. PCR or immunofluorescence tests may be used to distinguish *B. anthracis* from related bacteria.

PATHOLOGY

Unlike other *Bacillus* spp., *B. anthracis* possesses three virulence factors: a capsule, lethal toxin, and oedema factor. PA acts as the carrier component of the two bipartite toxins, combining with either lethal factor (LF) or oedema factor (EF) to produce the respective toxins, which act synergistically. EF interferes with the cAMP system of cells leading to an accumulation of fluid in the extracellular space, while LF is cytotoxic, especially to macrophages. The combined effect is tissue destruction, impaired immune response, and a rapidly rising level of toxin and organisms in the affected site. If the infection spreads beyond the initial site, these phenomena affect multiple organs giving the classic picture of septicemic anthrax with high levels of circulating toxins and organisms (up to 10^8/mL). The cause of the characteristic coagulation deficit and sharp fall in platelets seen in the final stages of clinical disease is not fully understood.

DISEASE

Disease may present with one of three clinical syndromes (cutaneous, gastrointestinal, or inhalational) that reflect the route of inoculation. Haemorrhagic meningitis is a fairly common occurrence in late disease, but may be the presenting symptom.

Inhalational anthrax

Inhalational anthrax results from the inhalation of *B. anthracis* spores, which may occur while working with contaminated animal products such as wool, hides, or bone meal. Inhalational anthrax also has occurred from deliberate and accidental release of weaponized spore preparations. Spores are phagocytosed by alveolar macrophages and transported to mediastinal lymph nodes where they germinate, multiply, and release toxins causing haemorrhagic necrosis. The incubation period for inhalation anthrax is estimated to be 1 to 7 days, although it may be significantly longer. Early symptoms are entirely nonspecific but experience from the 2001 bioterrorism outbreak in the US (where exposed individuals were identified early) shows that early disease may respond to antibiotics. Without treatment, the prodromal phase is followed by a fulminant bacteraemic phase with an almost universally fatal outcome.

Gastrointestinal anthrax

Gastrointestinal anthrax follows the ingestion of undercooked infected meat and tends to occur in point source outbreaks. Spores germinate in the alimentary tract epithelium causing necrotic ulcers similar to those seen in cutaneous disease. As with inhalational anthrax, mesenteric lymph nodes may be become enlarged and haemorrhagic. The case-fatality rate of established gastrointestinal anthrax is estimated to range from 12 to 50%.

Cutaneous anthrax

Cutaneous anthrax is the most common form of the disease. The incubation period is usually 5 to 7 days with a range of 1 to 12 days. Infection usually occurs on an exposed area of skin and begins as a small, painless papule that enlarges and ulcerates to form a painless necrotic ulcer with a black, depressed eschar, from which anthrax takes its name (άνθρακας [ánthrakas] is Greek for coal). Toxin release results in severe localized oedema and regional lymphadenopathy, and lymphangitis is usually present. With antibiotic therapy and surgery, the case fatality rate is less than 1%.

Injectional anthrax

Injectional anthrax is an additional anthrax syndrome that has been proposed following the Scottish outbreak to reflect the increased severity and mortality of anthrax acquired by this specific route. Individuals with injectional anthrax are often diagnosed late often because of the unfamiliar clinical presentation and lack of evident clustering, and they demonstrate a notably severe course of disease (case fatality rate of about 37%). The severity may result from the size and depth of the inoculum, the higher rate of septicaemia and the baseline health of those affected.

Diagnosis

The diagnosis of anthrax may be challenging because a high index of suspicion is required in order for the correct tests to be selected. A detailed history focussing on potential exposures is critical. These exposures may seem tenuous, as in the case of an American man and his child, who acquired anthrax from untreated hides that they were using to make drums. The organism may be cultured from blood and tissue samples prior to antibiotic administration. PCR is available at specialist diagnostic laboratories and may be performed on a variety of samples (for example, plasma, skin, deep tissue, ascites, and bronchoalveolar lavage). Retrospectively, a diagnosis of anthrax can be made on serological testing of paired sera, although this test is not widely used.

Anthrax is extremely uncommon in the developed world but should be considered in individuals with exposures to animal carcasses and tissues and those with specific exposures to substances that may be contaminated with these, including paraphernalia associated with injection drug use. The clinical presentation varies with the site of entry of the anthrax spores and the resulting infection may carry a very high mortality, particularly for the gastrointestinal and respiratory forms of disease. Individuals with anthrax pose a low risk to contacts but the formation of very hardy spores mean that tissue samples and clothing or bed linen pose an infection risk and need to be sterilized prior to disposal to prevent onward transmission.

QUESTIONS

1. The most common syndrome associated with anthrax infection is:
   a. Soft-tissue infection
   b. Pneumonia
   c. Abdominal symptoms
   d. Meningitis
   e. Hepatitis

2. Which of the following may be disposed of in the orange clinical waste stream:
   a. Placentas
   b. Flowers
   c. Dental fillings
   d. Batteries
   e. Clinical gloves

3. Which of these diagnostic tests cannot be used to confirm a definite case of anthrax infection:
   a. Blood culture
   b. PCR of bronchoalveolar lavage fluid
   c. Serology of paired sera
   d. PCR of infected tissue
   e. Culture of CSF

4. Which of these exposures has not been associated with anthrax infection?
   a. Tanning leather hides
   b. Smoking heroin
   c. Injecting heroin
   d. Eating undercooked beef
   e. Swimming in freshwater

5. Anthrax is able to survive in the environment due to:
   a. Protective Factor
   b. Lethal Factor
   c. Poly-D-glutamic acid capsule
   d. Formation of endospores
   e. Antibiotic resistance

GUIDELINES

1. Infection Control Precautions during the Clinical Management of Drug Users with Probable or Confirmed Anthrax HPS (2010) Version 0.7 (www.documents.hps.scot.nhs.uk/giz/anthrax-outbreak/ic-management-anthrax-v0-7-2010-01-18.pdf).
2. Carucci JA, McGovern TW, Norton SA, et al. (2002) Cutaneous anthrax management algorithm. *J Am Acad Dermatol* **47**:766.

REFERENCES

1. Health Technical Memorandum 07-01 (2013) Safe management of healthcare waste. Department of Health. HMSO.
2. Health Protection Scotland. An outbreak of anthrax among drug users in Scotland, December 2009 to December 2010.
3. Hope V, Kimber J, Vickerman P, Hickman M & Ncube F (2008) Frequency, factors and costs associated with injection site infections: findings from a national multi-site survey of injecting drug users in England. *BMC Infect Dis* **18**(8):120.
4. Hope VD, Marongui A, Parry JV & Ncube F (2010) The extent of injection site infection in injecting drug users: findings from a national surveillance study. *Epidemiol Infect* **138**(10):1510–1518.
5. McGuigan CC, Penrice GM, Gruer L et al. (2000) Lethal outbreak of infection with Clostridium novii type A and other spore forming organisms in Scottish injecting drug users. *J Med Microbiol.* **51**: 971–977.
6. Taylor A, Hutchinson S, Lingappa J et al. (2005) Severe illness and death among injecting drug users in Scotland: a case control study. *Epidemiol Infect* **133**(2):193–204.
7. Brett MM, Hood J, Brazier JS, Duerden BI & Hahne SJ (2005) Soft tissue infections caused by spore forming bacteria in injecting drug users in the United Kingdom. *Epidemiol Infect* **133**(4): 575–582.
8. Cherkasskiy BL (1999) A national register of historic and contemporary anthrax foci. *J Appl Microbiol* **87**(2):192–195.
9. Ringertz SH, Hoiby EA, Jensenius M et al. (2000) Injectional anthrax in a heroin skin-popper. *Lancet* **356**:1574–1575.
10. Brachman P & Kaufmann A (1998) Anthrax. In Bacterial infections of Humans: Epidemiology and Control, 3rd ed (Evans A & Brachman P eds), p. 95. Plenum.

11. Quinn CP & Turnbull PCB (1998). Anthrax. In Topley and Wilson's Microbiology and Microbial Infection, 9th ed (Hausler WJ & Sussman M eds), p. 799. Edward Arnold.

12. Meselson M, Guillemin J, Hugh-Jones M et al. (1994) The Sverdlovsk anthrax outbreak of 1979. *Science* **266**(5188):1202.

13. Jernigan DB, Raghunathan PL, Bell BP et al. (2002) Investigation of bioterrorism-related anthrax, United States, 2001: epidemiologic findings. *Emerg Infect Dis* **8**:1019.

14. Kanafani ZA, Ghossain A, Sharara AI et al. (2003) Endemic gastrointestinal anthrax in 1960s Lebanon: clinical manifestations and surgical findings. *Emerg Infect Dis* **9**:520.

15. Sirisanthana T & Brown AE (2002) Anthrax of the gastrointestinal tract. *Emerg Infect Dis* **8**:649.

16. Doganay M, Almac A & Hanagasi R (1986) Primary throat anthrax: a report of six cases. *Scand J Infect Dis* **18**:415–419.

17. Lightfoot N, Scott R & Turnbull B (1990) Antimicrobial susceptibility of *Bacillus anthracis*. *Salisbury Med Bull Suppl* **68**:95.

18. Doğanay M & Aydin N (1991) Antimicrobial susceptibility of *Bacillus anthracis*. *Scand J Infect Dis* **23**:333.

19. Turnbull PC, Sirianni NM, LeBron CI et al. (2004) MICs of selected antibiotics for *Bacillus anthracis*, *Bacillus cereus*, *Bacillus thuringiensis*, and *Bacillus mycoides* from a range of clinical and environmental sources as determined by the Etest. *J Clin Microbiol* **42**:3626.

ANSWERS

MCQ	Feedback

1. The most common syndrome associated with anthrax infection is:
  a. Soft-tissue infection
  b. Pneumonia
  c. Abdominal symptoms
  d. Meningitis
  e. Hepatitis
	Cutaneous anthrax is the most common presentation of anthrax. In the Scottish outbreak, 93% of cases had soft tissue involvement, 55% gastrointestinal disease, 33% had central nervous system involvement, and 5% respiratory involvement.

2. Which of the following may be disposed of in the orange clinical waste stream:
  a. Placentas
  b. Flowers
  c. Dental fillings
  d. Batteries
  e. Clinical gloves
	Batteries and amalgam have dedicated waste streams, flowers are considered domestic waste, and placentas are anatomical waste, requiring incineration.

3. Which of these diagnostic tests cannot be used to confirm a definite case of anthrax infection:
  a. Blood culture
  b. PCR of bronchoalveolar lavage fluid
  c. Serology of paired sera
  d. PCR of infected tissue
  e. Culture of CSF
	Testing paired sera can give a very strong indication retrospectively that an infection was anthrax, but time is needed for an antibody response to develop and serology was therefore not used as a part of the case definition in the Scottish outbreak.

4. Which of these exposures has not been associated with anthrax infection?
  a. Tanning leather hides
  b. Smoking heroin
  c. Injecting heroin
  d. Eating undercooked beef
  e. Swimming in freshwater
	There has never been a recorded case of anthrax related to swimming in water: the likelihood of ingesting, inhaling, or inoculating enough spores into skin to cause infection is extremely low.

5. Anthrax is able to survive in the environment due to:
  a. Protective Factor
  b. Lethal Factor
  c. Poly-ᴅ-glutamic acid capsule
  d. Formation of endospores
  e. Antibiotic resistance
	Protective Factor and Lethal Factor are virulence genes that allow anthrax to cause aggressive disease. The capsule inhibits phagocytosis, allowing the organism to evade the immune system. The formation of spores allows anthrax to survive unfavourable environmental conditions in a state of dormancy.

ASPERGILLOSIS

CASE 2

Fred Pink[1]

[1]Barts Health NHS Trust, Department of Infection, Royal London Hospital, Whitechapel, London, UK.

A 31-year-old patient with acute myeloid leukemia (AML) is persistently pyrexial despite 5 days of meropenem and vancomycin. Her disease proved refractory to primary treatment with R-CHOP chemotherapy and was complicated by two episodes of neutropenic sepsis. She underwent allogeneic hematopoietic stem cell transplant 20 days ago. She had previously required nearly 4 weeks of broad-spectrum antibiotic treatment as an inpatient on the same hematology unit. Her bone marrow function had not recovered.

Blood cultures from her Hickman line had been negative to date. Her CRP was rising and she was slightly hypoxic so a chest X-ray was performed, which the team felt showed some haziness in the right upper lobe. Further blood cultures from the line and periphery were sent. Blood was also tested for galactomannan. A CT scan of the chest, abdomen, and pelvis was requested. The team was especially concerned as two other AML patients with long inpatient stays on the same 18-bed unit had behaved similarly and both were now being treated for invasive pulmonary aspergillosis. The team had not had any cases of invasive aspergillosis on the unit for several months.

INVESTIGATION OF THE CASE

The infection control team was aware that extensive emergency building work had been carried out over the past two months, including repairs to a storm-damaged roof. The hematology unit adjoins the affected part of the building. The hematology unit has 10 side rooms, all of which have en suite toilets and have highly efficient particulate air (HEPA) filters. The weather had been especially hot during the repair work and, although staff, visitors, and patients had been asked not to open windows of the unit facing the work area, open windows were found on several occasions. Additionally, a temporary domestic staff member had been disciplined after repeatedly propping doors open in the side rooms. All three affected patients were in side rooms; two faced the courtyard where much of the work had been taking place.

Respiratory specimens obtained by limited bronchoscopy were sent for fungal culture, as well as Ziehl-Neelsen stain, mycobacterial culture, Gram stain, and bacterial culture. *Aspergillus fumigatus* was isolated from one specimen.

Aspergillus species are ubiquitous in the environment, both indoors and outdoors. Healthy people who breathe in spores won't develop a subsequent infection. However, patients with underlying pulmonary disease or immunocompromised patients are much more prone to develop an infection upon exposure to *Aspergillus* spores.

Outbreaks of nosocomial aspergillosis are generally related to construction work or reno-vation without adequate infection control precautions and failure to control spread of contaminated dust or debris. Studies reporting outbreaks have investigated the relation-ship between *Aspergillus* spore counts and infection risk. However, until now, no exact relationship between number of spores and quantifiable infection risk among patients has been established. Molecular typing of *Aspergillus* spp. has been extensively reviewed and can be useful to determine a common source in an outbreak setting. Invasive aspergillosis rates of around 5% may be seen in acute leukemics, and rates are slightly higher for post-induction and consolidation aplasia patients. Rates far in excess of this warrant investigation.

The infection control team's initial hypothesis should link the increased aspergillosis case rate with the recent building work. They should establish that an adequate assess-ment of the increased risk of environmental contamination was made before building work commenced and that appropriate infection-control precautions were instituted and remained in place for the duration of the period of risk. Repeated opening of doors and windows reflects breaches of these arrangements and should be documented as part of regular compliance checks.

A timeline should be drawn up identifying the start of the period of risk, as well as any infection-control breaches and periods of more intense construction work. The time-line will help in compiling a list of patients at risk and clarify the need for enhanced precautions. Screening of remaining inpatients should be considered (and particularly at-risk outpatients). If not already in place, regular serum galactomannan testing could be considered in addition to investigation of patients reporting new, compatible symp-toms. Throughout this work the infection prevention and control team should work with contractors and with clinical, nursing, and other staff in the unit. The team should keep management informed and ensure that the local health protection unit is informed.

CLINICAL MANAGEMENT

Early diagnosis and treatment are essential in cases of invasive aspergillosis. Clinicians should have a low threshold for suspecting invasive fungal disease in bone marrow failure or neutropenic hemato-oncology patients who fail to respond to first- and second-line empiric antibiotic treatment. A high-resolution CT scan of the chest and galactomannan serum test should be performed. If the patient was fit and had suspicious imaging, a limited bronchoscopy or nondirected bronchoalveolar lavage could be attempted. Where there is clinical suspicion, most clinicians would institute antifungal treatment while awaiting further results. Failure to establish a pulmonary source should trigger wider investigation with imaging of the abdomen and sinuses as a minimum.

Research studies employ strict criteria for the diagnosis of invasive aspergillosis, the most popular being those of the European Organization for Research and Treatment of Cancer/Mycosis Study Group, which rates cases as proven, probable, or possible based on clini-cal, mycological, patient, and histopathological factors. In practice, diagnosis of invasive aspergillosis is difficult. Patients may not be fit for invasive investigations such as bronchos-copy or they may be severely thrombocytopenic despite platelet support and so unsuita-ble for biopsy. Very often, clinicians will have to base their diagnosis solely on symptoms of a lower respiratory tract infection and suspicious imaging in an immunosuppressed patient. A CT chest finding of a rounded, discrete lesion with the so-called halo sign would be highly suggestive of *Aspergillus* infection (Figure 2.1).

Microscopy of bronchoalveolar lavage specimens may demonstrate fungal hyphae. Alternatively *Aspergillus* species may be grown on subsequent culture. Increasingly, molecular tests using polymerase chain reaction (PCR) technology are being used to determine the presence of fungal genetic material in primary specimens.

Treatment

Treatment options for invasive aspergillosis include voriconazole and liposomal amphotericin B, and some units employ both agents or add in an echinocandin in refractory cases. The patient's renal and liver function should be monitored carefully when these treatments are instituted.

Remember that septic, severely immunosuppressed individuals may have polymicrobial infections. Some pathogens may be transient or not cultured, so empiric broad-spectrum antibiotic therapy should be maintained until some clinical response is achieved. As with so many infections in the immunosuppressed, rapid and effective treatment of the underlying hematological malignancy will have a very significant impact on outcomes from invasive aspergillosis. AML patients with adverse cytogenetics and those refractory to chemotherapy are at even higher risk of a poor outcome in invasive aspergillosis.

PREVENTION OF FURTHER CASES

In the UK, Guidance on Cancer Services—Improving Outcomes in Haematological Cancers: The Manual (NICE 2003) makes a range of infection control recommendations that depend in part upon the intensity and duration of the therapy patients receive (and thus how long and severely neutropenic they are likely to be). General measures to reduce the incidence of invasive aspergillosis in at risk patients fall into three main groups: strict adherence to infection control precautions, provision of an appropriate patient environment, and antifungal prophylaxis for higher-risk patients. The first two measures will also contribute to reducing all hospital-acquired infections.

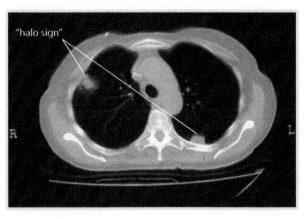

Figure 2.1. **The halo sign as seen on the CT chest scan, highly indicative of Aspergillus infection.** (Courtesy of Aspergillus & Aspergillosis Website, http://www.aspergillus.org.uk.)

Infection control practice

Although very widely practiced, and perhaps of value in reminding staff and visitors of the vulnerability of patients, there is a lack of firm evidence for barrier nursing. Therefore its role is unclear. Instead National Institute for Health and Care Excellence (NICE) guidance emphasizes basic infection control practices, such as handwashing. Compliance with infection control precautions is a key component in decreasing the risk of hospital-acquired infection.

Other practical measures include direct admission to specialist units, minimizing patient movements, and avoiding unnecessary stays in wards with little to no experience in caring for complex patients, such as the hematological population. A recent case-controlled study found that patients with hematological malignancies who were transferred more than five times from the hematology unit (including transfers for imaging) showed a sixfold increase in their risk of an invasive fungal infection.

Appropriate patient environment

Rather than placing all patients who are neutropenic in side rooms, it is recommended that patients who are receiving intensive therapy (such as induction chemotherapy) should be managed in a specialist hematology unit with strict adherence to infection control precautions. For example, single rooms with en suite facilities and laminar airflow and HEPA filtration should be available for patients undergoing high-dose therapy and for bone marrow transplant recipients. Good air filtration minimizes airborne microbiological contamination.

Patients with likely transient neutropenia, such as those managed on lower intensity regimens (for example, soral hydroxycarbamide), may be admitted to a bay in a general medical ward.

Standard single rooms appear to be sufficient for autologous bone marrow transplant patients if strict infection control practices are adhered to.

Antifungal prophylaxis

Current guidelines state that leukemia patients who receive induction chemotherapy, allogeneic bone marrow transplant recipients in their initial neutropenic phase, and allogeneic bone marrow transplant recipients who are experiencing graft-versus-host disease should receive appropriate antifungal prophylaxis. Antifungal agents with activity against *Aspergillus* include voriconazole, itraconazole, and posaconazole.

Reducing the risk of invasive aspergillosis during building work

Several outbreaks of invasive aspergillosis have been attributed to construction work on hospital sites, and the relationship between the two is well recognized. Appendix 3 of the Department of Health Building Note 00-09—Infection Control in the Built Environment—provides guidance on planning for construction or refurbishment activity in healthcare facilities. It stresses the need for infection-control teams to work with estates departments (or whoever is responsible for building work) and contractors to schedule work and decide on appropriate infection-control precautions. There is also an ongoing need for daily monitoring to ensure adherence to recommendations made by the infection-control team. There are three guiding principles in these recommendations:

1. Identify susceptible patient groups.
2. Protect susceptible patients from airborne fungal spores generated by construction work.
3. Minimize the dissemination of fungal spores during construction work.

Unless they also part of an at-risk group, most staff and visitors to the hospital are not likely to develop an invasive fungal disease. Patients who require mechanical ventilation on ventilators, dialysis, or long-term corticosteroid therapy will be at increased risk. Neutropenic patients, neonates on intensive care, and organ transplant recipients should be seen as high risk. Patients considered to be at very high risk include bone marrow transplants and those with aplastic anemia, persistent neutropenia, or severe combined immunodeficiency syndromes.

Areas may be ranked according to whether higher-risk patients might be exposed to fungal spores if proper precautions aren't in place: office space and plant rooms are deemed as low risk; general wards, radiology, and endoscopy as medium risk; intensive care units, operating theaters, pharmacy clean rooms, and cardiac catheter suites are classified as high risk.

Building and renovation work will fall into one of the four following categories:

- Type A includes removal of ceiling tiles for inspection work, minor electrical, and plumbing work.
- Type B involves short-term activity that doesn't generate a lot of dust, such as installing computer cabling.
- Type C involves the generation of moderate to high amounts of dust, which might include minor demolition work.
- Type D involves major demolition and construction work.

The Department of Health guidance includes a matrix weighting patient risk areas against the type of building work, which enables infection-control teams and estates departments (responsible for building work) to determine appropriate risk measures. These are recommendations to prevent all inpatients from developing a hospital-acquired infection.

For the hematology population, the following recommendations apply:

- Avoid nonemergency admissions during heavy construction periods.
- Place high-risk patients as far as possible from areas of demolition or construction.
- Seal off patient care areas with adequate and impermeable barriers and keep doors closed.
- Verify air filtration (HEPA filtration). Check possible plugging or leakage of air filters. Ensure that air pressure relationships are adequate compared with adjunct rooms. Aim for positive pressure in patient rooms and for negative pressure in in-house construction areas.
- Provide treatment in the patient's room if possible.
- Transport via an alternate route, schedule transportation during periods with minimal construction activity, minimize waiting times outside, and use appropriate face

masks for susceptible patients if it is necessary for them to leave their rooms and pass through potentially contaminated areas.

- Wet-clean wards thoroughly without raising dust.
- Monitor infections in patients who are at increased risk for invasive aspergillosis (IA).

Prior to starting work, contractors should be briefed by the infection-control team regarding risk management measures and the rationale for them. During the course of any work, clinicians should report any cases of suspected invasive fungal infection to the microbiology and infection-control teams. Clear arrangements should be in place for final cleaning of the site and removal of any screening materials. Although the emphasis is on airborne fungal spores, attention should be paid to the risks of aerosols arising from contaminated water, particularly through showers or taps.

Laboratory

Patient samples should be sent to the laboratory in CE-marked (conform to EU health, safety, and environmental requirements) leak-proof containers. Processing of specimens for fungal culture should take place in a microbiological safety cabinet. Many microbiology departments will have a dedicated laboratory space designed to process respiratory specimens such as bronchoscopy specimens.

EPIDEMIOLOGY

Aspergillus species are widely found in the environment; they can be isolated from soil, water, decaying vegetable matter, and unfiltered air. Building work tends to mobilize high numbers of their spores into the atmosphere. The most common species isolated is *A. fumigatus*, but *A. flavus*, *A. terreus*, and *A. niger*, among others, are known to cause disease.

Public Health England estimates that 4000–5000 cases of invasive aspergillosis occur in the UK each year, although diagnosis is often made late, if at all. Cases occur in the immunocompromised and those with underlying lung disease taking corticosteroids. Rates have increased over the past few decades, with a concomitant increase in the hematology and oncology patient population and from the use of immunosuppressive therapy and chemotherapy. High doses of corticosteroids used in the management of graft-versus-host disease have caused a shift in the onset of invasive aspergillosis, which often occurs many months after transplantation. The disease is more common in allogeneic hematopoietic stem cell transplant recipients as they tend to be profoundly immunosuppressed for extended periods of time. Invasive pulmonary aspergillosis is relatively more common in lung transplant patients compared to other solid organ recipients. Invasive aspergillosis carries a very poor prognosis. It is often diagnosed late, resulting in a delay in therapy. Mortality approaches 60% and may be higher in hematopoietic stem cell transplant recipients. Late diagnosis is associated with a poorer prognosis.

Approximately 40,000 cases of allergic bronchopulmonary aspergillosis (ABPA) in immunocompetent patients with underlying lung disease are seen in the UK each year. It affects around 1% of patients suffering from asthma and a higher percentage of cystic fibrosis patients. Globally the incidence of chronic pulmonary aspergillosis relates to the prevalence of lung cavitation in pulmonary tuberculosis patients. The World Health

Organization (WHO) has estimated that there may be nearly 380,000 cases globally in pulmonary tuberculosis patients alone. Most of these occur in China and India, where mortality even in patients treated for aspergillosis may still exceed 50% at five years.

BIOLOGY

Aspergillus species are molds. There are now more than 250 species categorized into several subgenera. Correct identification of some of the more unusual isolates requires the use of molecular diagnostic techniques. The majority of species reproduce asexually, producing spores (conidia) of around 3 μl diameter, which are only filtered out by HEPA filters of EU12 quality and above. The organism may be cultured from biopsy or other clinical sites and will usually grow within three days. Isolates grow well at body temperature (37°C). Interestingly, *A. fumigatus* may grow at significantly higher temperatures, up to 50°C. Susceptibility to antifungal agents varies by species; most notably *A. terreus* is resistant to amphotericin B, and so accurate speciation is important. Generally, growth on the plate is relatively rapid, with the colonies developing a characteristic color evident on the underside of the plate, for instance yellow in *A. flavus*, black with *A. niger*, and beige or brown with *A. terreus* (Figure 2.2).

Microscopy of bronchoalveolar lavage specimens may demonstrate fungal hyphae, alternatively *Aspergillus* species may be grown on subsequent culture. Increasingly, molecular tests based on polymerase chain reaction technology are being used to determine the presence of fungal genetic material in primary specimens.

PATHOLOGY

The route of *Aspergillus* infection is by inhalation of fungal spores, though they may directly inoculate tissues, especially in exposed wounds. It has been estimated that disease may develop as early as two weeks after inhalation of *Aspergillus* species spores. Once spores germinate, they produce fungal hyphae, the growth of which is increased by corticosteroids. In order to cause disease, the fungal spores must be able to adhere to pulmonary epithelia before tissue invasion by the hyphae.

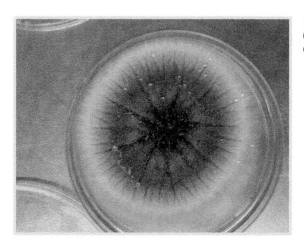

Figure 2.2. **Growth of *Aspergillus* species (*A. fumigatus*) on an agar plate.** (Courtesy of Jankaan published under CC BY-SA 3.0)

DISEASE

Aspergillus-related disease is most common in the immunosuppressed population, but it is not restricted to this diverse group of patients. For immunocompetent patients, without any lung pathology and no history of corticosteroid use, isolation of *Aspergillus* species from clinical samples, especially blood cultures, is highly likely to represent contamination. However, in patients with pulmonary cavitation (for example, post pulmonary tuberculosis) or underlying lung disease (cystic fibrosis or chronic obstructive pulmonary disease), potentially pathogenic *Aspergillus* species in pulmonary samples may reflect colonization or disease.

Aspergillus species may provoke an allergic response in the host, most commonly as allergic bronchopulmonary aspergillosis. This is a chronic condition that, if untreated, may ultimately lead to pulmonary fibrosis. It is classically characterized by central bronchiectasis in the context of asthma or cystic fibrosis, with elevated serum IgE and skin or antibody reactivity to *Aspergillus* species. An overactive immune response, predominantly driven by Th2 rather than Th1 cells, drives the production of excess interleukins that cause mucus plugging, inflammation of smaller airways, and atelectasis.

Other localized manifestations of *Aspergillus* disease include aspergillomas (fungal balls), particularly in lung cavities and sinuses, or wound infections. Invasive disease may affect a single organ system, but any invasion of vasculature will lead to disseminated disease. Invasive pulmonary aspergillosis is the most common site in the immunosuppressed, but invasive cerebral and sinus aspergillosis are well described and may occur together. If invasive disease is suspected at one site, imaging should be employed to identify other potential sites of infection elsewhere in the body.

Symptoms of invasive pulmonary aspergillosis typically include dyspnea, dry cough, and fevers. Imaging of the chest may demonstrate pulmonary infiltrates, pleural effusions, or cavitation. As already discussed, diagnosis is often difficult, but where there is clinical suspicion, treatment should be initiated promptly. Delays in treatment are associated with significantly worse outcomes.

QUESTIONS

1. Regarding invasive aspergillosis and hematopoietic stem cell transplant (HSCT) recipients, which of the following are true?
    a. NICE guidance recommends positive-pressure, HEPA-filtered side rooms for all patients undergoing HSCT and those who have received a HSCT.
    b. Recipients of allogeneic stem cell transplants are deemed lower risk for hospital-acquired and other infections than autologous HSCT patients.
    c. Refractory hematological malignancy is associated with poorer response to antifungal treatment of invasive aspergillosis.
    d. Invasive pulmonary aspergillosis is most likely to occur immediately after HSCT.
    e. Liposomal amphotericin B treatment is associated with renal dysfunction and deranged electrolytes.

2. Regarding invasive pulmonary aspergillosis, which of the following are true?
    a. Most cases are caused by *Aspergillus flavus*.
    b. Voriconazole is the first choice antifungal therapy.
    c. Diagnosis can only be made on the basis of positive culture.
    d. A positive culture of *Aspergillus* from bronchiolar lavage always indicates infection.
    e. It is more likely in a patient with HIV who takes antiretrovirals and whose CD4 count is 450 than in a patient taking 20 mg prednisolone each day as part of their rheumatoid arthritis treatment.

3. True or false: *Aspergillus* species:
    a. Are invariably sensitive to amphotericin
    b. Are primarily dermatophytes
    c. Isolated in clinical samples are always pathogenic
    d. Should always be processed in a safety cabinet in the microbiology laboratory
    e. Are molds

4. True or false: infection control measures to be taken in the event of a suspected cluster of *Aspergillus fumigatus* infections in a hematology unit must include:
    a. Identification of patients at risk
    b. Screening of staff members
    c. Assessment of existing infection control measures and evidence of any breaches, including whether isolation measures are available and used
    d. Contacting the local health protection unit
    e. Writing to all recent inpatients' families advising them of a potential outbreak

5. A new renal dialysis unit is scheduled to be built on the site of a former NHS staff recreation club, which will have to be demolished first. The site is next to the block containing the hospital's hematology inpatient and outpatient units. True or false: The infection control team should:
    a. Oppose the development as it is highly likely to put immunosuppressed patients at increased risk of *Aspergillus*-related conditions.
    b. Ensure that ventilation ducts, windows, and water pipes are isolated from areas where demolition is taking place.
    c. Insist that contractors use external waste disposal chutes when gutting the upper floors of the building to be demolished.
    d. Draw up an assessment of all at-risk groups on the hospital site.
    e. Discuss and plan measures to reduce exposure to *Aspergillus* and other potential pathogens with the contractor and clinical teams affected.

GUIDELINES

1. Guidance on Cancer Services—Improving Outcomes in Haematological Cancers: The Manual. (NICE, 2003).

2. Department of Health, Health Building Note 00-09: Infection Control in the Built Environment (especially appendix 3—IPC risk assessment during construction/refurbishment of a health-care facility), London 2013. (www.gov.uk/government/uploads/system/uploads/attachment_data/file/170705/HBN_00-09_infection_control.pdf.)

REFERENCES

1. Alberti C, Bouakline A, Ribaud P et al. (2001) Relationship between environmental fungal contamination and the incidence of invasive aspergilllosis in haematology patients. *J Hospl Infect* **48**(3):198–206.

2. Chabrol A, Cuzin L, Huguet F et al. (2010) Prophylaxis of invasive aspergillosis with voriconazole or caspofungin during building work in patients with acute leukaemia. *Haematol* **95**(6):996–1003.

3. De Pauw B, Walsh TJ, Donnelly JP et al. (2008) Revised definitions of invasive fungal disease from the European Organization for Research and Treatment of Cancer/Invasive Fungal Infections Cooperative Group and the National Institute of Allergy and Infectious Diseases Mycoses Study Group (EORTC/MSG) Consensus Group. *Clin Infect Dis* **46**(12):1813–1821.

4. Gayet-Ageron A, Iten A, van Delden C et al. (2015) In-hospital transfer is risk factor for invasive filamentous fungal infection among hospitalized patients with hematological malignancies: a matched case-control study. *Infect Control Hosp Epidemiol* **36**(3):320–328.

5. Oren I, Haddad N, Finkelstein R et al. (2001) Invasive pulmonary aspergillosis in neutropenic patients during hospital construction: before and after chemoprophylaxis and institution of HEPA filters. *Am J Hematol* **66**(4):257–262.

6. Napoli C, Marcotrigiano V & Montagna M (2012) Air sampling procedures to evaluate microbial contamination: a comparison between active and passive methods in operating theatres. *BMC Public Health* **12**:594.

7. UK Standards for Microbiology Investigations: Investigation of Bronchoalveolar Lavage, Sputum and Associated Specimens. *Public Health England*, Issue 2.5, Issued 02.06.2014.

8. Denning D, Pleuvry A & Cole D (2011) Global burden of chronic pulmonary aspergillosis as a sequel to pulmonary tuberculosis. *Bull World Health Organ* **89**:864–872.

9. Patterson T (2014) Aspergillus Species. In Mandell's Principles and Practice of Infectious Disease 8ed, pp. 2895–2910, Saunders.

ANSWERS

MCQ	Feedback

1. Regarding invasive aspergillosis and hematopoietic stem cell transplant (HSCT) recipients, which of the following are true?

a. NICE guidance recommends positive-pressure, HEPA-filtered side rooms for all patients undergoing HSCT and those who have received a HSCT.

    a. False. Many regimens have a significant outpatient component. NICE recommends that such patients are managed in specialist units with appropriate measures to avoid air contamination.

b. Recipients of allogeneic stem cell transplants are deemed lower risk for hospital-acquired and other infections than autologous HSCT patients.

    b. False

c. Refractory hematological malignancy is associated with poorer response to antifungal treatment of invasive aspergillosis.

    c. True

d. Invasive pulmonary aspergillosis is most likely to occur immediately after HSCT.

    d. False. More and more cases are seen many months after transplantation and relate to the use of corticosteroids for graft-versus-host disease.

e. Liposomal amphotericin B treatment is associated with renal dysfunction and deranged electrolytes.

    e. True

2. Regarding invasive pulmonary aspergillosis, which of the following are true?

a. Most cases are caused by *Aspergillus flavus*.

    a. False. *A. fumigatus* accounts for most cases.

b. Voriconazole is the first choice antifungal therapy.

    b. True. Other alternatives include liposomal amphotericin. Echinocandins have been used with reported success. Refractory disease may warrant dual antifungal therapy.

c. Diagnosis can only be made on the basis of positive culture.

    c. False. Obtaining invasive samples may not be possible, as for instance in profoundly neutropenic patients requiring platelet support.

d. A positive culture of *Aspergillus* from bronchiolar lavage always indicates infection.

    d. False. Unless the patient has underlying lung disease or is immunosuppressed, it is likely to represent a contaminant. However, if fungal hyphae were seen on initial microscopy or an isolate is repeatedly isolated, the patient should be investigated further including screening for acquired and genetic immunosuppressive conditions.

e. It is more likely in a patient with HIV who takes antiretrovirals and whose CD4 count is 450 than in a patient taking 20 mg prednisolone each day as part of their rheumatoid arthritis treatment.

    e. False

3. True or false: *Aspergillus* species:

a. Are invariably sensitive to amphotericin

    a. False. *A. terreus* is inherently resistant.

b. Are primarily dermatophytes

    b. False. Dermatophytes are fungi, such as *Trichophyton* species, which cause skin, hair and nail infections and are not normally associated with invasive disease. *Aspergillus* species may cause skin infections, particularly *A. flavus*, but are associated with invasive and disseminated infection in the immunocompromised.

c. Isolated in clinical samples are always pathogenic

    c. False. They are widespread in the environment.

d. Should always be processed in a safety cabinet in the microbiology laboratory

    d. True

e. Are molds

    e. True

MCQ	Feedback
4. True or false: infection control measures to be taken in the event of a suspected cluster of *Aspergillus fumigatus* infections in a hematology unit must include:	
a. Identification of patients at risk	a. True
b. Screening of staff members	b. False. *Aspergillus* species are ubiquitous, and there is likely to be an environmental reservoir.
c. Assessment of existing infection control measures and evidence of any breaches, including whether isolation measures are available and used	c. True
d. Contacting the local health protection unit	d. True
e. Writing to all recent inpatients' families advising them of a potential outbreak	e. False. *Aspergillus* species are ubiquitous, there is likely to be an environmental reservoir.
5. A new renal dialysis unit is scheduled to be built on the site of a former NHS staff recreation club, which will have to be demolished first. The site is next to the block containing the hospital's hematology inpatient and outpatient units. True or false: The infection control team should:	
a. Oppose the development as it is highly likely to put immunosuppressed patients at increased risk of *Aspergillus*-related conditions.	a. False. The development is likely to go ahead in any case. They should focus on working with the hospital architect, management, and contractors to put proper infection control measures in place and ensure the layout of the unit meets operational and infection control standards.
b. Ensure that ventilation ducts, windows, and water pipes are isolated from areas where demolition is taking place.	b. True.
c. Insist that contractors use external waste disposal chutes when gutting the upper floors of the building to be demolished.	c. False. Disposal chutes disperse fungal spores and other contaminants over a wide area. Waste should be removed in sealed bags along a designated route that is screened off as far as possible from the hospital's clinical areas.
d. Draw up an assessment of all at-risk groups on the hospital site.	d. True.
e. Discuss and plan measures to reduce exposure to *Aspergillus* and other potential pathogens with the contractor and clinical teams affected.	e. True.

BURKHOLDERIA CEPACIA COMPLEX (BCC) IN CYSTIC FIBROSIS

CASE 3

Elizabeth Sheridan,[1] Dervla Kenna,[2] Jane Turton,[2] Emma Lake[3]

[1]Poole Hospital NHS Trust, UK.
[2]Antimicrobial Resistance and Healthcare Associated Infections (AMRHAI) Reference Unit, Public Health England, UK.
[3]Cystic Fibrosis Trust, UK.

A PATIENT'S PERSPECTIVE

Cross infection is a big issue for me and for everyone with cystic fibrosis (CF). With the discovery of new superbugs such as *Mycobacterium abscessus*, I worry that the issue is only going to get bigger. I worry about how the cystic fibrosis specialist centres and clinics will cope with having to implement stricter guidelines and the impact this will have on our community. Everybody affected by cystic fibrosis has their own feelings about it. I know that technology has helped; people with cystic fibrosis can contact each other through text, email, social media, and forums, but it isn't the same is it? Lots of my friends have cystic fibrosis and it's not nice being told you can't see your friends. I understand the reasons for cross-infection guidelines, but that doesn't mean I have to like them. Sometimes I just want to stamp my feet and shout: "it's not fair."

Fair it might not be, but thanks to those pesky bugs, it's something I would rather not put to the test. Even if I am tempted to mix with others (and thanks to peer pressure I have been), I would hate to think I had caught something, or worse given something to someone else with cystic fibrosis. After all, even if you shared the same bugs as somebody else you can only rely on your last sputum sample; what if you have picked something new up or developed a new strain in the meantime? It is even more important to be careful if you are in hospital. It is likely that people are in hospital for IVs because they are growing new bugs, are colonised, or have suffered a flare up, and if these bugs have resulted in hospitalization they are definitely worth avoiding.

It can be difficult because it can make you look unsociable or like you are "selling out" (as teens tend to say). I really think that peer pressure from fellow patients, or rather, inmates, does make it harder, initially. My first personal experience of this was when segregation was first introduced on the ward. It was very difficult because there were some individuals who weren't so keen on the idea and would just not stay in their rooms. They would stand in the doorway or come into my room and chat and it was awkward to say, "go away I don't want your germs."

I don't really leave my hospital room unless I am going out with family and friends, and if I do come across someone else with cystic fibrosis during my travels I make sure that I speak to them at a good distance. I also just rely on good old common sense: if someone nearby is coughing, cystic fibrosis or no cystic fibrosis, I move away from them (although obviously discreetly; not running away at 50 mph with a look of horror!!).

Obviously this really sucks, especially if you are bored and there is no one to speak to. If someone else I know with cystic fibrosis is on the ward, I normally end up spending a lot of time on the phone talking to them or texting them.

I know it is hard. It is really strange that I work with eight other people with cystic fibrosis and yet have only ever waved and shouted at one of them from across a large room. But we still talk all the time and we all get on really well. It almost doesn't matter that we can't go down the pub together for a pint after work.

Over the last year or so I have found myself becoming more paranoid about the risks of picking up an infection, as a person with CF, I hear about new bugs or strains of bugs that we just don't know enough about how they are spread. I will no longer shake a person's hand (CF or not) and have developed somewhat of a phobia around air conditioning and the spores it can spread. It is becoming harder to not let this anxiety stop me from living life to the full.

"Maybe we should get the CF teams to invest in some NASA spacesuits!"

Emma Lake, Cystic Fibrosis Trust

The precautions described by Emma have only been put into place after it became evident that certain strains of the *Burkholderia cepacia* complex (Bcc) not only cause life-threatening lung damage in CF, but may be transmissible between patients. Although taking these precautions can have a number of adverse social, psychological, and lifestyle consequences for people with CF, they are essential in preventing acquisition of virulent strains of these organisms.

PATHOLOGY OF CYSTIC FIBROSIS

CF is a multisystem, inherited genetic disorder caused by an abnormality in sodium and chloride ion transport across cell membranes. In the lung, this abnormality results in persistent mucus hypersecretion and reduced mucociliary transport. Patients are predisposed to persistent and chronic lung infection, leading to bronchiectasis and a progressive decline in lung function as well as repeated acute episodes of infective exacerbation.

BIOLOGY OF THE ORGANISM

The Bcc consists of Gram-negative, oxidase-positive bacilli and currently comprises 20 closely related species, formerly known as genomovars (Table 3.1). First described as a cause of onion rot by Burkholder in 1950, these organisms are typically environmental in origin and have been isolated from soil (most usually associated with plant roots) and from water sources such as rivers. Some Bcc strains and species form symbiotic relationships with their plant hosts and have been used in the US as biocontrol and bioremediation agents, because they are able to produce antifungal agents and have the capacity to break down pollutants. However commercial use of these agents was reviewed in the US in the late 1990s, when it was discovered that some biocontrol strains

Table 3.1 **Species of the *Burkholderia cepacia* complex**

Species
Burkholderia cepacia (genomovar I)
Burkholderia multivorans (genomovar II)
*Burkholderia cenocepacia IIIA, IIIB, IIIC, and IIID (genomovars IIIA–IIID)
Burkholderia stabilis (genomovar IV)
Burkholderia vietnamiensis (genomovar V)
Burkholderia dolosa (genomovar VI)
Burkholderia ambifaria (genomovar VII)
Burkholderia anthina (genomovar VII)
Burkholderia pyrrocinia (genomovar IX)
Burkholderia ubonensis
Burkholderia latens
Burkholderia diffusa
Burkholderia arboris
Burkholderia seminalis
Burkholderia metallica
Burkholderia contaminans
Burkholderia lata
Burkholderia pseudomultivorans
Burkholderia territorii
Burkholderia stagnalis

* *Burkholderia cenocepacia* is represented by four distinct subgroups, of which *B. cenocepacia* IIIA and IIIB are the most clinically significant.

were indistinguishable from their clinical counterparts and therefore posed a clinical risk to immune-compromised patients such as those with CF.

EPIDEMIOLOGY OF BCC

Approximately 3% of CF patients in the United Kingdom (UK) are estimated to be infected with members of the Bcc, with prevalence increasing with age. It is thought that most Bcc infections in CF patients are acquired independently from the environment, although siblings typically share the same strain, and certain strains have been associated with a more widespread cross-infection risk. Transient infections with these organisms occur rarely, with the majority of patients becoming chronically infected. Usually infection is with just one strain, although on rare occasions co-infection with another species may occur.

Although all of the Bcc species have been isolated from CF patients, in the UK, Europe, and the US, *B. cenocepacia* and *B. multivorans* (see Table 3.1) are the most commonly isolated species, these being found in approximately 70% of infected patients. In recent years, *B. multivorans* has become the most prevalent species in the UK, with most of the strains isolated having unique pulsed-field gel electrophoresis profiles consistent with independent acquisition, rather than patient-to-patient transmission.

The Bcc came to prominence in the 1980s with the emergence of the transmissible *B. ceno-cepacia* IIIA strain ET12, which was isolated from CF patients in both North America (Canada, primarily) and the UK. ET12 is associated with enhanced transmissibility and with increased morbidity and mortality. The increased virulence and transmissibility of ET12 results from the presence of a surface adhesin known as the cable pilus and also to a genomic island that carries additional virulence genes. Cable pili are associated with increased binding to respiratory epithelial cells, cytotoxicity, and the mediation of cell death, as well as with the enhanced induction of inflammatory cytokines, when compared with nonpiliated control strains.

Other notable transmissible *B. cenocepacia* strains include the PHDC and Midwest clones from the US. Other Bcc species have occasionally been associated with transmissibility, such as certain strains of *B. dolosa* and *B. multivorans*. Segregation policies have now successfully reduced the prevalence of strains such as ET12.

Members of the Bcc are also occasionally associated with healthcare-associated oppor-tunistic infections of the bloodstream (rather than pulmonary infections) in people with-out CF. For example, they are related to IV (intravenous) lines and contaminated fluids.

Transmission characteristics

Spread is by droplets within a 1 m radius (coughing), direct contact, contaminated surfaces, and equipment. Aerosol transmission is not documented. Patients can remain infectious indefinitely. In the case of transient carriage, three negative specimens over a year are needed to consider the patient free from Bcc. The organism survives well in a wet environment and has been known to survive for several years in distilled water.

DISEASE

The host response to infections with the Bcc is variable, ranging from asymptomatic carriage to severe pneumonia. Approximately 20% of patients will develop cepacia syndrome: fever, leucocytosis, bacteraemia, and a rapid deterioration in lung function leading to death in most cases. Cepacia syndrome is most frequently associated with ET12 but has also occurred with strains of *B. cepacia* and *B. multivorans*.

CLINICAL MANAGEMENT

Infections with members of the Bcc tend to become chronic, and there is little evidence for the efficacy of any specific antibiotic regimen, which is in part a consequence of the lack of randomized clinical trials for Bcc-infected individuals. Members of the Bcc are intrinsically resistant to aminoglycosides, with many strains also exhibiting intrinsic resistance to beta-lactams (with the exception of meropenem). There is variable resist-ance to the quinolones, with inducible resistance a problem. In general infective exacer-bation is managed by a combination of high dose IV and nebulized antibiotics depending on sensitivity patterns. Other adjunctive therapies are also frequently used, for example, intensification of airway clearance therapies and macrolide antibiotics for their immuno-modulatory properties.

Bcc and lung transplantation

Lung transplant is a potential treatment for some patients with CF lung disease but is associated with a poor outcome in patients infected with Bcc. Internationally, Bcc infection, especially with *B. cenocepacia*, is well recognised as a contraindication to lung transplantation. Therefore, preventing acquisition is particularly important to keep this life-saving option open for patients.

INFECTION CONTROL PRECAUTIONS TO PREVENT COLONIZATION WITH VIRULENT RESPIRATORY ORGANISMS IN CF

In addition to the Bcc, a wide range of organisms can cause problems in CF through respiratory colonization and infection. Around 80% of adult CF patients are infected with *Pseudomonas aeruginosa*, which damages lung function and decreases survival. Epidemic strains with increased transmissibility, virulence, and antimicrobial resistance are a particular source of concern. An example is the Liverpool epidemic strain (LES), which is named after the CF centre, where it was first observed in 1996. A recent study estimated that LES was found in 11% of >1000 CF patients whose isolates were received by the national reference laboratory between 2010 and 2012. Other organisms associated with cross infection among CF patients include *Staphylococcus aureus*, *Stenotrophomonas maltophilia*, *Achromobacter xylosoxidans*, *Pandoraea* spp., nontuberculous mycobacteria, and respiratory viruses, notably respiratory syncytial virus (RSV) and influenza. National and international surveillance of these organisms is important to provide an early warning of potential outbreaks. The same infection control principles for inpatients and outpatients as for Bcc apply for these pathogens, although not all need strict segregation. Surveillance culture and molecular typing are performed to look for transmissible *P. aeruginosa* strains and evidence of transmission. Patients with these strains may be seen in a separate clinic like those with Bcc. Also consider separate clinics for *P. aeruginosa*-colonized and noncolonized patients. Early, aggressive treatment of *P. aeruginosa* colonization improves outcome by delaying or preventing the onset of chronic infection.

As far as exposure to environmental pathogens and risk of colonization is concerned, people with CF are encouraged to carry on with normal travel and outdoor activities as much as possible. Potential additional risks include travel to rural areas of South East Asia (*B. pseudomallei*), swimming in poorly maintained pools and spas (*Pseudomonas*), and exposure to animal manure and compost (fungi, in particular *Aspergillus*, can be acquired from the environment). People with CF are therefore encouraged to seek advice from their respiratory doctor before travelling or taking up new leisure activities. There is, however, little evidence base for specific risks of acquisition so decisions are not always easy and must always weigh up risks against benefits.

Screening

Regular respiratory sampling for all patients, at each hospital visit as well as during exacerbations, is performed for a range of pathogens. The results are used to guide antibiotic treatment of infective exacerbations and to identify infection control issues. The specimen is generally a sputum sample, but a cough plate is an alternative if the person is unable to expectorate. In addition to more general media, the samples are cultured on Bcc selective media as the organism is not always easy to see on a mixed plate and could be missed. Bcc is identified using phenotypic and molecular methods. Suspected Bcc isolates should be sent to a reference laboratory for confirmation, given the importance of correct identification for patient prognosis and for infection control. Molecular typing of all strains locally or at the reference laboratory is recommended to inform the local and national picture of which Bcc strains are prevalent in the community. Multiplex polymerase chain reaction (PCR) for respiratory pathogens in sputum including Bcc is useful in screening.

Healthcare precautions

Precautions should be established by setting up separate outpatient clinics for those colonized with Bcc, particularly *B. cenocepacia*, to minimize risk of transmission to other patients via direct contact, or airborne and environmental contamination. Additional segregated clinics, or at least separate time slots within the same clinic, can also be arranged for those with known transmissible strains of any organism, for example, *Pseudomonas*, methicillin-resistant *Staphylococcus aureus* (MRSA), *Mycobacterium abscessus*. It is particularly important to prevent the patients from waiting together in the same area of the clinic and to ensure good environmental cleaning between patients. If admitted to hospital, colonized patients should be isolated in a single room with its own shower, and standard contact, droplet, and respiratory precautions should be implemented. Some centres make patients wear a surgical mask though this approach is unproven.

Chest physiotherapy should be performed away from other CF patients. In outpatient clinics it is recommended that the clinical team move between the patients who remain in designated rooms, thereby reducing the risks of cross infection through direct (patient to patient) and indirect (contamination of the environment) routes.

Respiratory equipment (especially nebulizers) and the environment

Colonization with Gram-negative bacteria through contaminated equipment can be prevented by using antibacterial filters, sterile water, decontamination, and drying (including in the home setting), and not sharing equipment. Care should be taken to protect patients from reservoirs of infection in the hospital environment, such as showers and drains, which may harbour these organisms, through strict hospital cleaning protocols.

Community precautions

Guidelines are issued to avoid accidental transmission of respiratory pathogens. These should be supplemented with patient education. The importance of good respiratory and hand hygiene is made clear to those with cystic fibrosis by their medical and nursing teams as well as support organizations. All those with CF are advised not to share vehicles, prepare food, or have any physical contact with others with CF; if in the same room, they are advised that they should sit at least 1 m apart. Not more than one person with CF is to attend organized indoor events (for example, campaigning for cystic fibrosis charities). Gatherings of people with CF, such as summer camps and support groups, popular in the past, are no longer held, as it has been agreed that the risk of Bcc transmission outweighs the benefits. Specifically for Bcc, colonized individuals must not knowingly attend any events where other people with CF may be present. In the US, this applies even to out-door events. (The US guidelines are stricter than those in most other countries.) In prac-tice, however, it is not possible to prevent people socializing, so providing patients with clear advice as to how to minimise risk is key.

QUESTIONS

1. Which of the following are important to prevent the spread of *Burkholderia cepacia* complex in cystic fibrosis patients?
   a. Segregating patients in separate clinics
   b. Cleaning nebulizer equipment thoroughly after a colonized patient has used it
   c. Isolating hospitalized patients in a single room
   d. Use of negative pressure isolation for colonized patients
   e. Ensuring colonized patients do not come into contact with noncolonized patients at social events

2. Which of the following organisms have associated epidemic strains that may cause problems for CF patients?
   a. *Pseudomonas aeruginosa*
   b. *Haemophilus influenzae*
   c. *Mycobacterium abscessus*
   d. *Burkholderia cenocepacia*
   e. *Aspergillus fumigatus*

3. Which of the following are benefits of surveillance screening for Bcc?
   a. Potential epidemic strains are identified
   b. Infection control takes place at the local level
   c. International spread of virulent strains is tracked
   d. The need to send isolates to the reference laboratory is reduced
   e. People with CF information are able look after their own health

4. Which of the following are virulence factors in the transmissible strain *B. cenocepacia* ET12?
   a. Cable pilus
   b. Exotoxin A
   c. Genomic islands carrying virulence factors
   d. Clumping factor A
   e. Alginate production

5. A CF adult with a good lung function asks for your advice on several sporting and adventurous activities she would like to take part in. Which do you think may benefit her health and which should she avoid?
   a. Her friend has a horse and she has heard that it might be a good physio-boosting exercise. Do you agree?
   b. She is planning to go travelling with friends to Thailand and Vietnam and has heard that *Burkholderia* infections associated with monsoon weather conditions may pose a problem in this part of the world. She is wondering whether this is the case?
   c. She has recently joined a gym and has been advised not to use the hot tub because there is a risk of infection. She asks if this is correct.
   d. What about using gym equipment? She has heard that this might be covered with contaminants from other users and has been told not to use it. Should she take this advice?
   e. She has also heard that scuba diving may be one of the activities on her trip abroad and has heard this is safe if you have good lung function. Do you agree?

GUIDELINES

The Cystic Fibrosis Trust (Cysticfibrosis.org.uk) publishes useful guidelines such as those listed below.

1. The *Burkholderia cepacia* complex (Sept 2004) Suggestions for Prevention and Infection Control, 2nd ed. The UK Cystic Fibrosis Trust Infection Control Group.

2. Antibiotic Treatment for Cystic Fibrosis, 3rd ed (May 2009) Report of the UK Cystic Fibrosis Trust Antibiotic Working Group.

3. Laboratory standards for processing microbiological samples from people with cystic fibrosis (Sept 2010) Report of the UK Cystic Fibrosis Trust Microbiology Laboratory Standards Working Group. Cystic Fibrosis Trust.

4. CF Trust, *CF Today* (summer 2006 ed) (www.cysticfibrosis.org.uk/media/82745/CF_Today_Summer_06.pdf).

5. CF Trust, Melioidosis and tropical travel advice: (https://www.cysticfibrosis.org.uk/~/media/documents/life-with-cf/publications/factsheets/factsheet-melioidosis.ashx).

REFERENCES

1. Horsley A & Jones AM (2012). Antibiotic treatment for *Burkholderia cepacia* complex in people with cystic fibrosis experiencing a pulmonary exacerbation. **17**,10:CD009529. (www.ncbi.nlm.nih.gov/pubmed/23076960).

2. Turton JF, Kaufmann ME, Mustafa N et al. (2003) Molecular comparison of isolates of *Burkholderia multivorans* from patients with cystic fibrosis in the United Kingdom **41**(12):5750–5754 (www.ncbi.nlm.nih.gov/pmc/articles/PMC308997/).

3. Govan JR, Brown AR & Jones AM (2007) Evolving epidemiology of *Pseudomonas aeruginosa* and the *Burkholderia cepacia* complex in cystic fibrosis lung infection. *Future Microbiol* **2**:153–164. (www.futuremedicine.com/doi/pdf/10.2217/17460913.2.2.153).

4. LiPuma JJ (2010). The changing microbial epidemiology in cystic fibrosis. *Clin Microbiol Rev* **2**:299–323 (www.ncbi.nlm.nih.gov/pubmed/20375354).

5. Cheung KJ Jr, Li G, Urban TA et al. (2007). Pilus-mediated epithelial cell death in response to infection with *Burkholderia cenocepacia*. *Microbes Infect* **9**(7):829–837 (www.ncbi.nlm.nih.gov/pubmed/17537663).

6. Martin K, Baddal B, Mustafa N et al. (2013) Clusters of genetically similar isolates of *Pseudomonas aeruginosa* from multiple hospitals in the UK. *J Med Microbiol.* **62**(Pt 7):988–1000 (www.ncbi.nlm.nih.gov/pubmed/23558134).

7. Gibson RL, Burns JL & Ramsey BW (2003) Pathophysiology and management of pulmonary infections in cystic fibrosis. *Am J Respir Crit Care Med* **168**:918–951 (www.ncbi.nlm.nih.gov/pubmed/14555458).

8. Bryant JM, Grogono DM, Greaves D et al. (2013) Whole-genome sequencing to identify transmission of *Mycobacterium abscessus* between patients with cystic fibrosis: a retrospective cohort study. *Lancet* **381**(9877):1551–1560 (www.ncbi.nlm.nih.gov/pubmed/23541540).

9. Coeyne T & LiPuma JJ (2002) Multilocus restriction typing: a novel tool for studying global epidemiology of *Burkholderia cepacia* complex infection in cystic fibrosis. *J Infect Dis* **15**; 185(10):1454–1462 (www.ncbi.nlm.nih.gov/pubmed/11992281).

10. Coenye T, Spilker T, Van Schoor A et al. (2004) Recovery of *Burkholderia cenocepacia* strain PHDC from cystic fibrosis patients in Europe. *Thorax* **59**(11):952–954 (www.ncbi.nlm.nih.gov/pubmed/15516470).

ANSWERS

MCQ	Feedback
1. Which of the following are important to prevent the spread of *Burkholderia cepacia* complex in cystic fibrosis patients?	
a. Segregating patients in separate clinics	a. Important: Spread at clinics has been documented and is likely to occur through droplets when a patient coughs as well as through contamination of the clinic environment. Patients often wait together in small spaces at an outpatient clinic and are likely to socialize, increasing the risk. Further reading CF Trust infection control guidelines (Guideline 1).
b. Cleaning nebulizer equipment thoroughly after a colonized patient has used it	b. Less important: This would reduce the risk but not eliminate it. Patients with Bcc should not share equipment with others, even if decontamination is attempted. Further reading CF Trust infection control guidelines (Guideline 1).
c. Isolating hospitalized patients in a single room	c. Important: This must have en suite facilities to prevent contamination spread by showers, sinks, etc. Respiratory and droplet precautions are needed. Further reading CF Trust infection control guidelines (Guideline 1).
d. Use of negative pressure isolation for colonized patients	d. Less important: Aerosol spread has not been documented, so this is not necessary as droplets will be contained in a side room with standard ventilation.
e. Ensuring colonized patients do not come into contact with noncolonized patients at social events	e. Important: People with cystic fibrosis should be aware of their colonization status. Any events (for example, fundraising, awareness raising, and support) should be organized in such a way as to minimize any contact between people with cystic fibrosis. Further reading CF Trust infection control guidelines (Guideline 1).
2. Which of the following organisms have associated epidemic strains that may cause problems for CF patients?	
a. *Pseudomonas aeruginosa*	a. Yes: A number of transmissible and virulent strains have been identified and active surveillance is essential to detect newly emergent ones before significant transmission has occurred. Further reading Martin et al., 2013 (Reference 6).
b. *Haemophilus influenzae*	b. No: This is a frequent colonizer of the lung early in the course of the disease in childhood with a range of nontypeable community strains being most commonly isolated. Patient-to-patient transmission is theoretically possible.
c. *Mycobacterium abscessus*	c. Yes: Most infections are independently acquired, but outbreak clusters associated with highly related strains have also been documented in the UK and worldwide. Further reading Bryant et al., 2013 (Reference 8).
d. *Burkholderia cenocepacia*	d. Yes: *B. cenocepacia* IIIA strain ET12 is particularly associated with poor outcome and transmissibility for CF patients, though other strains have also been implicated in outbreaks.
e. *Aspergillus fumigatus*	e. No: Infections with a variety of fungi including *Aspergillus* species occur, most likely from environmental exposure to spores.

MCQ	Feedback
3. Which of the following are benefits of surveillance screening for Bcc?	
a. Potential epidemic strains are identified	a. Benefit: The use of molecular methods such as multilocus sequence typing (MLST) and pulsed-field gel electrophoresis allow discrimination between strains of a species and also help to identify outbreak strains.
	Further reading Coeyne and LiPuma, 2002 (Reference 9).
b. Infection control takes place at the local level	b. Benefit: Surveillance within clinics helps clinical staff to monitor the existence and spread of transmissible strains, therefore permitting the introduction of additional infection-control measures.
c. International spread of virulent strains is tracked	c. Benefit: Portable sequence-based techniques such as MLST enable comparison of strain types across continents and provide useful information about successful clones.
	Further reading Coeyne et al., 2004 (Reference 10).
d. The need to send isolates to the reference laboratory is reduced	d Benefit: The reference laboratory can provide crucial information about the national prevalence of strains and species and is reliant on hospitals submitting isolates for this to be possible.
e. People with CF information are able look after their own health	e. Benefit: Accurate national and local surveillance provides both CF patients and their carers with a clearer picture of up-to-date health issues.
4. Which of the following are virulence factors in the transmissible strain *B. cenocepacia* ET12?	
a. Cable pilus	a. Yes: The presence of the cable pilus allows better adhesion to epithelial cells, with associated cytotoxicity and apoptosis and also with the induction of inflammatory cytokines.
b. Exotoxin A	b. No: Exotoxin A production is associated with virulence in *P. aeruginosa*.
c. Genomic islands carrying virulence factors	c. Yes: Additional blocks of DNA on the genome, often spread between bacteria by horizontal transfer, may carry virulence-associated genes. This is known to be to the case for ET12.
d. Clumping factor A	d. No: This is produced by *Staphylococcus aureus* and helps in the adhesion to platelets.
e. Alginate production	e. No: This is classicially associated with growth of *Pseudomonas aeruginosa* in the CF lung, where the organism grows as a thick, mucoid biofilm. However, some strains of the *B. cepacia* complex have been shown to produce an exopolysaccharide under certain conditions. In both cases they protect the organism from the host immune system.
	Further reading Cheung et al., 2007 (Reference 5).
5. A CF adult with a good lung function asks for your advice on several sporting and adventurous activities she would like to take part in. Which do you think may benefit her health and which should she avoid?	
a. Her friend has a horse and she has heard that it might be a good exercise. Do you agree?	a. Yes. The CF Trust does not feel that horseback riding poses any additional problems with regard to acquisition of infections (and it may be good physio), but mucking out stables is to be avoided due to an increased risk of *Aspergillus*.
	Further reading CF Trust, CF Today, summer 2006 ed (Guideline 4).
b. She is planning to go traveling with friends to Thailand and Vietnam and has heard that *Burkholderia* infections associated with monsoon weather conditions may pose a problem in this part of the world. She is wondering whether if this is the case?	b. Yes: *B. pseudomallei* infections have been known to affect CF patients travelling to Southeast Asia and northern Australia, among other places. The risk of acquiring infection is thought to be greatly increased during the monsoon season. Camping during the rainy season should therefore be avoided, as should outdoor activities in rural areas.
	Further reading CF Trusts, Melioidosis and tropical travel advice (Guideline 5).

MCQ	Feedback
c. She has recently joined a gym and has been advised not to use the hot tub because there is a risk of infection. She asks if this is correct.	c. Yes: Though swimming in a well-maintained, chlorinated pool is likely to be safe, and beneficial, hot tubs can be associated with a risk of *P. aeruginosa* infection if not correctly maintained. Further reading CF Trust, CF Today, summer 2006 ed (Guideline 4).
d. What about using gym equipment? She has heard that this might be covered with contaminants from other users and has been told not to use it. Should she take this advice?	d. No: There is deemed to be a slight risk of acquiring infections in this way, but the benefits of exercise are likely to outweigh the risk.
e. She has also heard that scuba diving may be one of the activities on her trip abroad and has heard this is safe if you have good lung function. Do you agree?	e. No: It would be very difficult to assess the risks (for example, maintenance of equipment may be inadequate), and there are no current guidelines for CF patients.

CAMPYLOBACTER JEJUNI INFECTION

John Holton[1,2]

[1]Department of Natural Sciences, School of Science & Technology, University of Middlesex.
[2]National Mycobacterial Reference Service - South, Public Health England.

A patient presented to Accident and Emergency (Emergency Room) with profuse bloody diarrhoea. The diarrhoea occurred more than 10 times a day and had started suddenly, although the patient admitted he had been "under the weather" for a few days. An initial diagnosis of inflammatory bowel disease was made and he was admitted to an open ward. Initial laboratory investigations included standard haematology and biochemistry and a specimen of diarrhoea was sent to the microbiology laboratory for culture of faecal gastro-intestinal bacterial pathogens and detection of relevant viruses by polymerase chain reaction (PCR).

Three days later the ward received a report from the laboratory that *Campylobacter jejuni* had been isolated from the faeces. The patient was then isolated and because of the severity of the attack was started on treatment with erythromycin. Two days later two additional patients on the ward developed severe diarrhoea, and they also were isolated and specimens sent to the laboratory. There was no evidence of direct contact between the index (initial) case and the new cases in that ward.

The following day a patient in a separate ward developed diarrhoea and subsequently a patient in a third ward also developed diarrhoea. In all cases, *C. jejuni* was isolated from the faeces. All cases excluding the initial case had been admitted to the ward more than 10 days prior to the onset of disease.

All cases excluding the index case had been admitted to the ward for ≥10 days.

INVESTIGATION OF THE CASE

When other patients on the same ward developed diarrhoea, the hospital held an outbreak meeting attended by the infection-control nurse, the hospital epidemiologist, the clinical microbiologist, the consultant in charge of the patient, the ward sister (head nurse), and a member of the administrative staff. The group determined that the index patient should have been placed in isolation with contact precautions, hand hygiene using soap and water for staff should have been emphasized, and bleach or disinfectants containing hydrogen peroxide for environmental decontamination should have been used on admission because the patient had an as yet undiagnosed cause for the diarrhoea.

The initial epidemiology of an index case who has not been isolated and two further cases of an uncommon infection acquired in the same ward is consistent with transmission from patient to patient by the feco–oral route. In this case, transmission could occur by the contamination of objects in the ward (fomites) or the hands of healthcare workers, or occur directly if the patients were in contact, for example, in a day room. The group mapped the relative bed position of the new cases to the index case and sought to find any evidence of direct or indirect contact, such as care by the same staff members, shared equipment, or shared bathrooms. They identified no likely points of transmission, however.

When two new cases appeared in two separate wards, the investigation widened to include the entire hospital. All patients with diarrhoea in the hospital should have microbiological investigation, but this often does not include *Campylobacter* because it is more frequently a community-acquired pathogen. In this case, all patients with diarrhoea in the hospital were investigated for *Campylobacter*. The outbreak team was expanded to include representatives from all affected wards. A search for an epidemiological link between all the cases was sought using a standardized questionnaire filled out by the medical and nursing teams responsible for each of the affected patients and for three patients admitted on the same date, who were the same age and on the same ward but who did not report diarrhoea and were not diagnosed with a *Campylobacter* infection. *C. jejuni* outbreaks are commonly linked to poultry, unpasteurized dairy products, and drinking water. Each patient was therefore questioned about visits to the cafeteria, foods and liquids consumed, hospital meals, and common visits to other departments, such as radiology.

The exposure that was more common among the patients with *Campylobacter* infections as compared to those who were not infected with *Campylobacter* was consumption of cannelloni provided by the kitchen, except for the index case.

Specimens of the cannelloni retained in the kitchen were retrieved and *C. jejuni* was detected by culture. All specimens of *C. jejuni* isolated from the patients and the cannelloni were sent to the reference laboratory for typing.

At the reference laboratory, identification of the isolates was by species specific PCR and typing by multilocus sequence typing (MLST). The most common MLST type in the UK is ST-21. In this case, all the specimens from the patients, including the index case, were ST-45. This finding suggested that the index case was part of the same outbreak, although the patient did not eat the cannelloni in the hospital. Further epidemiological investigations into past behaviour identified a supplier of chicken liver paté with a positive culture of ST-45 *Campylobacter* from chicken liver collected from that supplier's factory as the source of an extended regional outbreak. Frequently chicken liver paté is blended with other foods, and in this case it was included in the cannelloni.

CLINICAL MANAGEMENT

Symptomatic patients may become dehydrated and should be managed by fluid replacement. The infection is usually self-limiting, lasting about seven days, and does not require antibiotics. Antimicrobial agents decrease the duration of symptoms and bacterial shedding if administered early in the course of disease. Antibiotic therapy (azithromycin or ciprofloxacin) may be required if the condition is severe, with frequent episodes of bloody diarrhoea, high temperature, or failure to improve after a week of symptoms, or if the patient is deteriorating or is immunocompromised. Rates of antibiotic resistance have risen greatly in the past decade and high rates of resistance are now seen in many countries. Resistance to erythromycin is about 5% in many countries, whereas resistance to quinolones is about 20% and in some countries as high as 60%. Alternative agents are tetracycline or clindamycin.

PREVENTION OF FURTHER CASES

Isolation of known or suspected cases

A patient admitted with diarrhoea should be isolated immediately until an infectious cause is excluded. In this incident, following the initial breach in procedure, the standard admissions procedures were reinforced with reminders and education to the relevant staff. Enteric precautions should be applied (a single room, en-suite facilities, entry, and exit from the room require persons to decontaminate their hands). Although both alcohol gel and handwashing are options, handwashing with soap and water is preferred during undiagnosed enteric outbreaks in hospital as it is more active against *Clostridium difficile* and norovirus transmission. On entering the isolation room, staff should don gloves and a disposable gown or plastic apron. Direct contact with the patient or body fluids requires the use of gloves and a disposable gown or plastic apron. Finally bleach or disinfectants containing hydrogen peroxide for environmental decontamination should be used. Patients with acute diarrhoea should be routinely isolated until comprehensive microbiology and virology testing identify the infectious cause.

Control of infection in the food chain

Campylobacter is present in the guts of various animals reared and processed for food. Exposure to faecal matter is possible at all stages in the processing and preparation of meat, fruits, and vegetables. Therefore measures such as those listed in Table 4.1, are required to avoid contamination of products before consumption.

The World Health Organization (WHO) has established the Global Foodborne Infectious Network (GFN) for surveillance, control, and education related to enteric infections. On a national scale the US set up a population-based sentinel site surveillance system—the Foodborne Diseases Active Surveillance Network (FoodNet). FoodNet monitors gastrointestinal infections caused by a subset of foodborne pathogens including *Campylobacter*. National regulations in different countries have been introduced in relation to the whole food chain from farming to food production and preparation.

EPIDEMIOLOGY

Campylobacteriosis disease is a global zoonosis (a disease transmitted from animals to humans) because *Campylobacter* is part of the normal flora of many animals raised for food such as poultry, cattle, pigs, sheep, and shellfish. Campylobacteriosis can also be acquired from companion animals, including dogs or cats, and from petting zoos. It has even been postulated the tops of milk bottles could become contaminated with *Campylobacter* spp. by birds pecking their tops.

The most common route of transmission is believed to be food borne—from poorly cooked contaminated meats. Contamination can occur at any point during the food chain: at the slaughterhouse, during transport, or in the kitchen. Cross-contamination can occur in the kitchen or in places where uncooked food is stored. It also can occur if shared utensils are used in food preparation and in handling cooked food. For example, cross-contamination can occur by direct contact on the same work surface, such as chopping boards, or by indirect contact through use of contaminated equipment.

Table 4.1. **Control of Infections in the Food Chain**

Stage in the food-handling process	Method of infection control
Slaughterhouse	The design and construction of a slaughterhouse should permit cleaning and disinfection, minimize air-borne contamination, control pests, and allow adequate working space for hygienic processing.
	During production, processing, and distribution, the food should be protected against any contamination that would make it unfit for human consumption.
	Processed foodstuffs should have separate working and storage facilities from raw materials.
	The premises should have suitable temperature-controls, which should be monitored.
	There should be an adequate number of washbasins for the staff.
	Slaughterhouses approved for the slaughter of different animal species should have separate storage facilities, and operation on different animals should be spatially or temporally separated to prevent cross-contamination of separate slaughter lines.
	There should be separate facilities for different stages of the manufacturing process.
	There should be separate lockable facilities for storing meats not fit for consumption.
	The abattoir should undergo veterinary inspection.
Transport	Containers used for transporting foodstuffs should be clean and designed to permit adequate cleaning and disinfection.
	There should be separation between different foodstuffs if conveyed at the same time.
	Containers that carry different foodstuffs should be decontaminated between each conveyance.
Kitchen hygiene	Hands should be washed with soap and water before preparing food and after handling raw food.
	Raw and cooked food should be separated in the kitchen.
	Kitchen utensils should be kept clean.
Personal hygiene	Employees should wash their hands with soap and water after visiting the lavatory.
	Employees should wash their hands with soap and water after contact with pets or farm animals.
	Drinking water from nonpotable sources should be avoided.
Storing, cleaning, and cooking	Food should be stored below 8°C or frozen below 18°C.
	Fresh fruits and vegetables should be washed under running water.
	A food thermometer should be used to ensure that foods are cooked to a safe internal temperature (63°C for whole meats, 71°C for ground meats, and 73°C for poultry).
	Cooked food should be kept above 63°C.

Additional methods of transmission are ingestion of contaminated water or unpasteurized milk and dairy products. Feco–oral transmission can occur from an animal to a human or rarely person to person. The infectious dose of *C. jejuni* appears to be as low as 800 organisms, although in an immunocompromised person it can be much lower.

Most *Campylobacter* infections are thought to be sporadic, but large outbreaks from a common source can occur and have often been associated with contaminated drinking water. Disease occurs throughout the year, but the incidence varies. A peak in infections often occurs in the summer months, which may arise from the climate changes affecting *Campylobacter* prevalence in reservoirs and by affecting human behaviours, such as increased swimming in surface waters and barbecuing, where meats are often undercooked.

Between 2000 and 2012 in the UK, there were 698,122 reported cases of campylobacteriosis. Between 2001 and 2008 the prevalence decreased, but since 2008 has increased again. There is also a seasonality with higher numbers between May and July, although this peak seems to be occurring later in the year. In the UK in cattle, over 1000 samples collected over a one-year period demonstrated, using MLST, three clonal lineages (ST 61, ST 21, and ST 42) all at approximately 20–25% prevalence. About 26% of milk-related outbreaks of *Campylobacter* have been caused by faecal contamination. In 1997, an outbreak in a school involving eight children was reported and associated with damaged milk-bottle tops, presumably because birds pecked at them. However, as the delivery of milk bottles becomes rarer, this source of contamination will disappear, perhaps to be replaced by the increasing sale and consumption of raw milk. In Canada, a study of 40 dairy farms yielded 29 (72%) to be positive for *Campylobacter* spp. In the UK, a study of 56 cattle farms demonstrated 62% were positive for *Campylobacter*. In the US, studies have shown that 77% of retail chicken livers are contaminated with *Campylobacter* in both the internal tissues as well as on the surface.

In the US the incidence is about 13 per 100,000. Incidence in Canada is 40 per 100,000, in Australia 112 per 100,000, and in New Zealand 400 per 100,000. *Campylobacter* incidence is highest in males and young children (0–4 years) but infections occur in all age groups.

Typing methods include pulsed-field gel electrophoresis (PFGE), MLST, FlaA restriction fragment length polymorphism (RFLP). In the UK, identification is by reverse transcription polymerase chain reaction (RT-PCR) and typing by MLST, although the introduction of whole genome sequencing may improve outbreak detection and source identification, giving a more definitive understanding of the epidemiology.

BIOLOGY

The family Campylobacteraceae includes the genera *Campylobacter* and *Arcobacter*. These are taxonomically closely related to the family Helicobacteraceae, which contains many *Helicobacter* species, notably *H. pylori*, which was initially thought to be a *Campylobacter* species. Both families lay within the epsilonproteobacteria. There are over 20 species of *Campylobacter*, but *C. jejuni* causes most infections.

C. jejuni (NCTC11168) has been sequenced and demonstrates many homopolymeric hypervariable regions, mainly linked with surface structures, and unusual promoter regions. Because of the hypervariability, it is difficult to construct a definitive genome sequence. There are a predicted 1654 genes comprising 94.3% of the genome, which makes it currently the most gene-dense bacterium known. Unusually, the genome has very few insertion sequences as compared to other bacteria and only a very few organized operons.

Campylobacter spp. are spiral-shaped, motile, nonsporulating, Gram-negative rods. They are micro-aerophiles (5% oxygen : 10% carbon dioxide) and are oxidase and catalase positive. There is no simple gold standard protocol for culturing all *Campylobacter* spp., as growth conditions vary by species and appropriate culture conditions need to be applied. Most *C. jejuni* hydrolyze hippurate, which differentiates them from other *Campylobacter* spp. Culture may take up to three days and organisms are usually isolated on selective media, for example, Skirrow or charcoal-cefoperazone-deoxycholate agar (CCDA). Growth occurs between 37°C and 42°C and the organism is bile-resistant.

There are a large number of serotypes defined by the organism's somatic (O), flagellar (H), or capsular antigens. Many of the O antigens contain sialic acid and are structurally similar to host cell gangliosides. Two main typing schemes are used: the Penner scheme based on heat stable O antigens and the Lior scheme based on heat labile surface proteins.

Many surface proteins of *C. jejuni* are *N*- or *O*-glycosylated and are important for adhesion and colonization of the organism. *C. fetus* is covered by a protein S-layer which affects binding of complement (C3b) making the organism complement resistant whereas *C. jejuni* does not have an S-layer and is complement sensitive.

DISEASE

The most common *Campylobacter* species causing human infections are *C. jejuni* (80–90% of infections), *C. coli*, *C. fetus*, *C. lari*, *C. helveticus*, and *C. upsaliensis*. *C. jejuni* is a common cause of traveler's diarrhoea.

Some *Campylobacter* infections are asymptomatic. When symptoms occur, the incubation period is usually 3–10 days. Symptoms vary depending on the virulence of the strain, infectious dose, and susceptibility of the patient, as protective immunity to *Campylobacter* is well documented. Onset of symptoms is often abrupt but can begin with a prodrome of fever, headache, and myalgia followed by cramping abdominal pains and profuse watery diarrhoea, which may be blood-stained. In rare cases the condition may progress to toxic megacolon. Other species such as *C. fetus* either cause mild gastrointestinal disease or bacteraemia, commonly in immunocompromised individuals. In this case prolonged treatment is required, and the patient may develop an abscess or meningitis, cellulitis, or an endovascular device-related infection.

Late-onset postinfection complications such as reactive arthritis (2–5% of patients), irritable bowel syndrome (9–13% of patients), and Guillain-Barré syndrome (GBS, 0.1% of patients) are associated with antecedent *Campylobacter* infections. GBS is an ascending acute flaccid paralysis and *C. jejuni* is the most commonly identified antecedent infection of GBS preceding paralysis in about 30% of GBS patients. GBS symptom onset typically occurs 1–3 weeks after onset of *C. enteritis*. *C. jejuni* can also cause immunoproliferative small intestinal disease (IPSID; also called alpha-chain disease), a type of mucosa-associated lymphoid tissue (MALT) lymphoma. It is found mainly in the Middle East or Africa. It presents with chronic diarrhoea and wasting.

PATHOLOGY

C. jejuni invades cells of the gastrointestinal tract affecting the small intestine but can also involve the colon and rectum. The main adhesins of *C. jejuni* are the major flagella subunit FlaA; *Campylobacter* adhesion protein A (CapA); an outer membrane lipoprotein

(JlpA); an *O-* glycosylated major outer membrane protein (MOMP); two outer membrane *N*-glycosylated peptides (PEB1a and PEB3), which also act as transport proteins; and a fibronectin binding protein (CadF).

The adhesins bind to blood group antigens on the surface of gastrointestinal epithelial cells. *Campylobacter* invasive antigens are secreted through the flagella export apparatus into the host cell. The organism disrupts the tight junction and penetrates to the basolateral side of the cell where it may penetrate into the epithelial cell inducing cell death. A pro-inflammatory response is induced by activation of NFkB and MAPkinase. Virulence factors include flagella (motility and adhesion), the presence of a plasmid, pVir, and the production of toxins. The best characterized is the cytolethal distending toxin, which affects the host cell cycle by inducing apoptosis.

Histologically there is a nonspecific acute inflammatory reaction with neutrophils and mononuclear cells and loss of the epithelial layer as a result of cell death, producing ulcers. The similarity of the O antigens to host cell gangliosides (particularly O-19) is the basis for the autoimmune neurological conditions of GBS.

QUESTIONS

1. Which of the following are true statements about the epidemiology of *Campylobacter* gastroenteritis?
   a. It is a zoonosis.
   b. The organism can be acquired from companion animals.
   c. The main source of infection is milk.
   d. The infectious dose is believed to be approximately 10^2 organisms.
   e. The most frequent MLST type in the UK is ST 21 for *C. jejuni.*

2. True or false: A patient with acute *Campylobacter jejuni* gastroenteritis may present with which of the following?
   a. Watery diarrhoea
   b. Blood in the faeces
   c. Diplopia
   d. Myalgia
   e. Peptic ulcer

3. Which of the following are true concerning *Campylobacter jejuni*?
   a. It requires 21% oxygen for growth.
   b. It is oxidase negative.
   c. It can grow at 42°C.
   d. Its surface is covered by an S-layer.
   e. The Penner typing scheme is based on the O antigens.

4. Prevention of further cases of gastroenteritis involves which of the following?
   a. Separate processing lines for different meats
   b. Handwashing
   c. Vaccination of slaughterhouse staff
   d. Separate storage of cooked and uncooked food in a kitchen
   e. Veterinary inspection of carcasses

5. Which of the following are true about gastroenteritis caused by *C. jejuni*?
   a. The flagella is the sole adhesion of the organism.
   b. The bacterium utilizes a flagella export apparatus to secrete virulence factors.
   c. *C. jejuni* produces cytolethal distending toxin.
   d. *C. jejuni* produces vacuolating cytotoxin.
   e. Motility is a virulence factor.

GUIDELINES

1. UK Standards for Microbiology Investigations: Identification of Campylobacter Species, PHE 2014 (www.hpa.org.uk/ProductsServices/MicrobiologyPathology/UKStandardsForMicrobiol-ogyInvestigations/).

REFERENCES

1. Sheppard SK & Maiden MC (2015) The evolution of *Campylobacter jejuni* and *Campylobacter coli*. *Cold Spring Harb Perspect Biol* **7**:ii:a018119. (doi: 10.1101/cshperspect.a018119).

2. Olsen SJ et al. (2001) An outbreak of *Campylobacter jejuni* infections associated with food handler contamination: the use of pulsed-field gel electrophoresis. *J Infect Dis* **183**:164–167.

3. Gao B et al. (2014) Novel components of the flagellar system in epsilonproteobacteria. *MBio* **5**:e01349-14 (doi: 10.1128/mBio.01349-14).

4. Kuwabara S & Yuki N (2013) Axonal Guillain-Barré syndrome: concepts and controversies. *Lancet Neurol* **12**:1180–1188.

5. Wagenaar JA (2013) Preventing *Campylobacter* at the source: why is it so difficult? *Clin Infect Dis* **57**:1600–1606.

6. Shin E et al. (2015) First report and molecular characterization of a *Campylobacter jejuni* isolate with extensive drug resistance from a travel-associated human case. *Antimicrob Agents Chemother* ii:AAC.01395-15.

7. Kaakouysh NO et al. (2015) Global epidemiology of *Campylobacter* infection. *Clin Microbiol Rev* **28**:687–720.

8. Fitzgerald C (2015) Campylobacter. *Clin Lab Med* **35**:289–298.

9. Ramirez-Hernandez A et al. (2015) Adherence reduction of *Campylobacter jejuni* and *Campylobacter coli* strains to HEp2 cells by mannan oligosaccharides and a high molecular weight component of cranberry extract. *J Food Prot* **78**:1496–1505.

10. Heikema AP et al. (2015) *Campylobacter jejuni* capsular genotypes are related to Guillain-Barre syndrome. *Clin Microbiol Infect* ii:S1198-743X(15)00563-7 (doi: 10.1016/j.cmi.2015.05.031).

11. Guerry P et al. (2012) Campylobacter polysaccharide capsules: virulence and vaccines. *Front Cell Infect Microbiol* **2**:1–10.

ANSWERS

MCQ	Feedback
1. Which of the following are true statements about the epidemiology of *Campylobacter* gastroenteritis? a. It is a zoonosis. b. The organism can be acquired from companion animals. c. The main source of infection is milk. d. The infectious dose is believed to be approximately 10^2 organisms. e. One of the most frequent MLST types in the UK is ST 21 for *C. jejuni*.	*Campylobacter* is a zoonosis, which can be acquired from several different animal sources, including food-source animals and companion animals. Although milk can be a source, it is not the main route of infection in the UK. The infectious dose is considered to be more than 10^4, although in immuno-compromised individuals it can be much less. The three most common MLST types in the UK are 21, 42, and 61.
2. True or false: A patient with acute *Campylobacter jejuni* gastroenteritis may present with which of the following? a. Watery diarrhoea b. Blood in the faeces c. Diplopia d. Myalgia e. Peptic ulcer	a. True b. True c. False d. True e. False Typically, acute campylobacteriosis presents with the signs of abdominal pain, diarrhoea, and, if severe, blood in the faeces. Often the patients complain of a prodrome of myalgia. Diplopia is not part of the acute presentation but may occur some weeks following the gastroenteritis as an immunological complication called Miller–Fisher syndrome, or as an ascending paralysis (Guillain–Barré syndrome). Peptic ulcer is caused by *Helicobacter pylori*.
3. Which of the following are true concerning *Campylobacter jejuni*? a. It requires 21% oxygen for growth. b. It is oxidase negative. c. It can grow at 42°C. d. Its surface is covered by an S-layer. e. The Penner typing scheme is based on the O antigens.	Campylobacters are micro-aerobic requiring 5% oxygen and 10% carbon dioxide. *C. jejuni* is oxidase positive and can grow between 37–42°C. *C. jejuni* does not have an S-layer and is complement sensitive, whereas *C. fetus* does have an S-layer and is complement resistant.
4. Prevention of further cases of gastroenteritis involves which of the following? a. Separate processing lines for different meats b. Handwashing c. Vaccination of slaughterhouse staff d. Separate storage of cooked and uncooked food in a kitchen e. Veterinary inspection of carcasses	Contamination can occur at any stage in meat production from source (for example, *C. jejuni* is found in chickens commonly as part of their microbiome) to food preparation in the kitchen. The principles of preventing meat contamination is to stream the production process to prevent cross-contamination and personal hygiene particularly hand washing. Vaccines are being developed for poultry and sheep and a human vaccine is also in development.
5. Which of the following are true about gastroenteritis caused by *C. jejuni*? a. The flagella is the sole adhesion of the organism. b. The bacterium utilizes a flagella export apparatus to secrete virulence factors. c. *C. jejuni* produces cytolethal distending toxin. d. *C. jejuni* produces vacuolating cytotoxin e. Motility is a virulence factor.	*C. jejuni* has several adhesins including the flagella subunits FlaA, outer membrane proteins, MOMP and jlpA (a lipoprotein), CapA (campylobacter adhesion protein), and two outer membrane peptides, PEB1a and PEB3. Proteins are secreted via the flagella export mechanism. Virulence is associated with motility and the secretion of a cytolethal distending toxin. The vacuolating cytotoxin is produced by *Helicobacter pylori* not campylobacters.

CLOSTRIDIUM DIFFICILE INFECTION

Alicia Yeap[1] and Nandini Shetty[2]

[1]Specialist Registrar, Public Health England, UK.
[2]Consultant Microbiologist, Public Health England, UK.

A tertiary referral hospital in England was found to have an unexpectedly high rate of *Clostridium difficile* infection (CDI). It was noted that the quarterly laboratory-confirmed cases that had been reported to the national surveillance program from this hospital had exceeded the target trajectory when compared to the predefined target set for the whole year. National targets for each hospital are set with the aim of reducing CDI rates and are partly determined from local CDI rates from preceding years.

INVESTIGATION OF THE INCIDENT

In order to gain a better understanding of the local epidemiology of CDI at this hospital, a prospective survey (from the patients' charts) of all CDI cases occurring over the three-month period using a simple questionnaire was conducted. A case of CDI was defined as a positive stool glutamate dehydrogenase (GDH) enzyme-linked immunoassay (EIA) result and a positive stool toxin EIA result, in accordance with national guidance for *C. difficile* testing.

In this time period, 34 cases of CDI were identified, with 38% occurring in patients with underlying haematology or oncology conditions. Clustering of 4–6 cases (spatially and temporally related) was noted across three specialty wards (haematology, oncology, and surgery), where many of these patients were placed. Of these cases, 76% were considered to be hospital-acquired infections, as defined by a period of more than 48 hours between admission to hospital and positive stool *C. difficile* test results (according to the UK Department of Health).

All cases had exposure to antibiotics prior to onset of diarrhoea, and 29% were receiving proton-pump inhibitor (PPI) therapy. Of the nonhaematology and oncology patients, 24% had severe disease, as defined by the presence of one or more of the following: fever greater than 38.5°C, leucocytosis greater than 15×10^9 cells/L, serum creatinine greater than 1.5 times the baseline value, and the presence of clinical or radiological evidence of severe colitis, or both. None of the cases needed surgical intervention such as colectomy or admission to the intensive care unit as a consequence of CDI. Severity was not assessed for the haematology and oncology patients because of the high frequency of neutropenia and alternative causes for fever and colitis in this group. Stool *C. difficile* test results were available for 71% of cases within 24 hours of onset of symptoms and 82% within two days of onset of symptoms. According to hospital policy, all patients with diarrhoea are routinely isolated. Some patients may be deisolated if no pathogens are identified and the diarrhoea is thought to be food related. This investigation revealed that some patients with diarrhoea were not isolated and no reason was documented in the patient's chart.

Ribotyping of isolates revealed three small clusters of 002 and 015, corresponding to the three specialty wards with a high incidence of CDI. Each cluster had two to four cases and the remaining cases belonged to a wide range of ribotypes.

CLINICAL MANAGEMENT

All positive results were telephoned to the clinical teams in charge of the patients' care by the microbiologists and clinical management discussed in detail. The cases would then be reviewed on the CDI ward round on the same or following day.

According to National Guidance in the UK, there are multiple clinical definitions of CDI (Table 5.1).

Following national guidance, oral metronidazole was advised for mild and moderate disease and oral vancomycin for severe disease. If oral administration was not possible, a combination of intravenous metronidazole and nasogastrically administered vancomycin was given. One case was given fidaxomicin because of a history of previous CDI relapse. Concomitant antibiotics and proton-pump inhibitors (PPIs) were stopped if possible and hydration was addressed. One patient with severe disease required a surgical review.

PREVENTION OF FURTHER CASES

Infection control nurses were notified of all new positive *C. difficile* results daily by the microbiology laboratory. The nurses worked closely with staff on affected wards to ensure that appropriate infection-control precautions were taken. Patients with CDI were isolated in side rooms until diarrhoea resolved. If side rooms were not available, an alternative strategy of grouping patients in a bay with the door closed was adopted. Staff members wore disposable gloves and aprons during contact with CDI patients and handwashing with soap and water was emphasized.

Table 5.1. **Types of clinical CDI according to UK National Guidance**

Type of Clinical CDI	Definition
Mild CDI	Not associated with a raised white cell count (WCC); it is typically associated with <3 stools of types 5–7 on the Bristol Stool Chart per day.
Moderate CDI	Associated with a raised WCC that is <15 × 10^9/L; it is typically associated with 3–5 stools per day.
Severe CDI	Associated with a WCC >15 × 10^9/L, or an acute rising serum creatinine (that is, >50% increase above baseline), a temperature of >38.5°C, or evidence of severe colitis (abdominal or radiological signs). The number of stools may be a less reliable indicator of severity.
Life-threatening CDI	Includes hypotension, partial or complete ileus or toxic megacolon, or computerized tomography (CT) scan evidence of severe disease.

Signs were placed on the isolation room doors to remind staff of the use of contact precautions, that is, appropriate personal protective equipment (PPE) and hand hygiene (Figure 5.1). PPE should include gloves and gowns or aprons; masks are not needed. Enhanced environmental cleaning was undertaken, which involved daily cleaning and

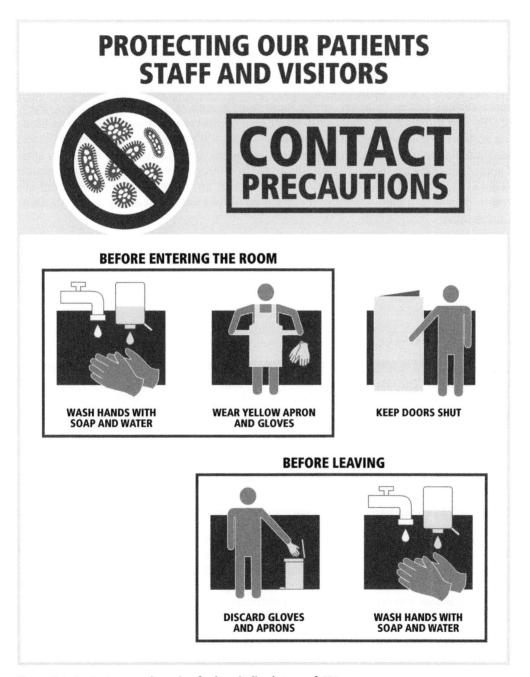

Figure 5.1. Contact precautions sign for hospitalized cases of CDI.

disinfection of surfaces and equipment with special attention to high-touch surfaces and cleaning of toilets 3–4 times a day. Disinfectants, such as chlorine-releasing agents, which are effective against *C. difficile* spores, and the terminal cleaning of curtains and mattresses upon discharge of patients were also part of the regimen to prevent further cases. Ward audits of hand hygiene and environmental cleaning were reviewed and problems identified from these audits addressed. This is the recommended standard of care to prevent spread in the hospital environment.

Daily CDI ward rounds were implemented in response to the unexpectedly high rate of CDI. These ward rounds were carried out by a team that included a microbiologist, an infection-control nurse, and a pharmacist. The team reviewed all CDI cases within a day of diagnosis and gave further advice on infection-control measures and clinical management to ward staff and clinical teams. These ward rounds also served to raise awareness of CDI and to educate hospital staff about general preventive measures, such as limiting antibiotic exposure and using narrow-spectrum antibiotics when possible. Antibiotic stewardship strategies including restriction of cephalosporin and broad-spectrum antibiotics, providing an indication for all antibiotic prescriptions with dose and duration clearly marked, and de-escalation to narrow-spectrum, targeted therapy once culture results became available were also measures employed to decrease the burden of CDI in the hospital.

As hematology and oncology patients had been identified as a high-risk group for CDI, microbiologists increased their attendance at multidisciplinary team meetings in which the care of these patients was discussed, and their participation contributed to improved antibiotic stewardship for this patient group.

EPIDEMIOLOGY

The incidence of healthcare-associated CDI in England has been steadily declining in recent years. The national mandatory surveillance program from NHS Hospitals in England showed that counts and rates of hospital-associated CDI cases both decreased by 16.4% (from 1322 to 1105 cases and from 15.4 to 12.8 cases per 100,000 bed-days, respectively) over the same time period (April–June 2015 to April–June 2016). There is some variation in CDI incidence among geographical regions, but the decline in incidence has been observed in all regions. Prior to 2007, there had been a steady increase in CDI incidence, leading to the institution of mandatory enhanced surveillance and a national ribotyping network service. Scotland, Wales, and Northern Ireland have seen similar trends in CDI rates and have implemented similar surveillance mechanisms.

Ribotypes 001, 002, 012, 014, 015, 018, 027, and 078 are the predominant strains found in Europe, with marked differences in distribution across countries. With the exception of 012 and 018, these are the most common strains of *C. difficile* in England. Previously, ribotype 106 was a predominant strain in England, despite rarely being found elsewhere, but this strain is now no longer in the top 10 most common strains of *C. difficile* in the country for unclear reasons. Ribotypes 001 and 027 are also becoming less common in England.

There has been much interest in ribotype 027, also known as North American pulsed-field type 1 (NAP1) or restriction endonuclease analysis (REA) group BI. This strain has caused several large and highly publicized hospital outbreaks of CDI. The first of these was described in Quebec, Canada in 2003, followed shortly after by two consecutive outbreaks in Stoke Mandeville Hospital in England. Severity of disease, case fatality, and transmissibility were noted to be markedly increased in these early outbreaks, though these findings have not been consistently reported since in subsequent outbreaks of this strain in other parts of the world.

C. difficile is the underlying cause of 10–30% of cases of nosocomial diarrhoea. Older age is a major risk factor for CDI, with most CDI cases occurring in people aged 65 years and older. Women are slightly more likely to develop CDI than men, though the reasons for this gender difference are not known. People with severe underlying comorbidities, those with prolonged admission to the hospital, and critical-care patients are also more likely to develop CDI. Specific underlying conditions such as abdominal surgery, malignancies, renal failure, and inflammatory bowel disease have been associated with an increased risk of disease.

The most important risk factor for developing CDI is the prior use of antibiotics (within the previous 90 days), especially broad-spectrum agents that significantly alter normal bowel flora. Most classes of antibiotics have been implicated as risk factors for CDI. Historically, CDI was first described in people who developed diarrhoea and pseudomembranous colitis after taking clindamycin. Second- and third-generation cephalosporins and fluoroquinolones were the next in line to be incriminated. Their broad spectrum of activity provides biological plausibility for a role in causality. The use of these classes of antibiotics has also been temporally associated with several outbreaks and rises in CDI rates. Similarly, the introduction of restrictions on the prescriptions of these antibiotics in hospital settings has been reported to coincide with significant reductions in CDI rates, though often with the concurrent implementation of other infection control measures. Co-amoxiclav and piperacillin-tazobactam are currently the two most commonly prescribed antibiotics in CDI cases in England, which probably reflects changes made to hospital formularies to reduce the use of cephalosporins and fluoroquinolones. Associations with specific classes of antibiotics can be confounded by many other factors such as age, comorbidities, and hospitalization, but attempts have been made to categorize antibiotics as high risk or low risk for the development of CDI. CDI can occur even after a single dose of antibiotics.

With regard to other drugs, gastric acid suppressants such as proton-pump inhibitors and H2 receptor antagonists have been shown in some studies to be associated with increased acquisition of CDI, CDI recurrence, and refractory disease, whereas other studies have failed to demonstrate these associations. Some chemotherapeutic agents, bowel prokinetic drugs, and laxatives have also been linked with an increased risk of CDI.

Although CDI is often considered to be a hospital-acquired infection and most interventions for prevention and control have thus far targeted hospital settings, it is currently estimated that 25–30% of CDI cases may be acquired in the community. There are some indications that the rate of community-acquired infections may be rising, with different strain distributions implicated. Definitions for community acquisition vary widely, with thresholds for previous hospitalization ranging from four weeks to a year. The true incidence of community-acquired infection is also confounded by variations in *C. difficile* assay characteristics and by the possibility of detection of carriage in the absence of testing for other community-acquired diarrheal pathogens. The characteristics of community-acquired CDI have yet to be fully elucidated, but studies so far suggest that although the same risk factors are likely to remain relevant in some cases, there may be a significant proportion of cases who have not received prior antibiotics nor had recent

hospitalization. Cases of community-acquired CDI are younger than those with hospital-acquired disease.

C. difficile is ubiquitous in the environment and toxigenic strains known to cause human
disease have been isolated from water sources, domestic animals and those raised for
food, and food products of both meat and vegetable origins. There have not been any
recognized outbreaks of CDI linked to these potential sources so far. The predominant
strains isolated from food animals such as cattle and pigs and food products have been
ribotype 078 and 027, strains recognized to cause human disease. Several large surveys
have investigated the prevalence of *C. difficile* in food products and have reported finding
C. difficile in 0–42% of samples tested. More work is needed to determine the role of these
potential sources in the evolving epidemiology of CDI.

BIOLOGY

C. difficile is an obligate anaerobic Gram-positive bacillus, measuring 2–17 μm in length.
It is closely related to *C. sordellii*. *C. difficile* produces spores that can survive for long periods of time in the environment and are highly resistant to adverse conditions, such as
desiccation and extremes of temperature. The spores are also resistant to alcohol and may
not be removed adequately from hands by alcohol hand rubs. The main mode of transmission is by the ingestion of spores from a contaminated environment. Spores are relatively resistant to gastric acid and, when ingested, pass through the stomach and convert
into vegetative forms in the small bowel. These forms grow and replicate within the lumen
of the large bowel. The role of adhesion factors on the organism is unclear. Disruption of
normal bowel flora is necessary for *C. difficile* to establish itself within the colon, which
usually occurs through exposure to antibiotics. *C. difficile* produces ρ-cresol, a bacterio-
static substance that may be inhibitory to other gut microbes and gives the organism its
characteristic smell.

Disease is mediated by two large toxins, toxin A (205 kDa) and toxin B (308 kDa). These
toxins bind to specific carbohydrate receptors expressed on bowel epithelium and
enter the cytosol of these cells. Here, both toxins glucosylate small guanosine triphos-
phate-binding proteins, Rho, Rac, and Cdc42. This interferes with signalling pathways,
induces pore formation, and causes derangement of the actin cytoskeleton, which
results in the loss of cellular structural integrity and leaking of tight junctions between
enterocytes, which in turn increases permeability and fluid secretion across the bowel
wall. These toxins also cause apoptosis of enterocytes. Toxin A directly activates neutrophils, promotes neutrophil chemotaxis, and stimulates the release of pro-inflammatory cytokines such as tumour necrosis factor α and interleukin-1, which contributes
to mucosal inflammation. Toxin B is much more potent than toxin A, despite lacking
some of its actions. Some isolates of *C. difficile*, typically ribotype 017, which is more
commonly found in Asia, have been found to produce toxin B only and cause disease of
similar severity to those that produce both toxins. Toxin production can be up-regulated
in the presence of antibiotics. The presence of antitoxin antibodies has been shown to
be protective against disease.

Binary toxin, also known as *C. difficile* transferase, is a third toxin produced by 6–12.5%
of strains of *C. difficile*. It modifies actin by ribosylating adenosine diphosphate, a

mechanism of action common to other clostridial toxins. Binary toxin has been postulated to increase severity of disease but is unlikely to be able to cause disease in the absence of other toxins.

Ribotype 027 has been shown to generate more toxin *in vitro* than other strains. This strain harbours an 18-base-pair deletion in the *tcdC* gene, a negative regulatory gene for toxin production. It also has a single base-pair deletion at position 117, which results in truncation of the *tcdC* gene. Ribotype 027 also produces binary toxin, and an increased sporulation capacity has been described. Fluoroquinolone resistance is common in this strain. The significance of these findings remains controversial as several studies have not demonstrated increased virulence or transmissibility.

DISEASE

Asymptomatic colonic carriage occurs in up to 3% of adults in the community and up to 20% of hospitalized adults. Studies of carriage in children under the age of one year have shown acquisition rates of up to 63%, with toxigenic strains being common. There have, however, been relatively few reports of disease in infants and young children. Possible reasons for this may be the absence of cell surface receptors for *C. difficile* toxins in this age group or the presence of protective maternal antibodies.

The most common presentation of CDI is diarrhoea, which is watery in nature in most cases but can also be mucoid. The onset of diarrhoea can vary from a few days to 10 weeks after antibiotic exposure. Visible blood in the diarrheal stools is rare. A characteristic odour may be present. Fever occurs in up to 66% of patients with CDI and abdominal pain in up to 33%. Leucocytosis has been noted in up to 60% of cases. Symptoms may resolve spontaneously in a small minority. Dehydration and consequent acute renal dysfunction may result.

Markers of severity of disease have been extensively investigated and the most well-recognized of these are the presence of leucocytosis of greater than 15×10^9 cells/L, a rise in serum creatinine to more than 1.5 times the premorbid value, and the presence of signs of colitis. Other severity markers are fever greater than 40°C, hypoalbuminemia, and age greater than 70 years. Predictors of poor outcome include raised serum lactate levels, older age, comorbidities, and admission to critical care. Fulminant disease occurs in less than 5% of CDI cases and 30-day all-cause mortality may be up to 20%. Relapses of CD are common and can occur in about 20% of cases after the first episode and in 50% of cases after the second episode.

Pseudomembranous colitis can be diagnosed on endoscopy, though the procedure carries a higher risk of perforation if this condition is present. Colitis varies in extent of gut involvement and severity. Inflammation typically does not involve the small bowel. Diarrhoea may be absent in patients who present with paralytic ileus or toxic megacolon (Figure 5.2). Colonic volvulus, perforation, and peritonitis may ensue.

Reactive arthritis, with or without features of Reiter's syndrome, has been linked to CDI, in association with human leukocyte antigen (HLA) B27. Other extracolonic manifestations that have been described, albeit rarely, are bacteraemia, splenic abscesses, and osteomyelitis.

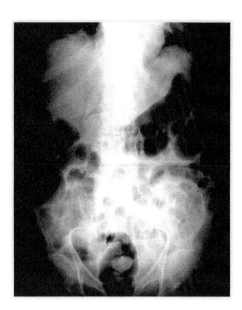

Figure 5.2. **Abdominal X-ray demonstrating dilated bowel loops consistent with toxic megacolon, most marked in the right upper quadrant.** (Courtesy of University College Hospital, London.)

Diagnosis of CDI is made by the combination of a compatible clinical picture with fecal tests that demonstrate the presence of the organism and toxin production. Anaerobic culture of the organism and cell culture-based techniques for detecting the cytopathic effects of toxins are considered to be the gold standard diagnostic tests, but these methods require considerable expertise and time. Toxin EIAs are simple and quick to perform and hence have been extensively used, but they generally suffer from a considerable lack of sensitivity and are no longer recommended as standalone tests. A two-step testing algorithm is now the recommended method for CDI diagnosis. A GDH, EIA, or toxin gene nucleic acid amplification test (NAAT) should be performed as the screening test, with a negative result indicating that CDI is unlikely and a second test is not required. If the screening test is positive, a sensitive toxin EIA test should be carried out and is diagnostic of CDI, if positive. Some laboratories use a NAAT assay with primers specific for the toxin B gene as a confirmatory test. The sensitivity and specificity of this polymerase chain reaction (PCR) assay has been reported as 93–97% and 93–96%, respectively, compared to toxigenic anaerobic culture or cytotoxicity (EIA) assay. If the confirmatory test is negative, the patient may be a carrier with transmission potential and isolation measures may be appropriate.

Oral metronidazole is the current treatment of choice for mild to moderate disease. Oral vancomycin may be more effective for severe disease. A new oral antibiotic, fidaxomicin, has been shown to be at least equivalent to oral vancomycin in efficacy and may result in fewer recurrences. Intravenous metronidazole with or without the addition of intracolonic vancomycin has been used where oral treatment is not possible. Surgery may be required for extensive severe disease. Intravenous immunoglobulin has been used for severe or recurrent disease. Other treatment strategies for recurrent disease include pulsed and tapered vancomycin regimens, probiotics, and fecal transplants.

Recurrence or reinfection is thought to occur in 20–30% of CDI cases, with multiple relapses occurring in 5–8% percent of cases. Older age, female gender, PPI use, recent abdominal surgery, chronic renal failure, and community-acquired disease have been associated with recurrences. Concomitant antibiotics increase the risk of recurrence and treatment failure. Treatment failure may be difficult to identify as a response may only be seen after three to five days of treatment and residual diarrhoea, despite improvement, is not uncommon. Treatment failure has been estimated to occur in up to 50% of CDI cases and may be more common with oral metronidazole, though this does not seem to be related to metronidazole resistance. However, reduced susceptibility to metronidazole has been noted in historically epidemic strains, and there are concerns that the levels of metronidazole achievable in the colon may be inadequate to treat strains with higher minimum inhibitory concentrations, especially if colonic inflammation is reduced. Raised minimum inhibitory concentrations (MIC) to vancomycin are of less concern because stool concentrations of vancomycin with oral therapy remain very high regardless of the degree of colonic inflammation.

PATHOLOGY

Nonspecific inflammatory changes are common with CDI. Colitis most often involves the rectosigmoid colon but may only affect the ascending colon in some cases. In classic pseudomembranous colitis, white or yellow plaques of fibrinopurulent debris are seen loosely attached to the bowel wall. The mucosa may initially be normal but becomes oedematous and hyperemic with progression of disease. Plaques may coalesce, forming pseudomembranes.

Microscopically, the disease is characterised by foci of degenerating bowel epithelium and crypts with overlying eruptions of fibrin, mucin, neutrophils, and necrotic cellular debris. Glands are distended with mucin, and superficial erosions and ulceration develop with disease progression. There is a histological grading system to delineate severity of disease.

QUESTIONS

1. Choose the correct statement about tests for CDI from the list below:
   a. A positive GDH EIA is diagnostic of CDI.
   b. A positive toxin NAAT is diagnostic of CDI.
   c. Toxin EIAs should not be used for CDI diagnosis.
   d. Cytotoxin assays using cell culture can be used in CDI diagnosis.
   e. *C. difficile* may be found incidentally on cultures for other diarrheal bacterial pathogens.

2. Of the following list of hospital infection-control precautions, which option is inappropriate for CDI?
   a. Cases should be isolated in single rooms.
   b. Masks should be worn during patient contact.
   c. Handwashing with soap and water should be performed after patient contact.
   d. Cases should be isolated in cohort areas.
   e. Gloves and aprons should be worn during patient contact.

3. The following are risk factors for developing CDI, except:
    a. Older age
    b. Raised serum lactate
    c. Chemotherapy
    d. Prior use of antibiotics
    e. Inflammatory bowel disease

4. *C. difficile* ribotype 027 has been associated with the following, except:
    a. Production of binary toxin
    b. Reduced sporulation capacity
    c. Fluoroquinolone resistance
    d. Increased severity of disease
    e. Presence in food animals and food products

5. The following statements about community-acquired CDI are true, except:
    a. Prior antibiotic exposure may not be a risk factor.
    b. About 25% of CDI may be acquired in the community.
    c. Patients with community-acquired CDI are younger than those with hospital-acquired CDI.
    d. Strains causing human disease have been found in food animals.
    e. Strains causing human disease have not been found in food products.

GUIDELINES

1. Sartelli M et al. (2015) WSES guidelines for management of *Clostridium difficile* infection in surgical patients. *World J Emerg Surg* **10**:38 (doi: 10.1186/s13017-015-0033-6).

REFERENCES

1. Burnham CA & Carroll KC (2013) Diagnosis of *Clostridium difficile* infection: an ongoing conundrum for clinicians and for clinical laboratories. *Clin Microbiol Rev* **26**(3):604–630.
2. Freeman J, Bauer MP, Baines SD et al. (2010) The changing epidemiology of *Clostridium difficile* infections. *Clin Microbiol Rev* **23**(3):529.
3. Gould LH & Limbago B (2010) *Clostridium difficile* in food and domestic animals: a new food-borne pathogen? *Clin Infect Dis* **51**(5):577–582.
4. Department of Health (2013) Updated guidance on the diagnosis and reporting of *Clostridium difficile*. London: Department of Health.
5. Department of Health, Health Protection Agency (2008) *Clostridium difficile* infection: how to deal with the problem. London: Department of Health.
6. Valiente E, Cairns MD & Wren BW (2014) The *Clostridium difficile* PCR ribotype 027 lineage: a pathogen on the move. *Clin Microbiol Infect* **20**(5):396–404.
7. Voth DE & Ballard JD (2005) *Clostridium difficile* toxins: mechanism of action and role in disease. *Clin Microbiol Rev* **18**(2):247–263.
8. Malnick S & Melzer E (2015) Human microbiome: from the bathroom to the bedside. *World J Gastrointes Pathophysiol* **6**:79–85.
9. Monaghan TM et al. (2015) Pathogenesis of *Clostridium difficile* infection and its potential role in inflammatory bowel disease. *Inflamm Bowel Dis* **21**:1957–1966.

ANSWERS

MCQ	Feedback
1. Choose the correct statement about tests for CDI from the list below: a. A positive GDH EIA is diagnostic of CDI. b. A positive toxin NAAT is diagnostic of CDI. c. Toxin EIAs should not be used for CDI diagnosis. d. Cytotoxin assays using cell culture can be used in CDI diagnosis. e. *C. difficile* may be found incidentally on cultures for other diarrheal bacterial pathogens.	Cell culture-based cytotoxin assays can be considered for use in the second stage of the recommended CDI testing algorithm; however, they are more time-consuming and require more expertise to perform and interpret than toxin EIAs, and these factors should be considered by laboratories when choosing CDI tests. GDH EIAs and toxin NAATs are recommended as screening tests but lack specificity if used as standalone tests. Toxin EIAs should not be used alone but have a role in the second stage of the recommended testing algorithm—a positive sensitive toxin EIA result with a positive screening test or a screening test followed by a real-time PCR for toxin B is strongly suggestive of CDI. *C. difficile* requires strict anaerobic conditions to grow and selective agar is needed to identify it, hence it is not usually found on standard cultures for other diarrheal bacterial pathogens. Its name derives from the relative difficulty in isolating this organism historically. **Further reading** Burnham CA, Carroll KC (2013) Diagnosis of *Clostridium difficile* infection: an ongoing conundrum for clinicians and for clinical laboratories. *Clin Microbiol Rev* 26(3):604–630. Department of Health (2013) Updated guidance on the diagnosis and reporting of *Clostridium difficile*. London: Department of Health.
2. Of the following list of hospital infection control precautions, which option is inappropriate for CDI? a. Cases should be isolated in single rooms. b. Masks should be worn during patient contact. c. Handwashing with soap and water should be performed after patient contact. d. Cases should be isolated in cohort areas. e. Gloves and aprons should be worn during patient contact.	There is no evidence that inhalation of infectious aerosols is a route of transmission of *C. difficile*. Isolation of hospitalized cases is indicated to limit transmission to other patients—single rooms should be used or cohort isolation if numbers of cases are large, for example, in outbreaks. Contact precautions include wearing gloves and aprons and washing hands with soap and water to remove spores that are resistant to alcohol hand rubs. **Further reading** Department of Health, Health Protection Agency (2008) *Clostridium difficile* infection: how to deal with the problem. London: Department of Health.
3. The following are risk factors for developing CDI, except: a. Older age b. Raised serum lactate c. Chemotherapy d. Prior use of antibiotics e. Inflammatory bowel disease	Raised serum lactate predicts poor outcome with CDI but is not a risk factor for developing CDI. Older age, chemotherapy, prior use of antibiotics, and inflammatory bowel disease have all been found to be risk factors for developing CDI. **Further reading** Freeman J, Bauer MP, Baines SD et al. (2010) The changing epidemiology of *Clostridium difficile* infections. *Clin Microbiol Rev* 23(3):529.
4. *C. difficile* ribotype 027 has been associated with the following, except: a. Production of binary toxin b. Reduced sporulation capacity c. Fluoroquinolone resistance d. Increased severity of disease e. Presence in food animals and food products	Increased sporulation capacity has been described for ribotype 027 and is postulated to increase its transmissibility compared to other strains of *C. difficile*. The other answers describe characteristics that have been associated with ribotype 027, though it should be noted that increased disease severity has not been a consistent finding. **Further reading** Valiente E, Cairns MD & Wren BW (2014) The *Clostridium difficile* PCR ribotype 027 lineage: a pathogen on the move. *Clin Microbiol Infect* 20(5):396-404. Freeman J, Bauer MP, Baines SD et al. (2010) The changing epidemiology of *Clostridium difficile* infections. *Clin Microbiol Rev* 23(3):529.

MCQ	Feedback

5. The following statements about community-acquired CDI are true, except:
  a. Prior antibiotic exposure may not be an important risk factor.
  b. About 25% of CDI may be acquired in the community.
  c. Patients with community-acquired CDI are younger than those with hospital-acquired CDI.
  d. Strains causing human disease have been found in food animals.
  e. Strains causing human disease have not been found in food products.
	Strains causing human disease have been found in food products. The other statements are true.

AN OUTBREAK OF *CRYPTOSPORIDIUM* SP. ASSOCIATED WITH A PUBLIC SWIMMING POOL

CASE 6

Christina J. Atchison[1] and Rachel M. Chalmers[2]

[1]North West London Health Protection Team, Public Health England, UK.
[2]Cryptosporidium Reference Unit, Public Health Wales Microbiology ABM, Singleton Hospital, Swansea, UK.

On September 22, 2010, an Environmental Health Officer (EHO) from the local city council notified the Greater Manchester Health Protection Unit (GMHPU) that 11 children from a local swimming club were experiencing gastrointestinal symptoms (diarrhoea and abdominal cramps). The swimming club was based at a local leisure centre with a 25m swimming pool. The club, with over 200 members aged 4–20 years old, trained four evenings a week and regularly competed in competitions. At other times, the pool was used by members of the public, swimming classes for babies, toddlers, and adults with learning difficulties, and a number of local schools.

INVESTIGATION OF THE OUTBREAK

A questionnaire was completed and stool samples were collected from the 11 children. In addition, other cases were sought by requesting local GPs, paediatricians, walk-in centres, and hospital microbiologists to report to GMHPU suspected cases of diarrhoea, vomiting, or both, that were potentially linked to the swimming pool, and to obtain a stool specimen if appropriate. Initial review of the questionnaires suggested that illness was linked to attendance at a training session at the pool on September 13. By September 28, three cases had submitted stool samples to the local microbiology laboratory and they tested positive for *Cryptosporidium*. The tests detected oocysts using fluorescent microscopy with auramine-phenol stain. These specimens were typed as *Cryptosporidium hominis* by real-time polymerase chain reaction (PCR) at the National *Cryptosporidium* Reference Unit, Public Health Wales, Swansea. An outbreak was declared and an Outbreak Control Team (OCT) was convened with representatives from GMHPU, the City Council Environmental Health Department, regional epidemiology and microbiology services, the national *Cryptosporidium* Reference Unit, and local Public Health. An outbreak investigation followed.

A retrospective cohort study was conducted to identify risk factors for infection in members of the swimming club. A list of 129 active members of the club was obtained. EHOs from the City Council conducted face-to-face and telephone interviews. The questionnaire was designed to collect information on the nature of illness and key exposures (including swimming at the pool on specified training dates, swimming at competitions, eating or drinking at events, and swimming at other pools). The questionnaire was piloted (tested) by staff and members of the public who were not cases, and were not connected to cases, at the local leisure centre. A descriptive analysis of the data collected was performed to examine the characteristics of cases in time, place, and person. To assess risk factors for cryptosporidiosis, adjusted odds ratios (aOR) with 95% confidence intervals (95% CI) and *p* values were calculated.

The response rate among active members of the swimming club was 78%. From the respondents, 48 probable cases of cryptosporidiosis and 53 noncases (persons without symptoms) were identified. A probable case of cryptosporidiosis was defined as a member of the club with illness onset in September 2010 and having one or more of the following symptoms: diarrhoea ($\geq$3 loose stools per day), abdominal cramps, loss of appetite, fever, nausea, or vomiting, without laboratory confirmation of *Cryptosporidium* infection. Individuals with a history of foreign travel in the two weeks before illness onset were excluded from the study. Cases had a median age of 9 years (range: 4–26 years old) and noncases had a median age of 9 years (range: 4–17 years old). Cases reported a median duration of illness of 3 days (range: 1–9 days). Self-reported symptoms included diarrhoea (77%), vomiting (75%), loss of appetite (71%), abdominal pain (67%), nausea (67%), and fever (52%). The epidemic curve suggested a point source outbreak (Figure 6.1). The first case was a potential source as he had the earliest illness onset. Later cases may indicate secondary spread within households or schools.

The risk of illness was greater in members of the swimming club who attended the training session at the pool on September 13. The association was strong and highly significant statistically (aOR 28; 95% CI 8 to 99, *p* value <0.0001). This finding supports the hypothesis that the training session on September 13 was a significant exposure and is consistent with a point source exposure as illustrated by the epidemic curve (see Figure 6.1). Illness was not associated with eating any food items, use of toilets, changing rooms, or showers.

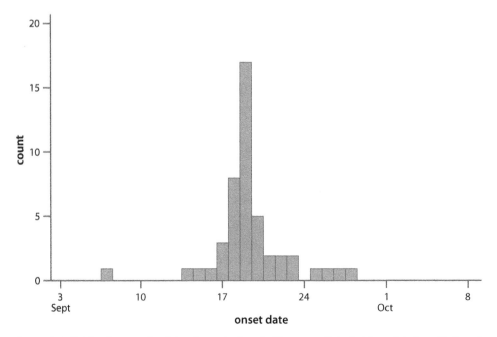

Figure 6.1 **Epidemic curve by date of onset of probable cases.** (Adapted from data from McCann R, Jones R, Snow J et al. [2014] *Epidemiol Infect* **142**:5155 [doi: 10.1017/S0950268813001143].)

Note that recall bias can be an issue in outbreak investigations in which data on exposure are collected retrospectively. Problems arise because the quality of the data is determined to a large extent on the person's ability to accurately recall past exposures, which may differ among cases and noncases and lead to bias. For example, in this investigation, an individual with cryptosporidiosis (case) may have reported his or her experience of attending the swimming pool differently from an individual without symptoms (noncase); that is, cases may tend to have a better recall on past exposures than noncases. Recall bias may result in either an underestimate or overestimate of the association between exposure and outcome.

The two main study designs used in the investigation of outbreaks are the retrospective cohort design, used in this investigation, and the case-control design. In the retrospective cohort design, all members of a defined cohort are included in the study and information on their exposure to different factors is investigated retrospectively. A higher incidence of disease in the exposed group suggests an association between that factor and the disease outcome. This study design is generally a good choice when dealing with an outbreak in a relatively small, well-defined source population (that is, the population from which cases arose), particularly if the disease being studied was fairly frequent, as was the situation in this outbreak. The case-control design is a better choice when the source population is large and ill defined, or when the disease outcome is rare. Exposures in cases are compared to exposures in a sample of the noncases (that is, the controls) drawn from the same at-risk population. If the cases have substantially higher odds of exposure to a particular factor compared to the control subjects, it suggests an association.

An environmental inspection of the swimming pool took place on September 28 and revealed that the pool had consistently met appropriate chlorination standards and disinfection procedures and it employed medium-rate sand filtration. Pool water testing for free chlorine, combined chlorine, and pH was performed three times a day and the pool was tested monthly for microbiology. Standards of cleanliness and repair were satisfactory. The pool water filtration system used two multigrade sand filters and automatic coagulant dosing with polyaluminium chloride. The pool water turnover rate was 2.5 hours. Backwashing occurred once a week, on Wednesdays at noon. Pool logs indicated that no faecal or vomiting incidents had been reported at the pool since the beginning of September 2010. The pool water was not tested for *Cryptosporidium* at the time of the inspection as the pool was well managed, the suspected contamination event had taken place two weeks earlier, there were no cases outside the swimming club members, case numbers had declined, and the turnover rate meant that the pool would have been capable of removing oocysts within six turnovers (15 hours).

CLINICAL MANAGEMENT

Thirty-five cases reported taking time off work or school as a result of illness and two cases were hospitalized. As there is no specific treatment for cryptosporidiosis, rehydration with oral or intravenous fluids was recommended as appropriate based on clinical assessment.

Cryptosporidium infection can be diagnosed by detection of oocysts, oocyst antigens, or sporozoite DNA in stool specimens. Histological examination of intestinal biopsies is also possible but usually only infrequently used for example, in patients with profound T-cell immune deficiency. The most commonly used method is microscopic examination of

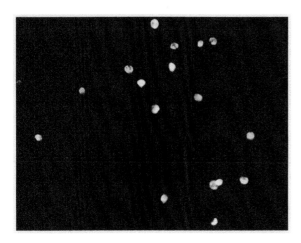

Figure 6.2. **Cryptosporidium oocysts: auramine-phenol staining, with ×50 objective.** (Courtesy of Cryptosporidium Reference Unit.)

stool (sometimes preserved in formalin and concentrated). Examination of stained faecal smears is necessary because the oocysts are too small (4–6 μm) to be accurately identified by unstained examination. In the UK the recommended microscopy method is examination of faecal smears stained with auramine-phenol (AP) (Figure 6.2) or a modified Ziehl–Neelsen (mZN) stain (Figure 6.3). In the UK, stools are submitted fresh, without formalin, so concentration is not required routinely prior to staining. Where concentration is required, modified methods should be used to limit oocyst losses and to prevent interference with the adhesion of oocysts to slides and interference with staining. Microscopy requires a skilled operator and can be time consuming. Enzyme immunoassays (EIAs) for diagnosing cryptosporidiosis are also available. They are relatively simple to perform and have high sensitivity and specificity, but positive reactions need to be confirmed by a sensitive method such as immunofluorescence microscopy. PCR methods are the gold standard for *Cryptosporidium* detection and are used in reference laboratories to confirm the diagnosis and differentiate between species. In exceptional circumstances, for maximum sensitivity, immunomagnetic separation with immunofluorescence microscopy can be used, as the method can detect as few as two organisms per gram of stool. Other

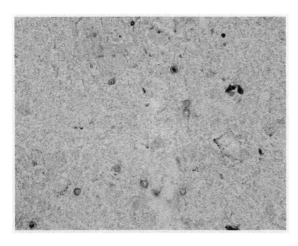

Figure 6.3. **Cryptosporidium oocysts: modified Ziehl-Neelsen staining, with ×50 objective.** (Courtesy of Cryptosporidium Reference Unit.)

samples occasionally examined in high-risk patients include bile, in cases of cholangitis, and bronchoalveolar lavage, in cases in which pulmonary cryptosporidiosis is suspected.

UK guidance states that all stool samples from community cases of diarrhoea should be tested for *Cryptosporidium*. However, laboratories have varying criteria for selecting stools for testing. In addition, examination for *Cryptosporidium* may not necessarily be included in a request for "ova, cysts, and parasites," as the methods of examination for the two tests differ. Therefore, clinicians are advised to specify *Cryptosporidium* on the request form to ensure appropriate testing is carried out. *Cryptosporidium* should be suspected in all patients who present with acute gastroenteritis, especially young children, or if the symptoms are prolonged (>3 days).

PREVENTION OF FURTHER CASES

Cryptosporidiosis is highly infectious person to person, as large numbers of oocysts are excreted and the infectious dose is low. To avoid spreading the illness, meticulous personal hygiene is required. Guidelines for the prevention of person-to-person spread following gastrointestinal illness should be followed.

Cases were advised to perform frequent and thorough handwashing, to properly dispose of excreta, and to wash soiled materials such as clothing or bedding at a high temperature. Other advice included avoidance of towel sharing and staying home from work (particularly for food handlers and staff of healthcare facilities), school, or nursery until 48 hours after symptoms had stopped.

Because there is no efficacious disinfectant residual in swimming pools, and pool water treatment may not adequately treat or remove *Cryptosporidium* oocysts, cases were advised not to use swimming pools for two weeks after diarrhoea had stopped because oocysts can still be shed during this time.

Local GPs were informed about the outbreak and reminded about the need to exclude patients from school, work, or swimming. The swimming club put this information on its website and put up posters at all of its pools. Local newspapers and the city council magazine were used to raise public awareness of the two-week, no-swim rule. Educational material in the form of leaflets, posters, and web-based information was prepared for all local pool operators and swimming teachers in a campaign to raise awareness. These materials were distributed before the annual autumn peak for cryptosporidiosis in the UK.

Swimming pool operators were urged to conduct filter backwashing at the end of the day, following pool closure, rather than at noon.

Most water disinfection processes do not kill *Cryptosporidium*. It is resistant to normal chlorine levels used for swimming pool disinfection. Therefore, filtering is required to remove the parasite. Swimming pool outbreaks are often associated with mismanagement and inadequate pool water filtration. Swimming pool operators should ensure that filters are operating well and with coagulant, that there is sufficient water replacement (particularly in periods of high bather load), that they conduct filter backwashing appropriately and after the pool has closed at night, and that they encourage preswim showering, and that babies and toddlers wear special swim nappies (diapers). UV light, which inactivates *Cryptosporidium* oocysts, plays an adjuvant role in *Cryptosporidium* disinfection in a minority of newer pools. Risks to public health can be minimized by ensuring swimming pool construction, engineering, management, procedures and pool water circulation, and treatment and disinfection are optimal and in accordance with current guidelines.

As medical treatment options are limited, prevention and risk reduction are the most important interventions. General precautions against *Cryptosporidium* infection include handwashing prior to eating or preparing food and after contact with animals, adequate treatment of nonpotable water, and thorough washing of fruit and vegetables prior to consumption. There is no clear consensus on whether prewashed, packaged fruit and vegetables should be washed again at home as this could risk cross-contamination with other organisms. The Food Standards Agency (FSA) advises consumers to follow the instructions on the packaging in this regard. Suitable handwashing facilities should be provided and used at open farms. In England, the chief medical officer advises that immunocompromised individuals with T-cell deficiencies should boil all drinking water (including bottled water) to reduce the risk of infection.

EPIDEMIOLOGY

Cryptosporidium is one of the most common human enteropathogens worldwide, but it is most common in less developed countries. Reported cases in industrialized countries are rising because of the leading role that *Cryptosporidium* plays as a cause of waterborne outbreaks. In recreational waters, *Cryptosporidium* is the leading microbial cause of outbreaks in both the UK and the US. The true global burden of cryptosporidiosis is not known. The greatest burden is in children living in developing countries. However, the burden is difficult to quantify as estimates vary widely as a result of differences in study design, study populations, and diagnostic methods used. A recent multicentre study investigating the etiology of moderate-to-severe paediatric diarrhoeal disease in sub-Saharan Africa and south Asia found that most cases resulted from four pathogens: rotavirus, *Cryptosporidium*, enterotoxigenic *Escherichia coli* producing heat-stable toxin, and *Shigella*. *Cryptosporidium* was the second most common pathogen in infants (0–11 months old). Within the European Union, *Cryptosporidium* is a notifiable organism and laboratory-confirmed case reports are collected through the European Surveillance System (TESSy). However, *Cryptosporidium* is statutorily notifiable in only some European countries (for example, the UK, Germany, Ireland, and Sweden), and it is likely that there are substantial differences in ascertainment between countries because of variations in access to health care, requests for and submission of samples, laboratory testing, and notification. In the UK, *Cryptosporidium* is the most common protozoal cause of acute gastroenteritis, with notification rates between 5.8 and 9.1 per 100,000 population reported from 2007 to 2011. However, *Cryptosporidium* is known to be under diagnosed and underreported. One UK study estimated an annual incidence of *Cryptosporidium* in the community of 69.5 cases per 100,000 UK population.

The distribution of *C. hominis* and *C. parvum*, the major species infecting humans, varies geographically. *C. hominis* (found mainly in humans) predominates in most parts of the world, particularly in developing countries, and *C. parvum* (found commonly in humans and animals) is most frequent in the Middle East. Both species are common in Europe.

Cryptosporidium is transmitted by the faecal–oral route either by direct contact with an infected human or animal or their faeces, or indirectly through contaminated food or water. Contamination of food crops and source water with faeces from livestock is an important mechanism of zoonotic transmission for *C. parvum*, as well as children's farm visits or exposure to animal dung during outdoor recreation. Sporadic infections occur as well as outbreaks of disease. *Cryptosporidium* spreads proficiently in settings without adequate sanitation and hygiene. The epidemiology of the parasite can be explained by a number of key intrinsic characteristics. Oocysts are immediately infectious on excretion in the faeces, are shed in high numbers (up to 10^9 per stool), and can be passed for as long as two months after diarrhoea has stopped. The infectious dose is low (as few as 10 oocysts for some *C. hominis* and *C. parvum* strains) and oocysts can remain infectious in the environment for ≥ 6 months if kept moist (for example, in damp soil) and resist disinfection (including normal chlorine-based disinfection). Oocysts can survive in properly chlorinated recreational water for ≥ 10 days. In addition, the extended incubation period (mean: 7 days; normal range: 3–12 days) allows transmission to continue for days before public health officials detect an outbreak.

Young children and persons living with HIV/AIDS in developing countries are disproportionately affected. Age-related decreases in disease incidence in endemic settings suggest protective immunity induced by prior exposure. Thus those lacking protective immunity, including young children and immunocompromised individuals, are most at risk. In developing countries incidence of disease peaks in young children, with most experiencing an infection by 2 years of age. Exclusive and partial breastfeeding appear to afford some protection. Peaks of cryptosporidiosis occur during the warm rainy months. In the UK, cryptosporidiosis is widespread geographically and has a bimodal age distribution, with the greatest number of reported laboratory-confirmed cases occurring in children 0–4 years old and among adults 15–44 years old. *C. parvum* infections peak in spring and *C. hominis* peaks in late summer and autumn. Cases are more common during the second half of the year in the UK. From 2003 to 2013 there was an average of 2915 cases of cryptosporidiosis reported during the second half of the year compared to an average of 1267 cases reported during the first half of the year (Figure 6.4). A reduction in cases for the first half of the year has occurred compared to the previous decade (1992–2000). This reduction is thought to have occurred because improvements in drinking water supply and regulation have resulted in fewer *C. parvum* infections acquired from contaminated drinking water. In contrast, the number of cases in the second half of each year has shown a slight increase (see Figure 6.4). Some of this increase was attributed to identified outbreaks related to swimming pools in Majorca in 2000 and 2003. Infections in the latter half of the year are predominantly *C. hominis* and are thought to be related to holiday and foreign travel during the summer months and increased use of swimming pools.

Risk factors associated with sporadic infection include contact with ill persons and livestock (especially young ruminants), and travel abroad (especially to less industrialized countries). Outbreaks are commonly associated with waterborne transmission, with infections caused by contaminated drinking water and swimming pools a particular problem because the oocysts are resistant to chlorine-based disinfectants. As outbreaks propagate, secondary transmission from ill contacts becomes increasingly important, in some cases eventually outweighing water exposure as a risk factor. The largest outbreaks of cryptosporidiosis are associated with contamination of drinking water by sewage effluent or manure. Outbreak settings have included treated recreational water venues (communal swimming pools or water parks), untreated recreational waters (lakes, rivers),

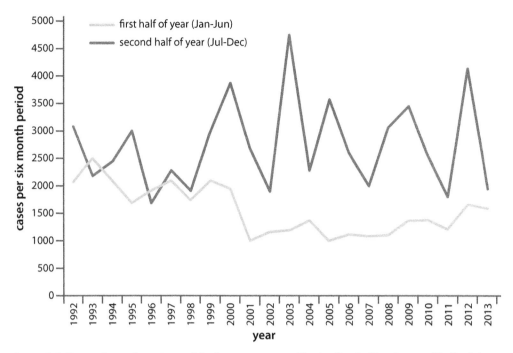

Figure 6.4. **Comparison of cryptosporidiosis cases reported in the first half and second half of the year in England and Wales, 1992 to 2013.** (Adapted from Public Health England Cryptosporidium Dataset [LabBase2] for England and Wales. Courtesy of Gastrointestinal, Emerging and Zoonotic Infections Department, Centre for Infectious Disease Surveillance and Control, Public Health England.)

institutions (day care centres and nurseries, schools, hospitals, prisons), farms (open farms, private farm open days, residential farm centres), and households and the general community (water or food consumption). Water is the most commonly reported vehicle of transmission in *Cryptosporidium* outbreaks. In the UK, in a review of 89 waterborne outbreaks of infectious intestinal disease (IID) involving 4321 cases, *Cryptosporidium* was implicated in 69%, making it the most common identified pathogen in this setting. It is the parasite's ability to survive in the environment and its resistance to chlorine disinfection that supports transmission through drinking water and recreational waters. In a European review, *Cryptosporidium* was found to be the most commonly identified causative organism in drinking water outbreaks, most of which were attributed to chronic filtration failures or to livestock and rainfall in the catchment area with inadequate or no filtration of water sources. Outbreaks in the 1980s and 1990s in the UK of *Cryptosporidium* from drinking water led to the development of regulatory requirements for *Cryptosporidium* in public drinking water. These regulations led to a substantial reduction in *Cryptosporidium* cases, especially in the first half of the year (see Figure 6.4), and fewer reported drinking water (from water mains) outbreaks. While mains drinking water quality is now highly regulated in industrialized countries, treated recreational water venues remain highly variable, and these are settings where *Cryptosporidium* outbreaks occur. Since 2001, swimming pools have been the most common setting for outbreaks of waterborne infectious intestinal disease in England and Wales, with *Cryptosporidium* as the leading cause. In the UK, the publication Swimming Pool Water: Treatment and Quality Standards for Pools and Spas and the Pool Water Treatment Advisory Group (PWTAG) Code of Practice provide guidance and set out quality standards for treated recreational waters. They are viewed as best practice, and swimming pool operators would be, and indeed have been, prosecuted for causing an outbreak by failing to follow this guidance.

Risk factors differ for the two species (*C. parvum* and *C. hominis*) that account for most disease in humans. Therefore, typing to species level yields useful epidemiological information, particularly in outbreaks, in terms of likely sources and routes of transmission. *C. hominis* infections tend to be associated with foreign travel. They also occur in children ≤1 year old and in adults, especially females aged 15–44 years old. Other contributing factors are changing children's diapers or swimming in toddlers' pools. *C. parvum* is associated with farm animal contact.

BIOLOGY

Cryptosporidium is an apicomplexan oocyst-forming protozoan parasite that infects a broad range of hosts, causing gastrointestinal illness in humans and young animals. Approximately 30 species of *Cryptosporidium* have been identified to date. The major species infecting humans are *C. hominis* and *C. parvum*, and they account for >90% of human cases of cryptosporidiosis. Other species that can occasionally infect humans or that may be more common in some settings include *C. meleagridis*, *C. felis*, *C. canis*, and *C. cuniculus*.

C. hominis and *C. parvum* can be subtyped by sequencing a surface antigen coding gene (known as gp60) involved in parasite attachment and invasion. Sequence differences are used to distinguish subtype families (*C. hominis*: Ia, Ib, etc.; *C. parvum*: IIa, IIc, IId, etc.). Subtypes Ib, IIa, IIc, and IId are among those most commonly observed in humans. Greater genetic diversity is seen in developing countries, particularly in rural settings. Subtyping can provide useful information concerning source of transmission (for example, animal-to-human versus human-to-human). However, gp60 sequencing alone may underestimate parasite diversity, but there is currently no consensus on multilocus subtyping methods. Currently there is limited understanding of the correlation between subtype and clinical manifestations, and the protective role of homotypic or heterotypic subtype immunity.

DISEASE

Cryptosporidium was first recognized as a causative agent of human gastroenteritis in 1976. Debilitating disease, chronic infections, and high mortality identified in AIDS patients during the early 1980s led to the inclusion of cryptosporidiosis as an AIDS-defining illness. Symptoms reflect the parasite's predominant pathogenesis of the small bowel. A UK study of laboratory-confirmed *Cryptosporidium* cases who had sought medical attention for their symptoms found that the most common symptom was diarrhoea (98%), which was typically watery (81%) and lasted a median of 5 to 10 days. Other symptoms included abdominal cramps (60–96%), fatigue, nausea (35%), loss of appetite, fever (36–59%), and vomiting (49–65%). However, many infections are undiagnosed or unconfirmed as they are asymptomatic or mild and self-limiting. Symptoms can persist for up to a month. Over one-third of cases have relapsing symptoms, indicative of persistent infection. The differential diagnosis is usually other causes of infectious gastroenteritis. Severity of illness is influenced by age, nutritional status, and immune status of the host, and possibly by infecting species and subtype. In developed countries, most cases are in immunocompetent individuals who experience a self-limiting illness. Among children in developing countries

Cryptosporidium is one of the most important causes of persistent diarrhoea. Complications in this population include malnutrition and long-term impairment of cognitive and physical fitness. In immunocompromised individuals, particularly those with T-cell immune deficiencies, the risk increases for severe, protracted disease, which can result in malabsorption, weight loss, and death. At most risk from cryptosporidiosis are those with a CD4 count <50/mm^3; patients with a CD4 count >200/mm^3 are not at increased risk.

Extraintestinal manifestations may occur and are most common in children and immunocompromised persons. Biliary tract disease including sclerosing cholongitis, acalculous cholecystitis, and pancreatitis as a complication in those severely immunocompromised has a poor prognosis. Respiratory disease in the form of pulmonary infiltrates and respiratory distress has been observed, most often in children. Sinusitis has also been described. In advanced HIV, pneumatosis cystoides intestinalis is a rare complication associated with cryptosporidiosis in which gas-filled cysts develop in the gut wall. Rupture of the cysts can lead to pneumoretroperitoneum and pneumomediastinum.

Reported long-term effects of *Cryptosporidium* infection include seronegative reactive arthritis and relapse of inflammatory bowel disease. Possible association with irritable bowel syndrome is under investigation.

Treatment options are limited. Rehydration with oral or intravenous fluids may be required. The aim of treatment in immunocompromised patients is improvement of symptoms, with complete clearance of the parasite unlikely unless the underlying immune deficiency can be corrected. In HIV-infected individuals, resolution of symptoms relies on restoration of immune status using highly active antiretroviral therapy (HAART). HAART improves the CD4 count and restores a degree of immunity. In addition, protease inhibitors seem to have some antiparasitic activity, an effect enhanced with paromomycin. In the US, nitazoxanide, a thiazole compound, is licensed for use and available by regular prescription for disease in immunocompetent patients >1 year old. It has been shown to decrease severity and duration of symptoms in this patient group. Parasite clearance has also been demonstrated. There is less conclusive evidence of the benefits of nitazoxanide in immunocompromised individuals. In the subgroup with the most advanced HIV disease, it has been shown not to be effective. Nitazoxanide is not licensed in the European Union but is available in the UK on a named-patient basis.

PATHOLOGY

Cryptosporidium has a monoxenous life cycle completed within the gastrointestinal tract of a single host. Infection follows ingestion of the oocyst life-cycle stage, which is shed in faeces. Under conditions triggered in the intestine, the oocysts each release four motile infectious sporozoites in a process known as excystation. These actively probe, attach, invade, and become engulfed by host epithelial cells at the luminal surface. An asexual cycle follows, involving differentiation and, sequentially, trophozoites, Type I meront, and merozoite production. The parasite proliferates as six or eight merozoites are released to invade neighbouring epithelial cells, and they either develop into trophozoites, repeating the asexual cycle, or into Type II meronts. The sexual cycle is initiated following production of four merozoites by Type II meronts, which are then released to invade neighbouring host cells and differentiate into either macrogamonts (ova) or microgamonts. Microgamonts release microgametes (sperm) into the intestinal lumen. The microgametes attach and penetrate infected epithelial cells to fertilize the macrogamete, producing a zygote. Following meiosis, the zygote differentiates into four sporozoites as the oocyst matures and is released into the intestinal lumen. Sporulated oocysts are shed in the faeces, often in large numbers, and are immediately infectious for the next susceptible host. Autoinfection of the host can occur as sporozoites may be released directly into the intestinal lumen, and the life cycle continues.

The main site of infection is the small bowel, although infection may spread throughout the gastrointestinal tract and extraintestinal sites. In HIV infected individuals, more proximal small bowel infections generally cause more severe diarrhoea and reduce survival rates, compared to heavy infection of the colon that, in the absence of small intestine infection, can cause intermittent diarrhoea or even asymptomatic infection. Osmotic, inflammatory, and secretory aspects of diarrhoea have all been investigated for cryptosporidiosis, but the causative mechanisms have not been fully understood. They are, however, multifactorial, consisting of the effect of the parasite and its products on the epithelial layer and the immunological and inflammatory responses of the host, leading to impaired intestinal absorption and enhanced secretion. Loss of barrier integrity and epithelial cell injury have been reported from *in vitro* studies and altered bowel habits have been reported in patients following *Cryptosporidium* infections.

QUESTIONS

1. *Cryptosporidium parvum* is commonly found in which host or hosts?
   a. Humans only
   b. Humans and mammals
   c. Birds only
   d. Mammals and birds
   e. Humans and birds

2. With respect to *Cryptosporidium hominis* and *C. parvum*, which statement is incorrect?
   a. *C. parvum* is zoonotic.
   b. *C. hominis* can be transmitted via swimming pools.
   c. Both species are a cause of waterborne outbreaks.
   d. Both species may be transmitted anthroponotically.
   e. Both species may be transmitted zoonotically.

3. Which is the most common symptom of *Cryptosporidium* infection in humans?
   a. Abdominal pain
   b Vomiting
   c. Fever
   d. Diarrhoea
   e. Nausea

4. *Cryptosporidium* oocysts in stool samples can be identified by which stain?
   a. Modified Ziehl–Neelsen
   b. Iodine
   c. Crystal violet
   d. Trichrome
   e. Acridine orange

5. Which is not considered to be a general precaution against *Cryptosporidium* infection?
   a. Handwashing prior to eating
   b. Thorough washing of fruit before consumption
   c. Boiling all drinking water
   d. Handwashing after contact with animals
   e. Adequate treatment of nonpotable water

GUIDELINES

1. Advice on the response from public and environmental health to the detection of crypto-sporidial oocysts in treated drinking water (2000) PHLS Advisory Committee on Water and the Environment. *Commun Dis Public Health* **3**(1):24–27.

2. Guidance for the investigation of *Cryptosporidium* linked to swimming pools (2016) *Cryptosporidium* Reference Unit, Swansea.

3. Swimming pool water: treatment and quality standards for pools and spas (2017) Pool Water Treatment Advisory Group.

4. PWTAG Code of Practice. Pool Water Treatment Advisory Group (2016).

5. Preventing person-to-person spread following gastrointestinal infections: guidelines for public health physicians and environmental health officers (2004) *Commun Dis Public Health* **7**(4):362–384.

6. Guidance on infection control in schools and other childcare settings (2010) Health Protection Agency, London.

7. Guidelines for the investigation of zoonotic disease, England and Wales (2009). Health Protection Agency, London.

8. Cryptosporidiosis Outbreak Response & Evaluation (CORE) Guidelines (2009) Centers for Disease Control and Prevention, Atlanta.

9. Guidelines for safe recreational water environments (2006). Swimming Pools and Similar Environments, vol 2. WHO, Geneva.

10. Guidelines for Drinking-Water Quality, 4th ed (2011). WHO, Geneva.

REFERENCES

1. McCann R, Jones R, Snow J et al. (2014) An outbreak of cryptosporidiosis at a swimming club—can rapid field epidemiology limit the spread of illness? *Epidemiol Infect.* **142**(1):51–55 (doi: 10.1017/S0950268813001143. Epub 2013 May 14).

2. Shirley DA, Moonah SN & Kotloff KL (2012). Burden of disease from cryptosporidiosis. *Curr Opin Infect Dis* **25**(5):555–563 (doi: 10.1097/QCO.0b013e328357e569).

3. Kotloff KL, Nataro JP, Blackwelder WC et al. (2013) Burden and aetiology of diarrhoeal disease in infants and young children in developing countries (the Global Enteric Multicenter Study, GEMS): a prospective, case-control study. *Lancet* **382**(9888):209–222 (doi: 10.1016/S0140-6736(13)60844-2. Epub 2013 May 14).

4. Chalmers RM, Campbell B, Crouch N & Davies AP (2010) Clinical laboratory practices for detection and reporting of Cryptosporidium in community cases of diarrhoea in the United Kingdom. *Euro Surveill* **15**(48),(pii):19731.

5. European Centre for Disease Prevention and Control. Annual epidemiology report on communicable diseases in Europe 2011. [2013 (12 March 2104)]; Available from: http://www.ecdc.europa.eu/en/publications/surveillance_reports/annual_epidemiological_report/Pages/epi_index.aspx.

6. Tam CC, Rodrigues LC, Viviani L et al. (2012) Longitudinal study of infectious intestinal disease in the UK (IID2 study): incidence in the community and presenting to general practice. *Gut* **61**(1):69–77 (doi: 10.1136/gut.2011.238386. Epub 2011 Jun 27).

7. Jokipii L & Jokipii AM (1986). Timing of symptoms and oocyst excretion in human cryptosporidiosis. *N Engl J Med* **315**(26):1643–1647.

8. Chappell CL, Okhuysen PC, Langer-Curry R et al. (2006) *Cryptosporidium hominis*: experimental challenge of healthy adults. *Am J Trop Med Hyg* **75**(5):851–857.

9. Chappell CL, Okhuysen PC, Sterling CR et al. (1999) Infectivity of *Cryptosporidium parvum* in

healthy adults with pre-existing anti-C. parvum serum immunoglobulin G. *Am J Trop Med Hyg* **60**(1):157–164.

10. Chalmers RM, Elwin K, Thomas AL et al. (2009) Long-term *Cryptosporidium* typing reveals the aetiology and species-specific epidemiology of human cryptosporidiosis in England and Wales, 2000 to 2003. *Euro Surveill* **14**(2),(pii):19086.

11. Chalmers RM, Smith R, Elwin K et al. (2011) Epidemiology of anthroponotic and zoonotic human cryptosporidiosis in England and Wales, 2004–2006. *Epidemiol Infect* **139**(5):700–712 (doi: 10.1017/S0950268810001688. Epub 2010 Jul 12).

12. Chalmers RM & Katzer F (2013) Looking for *Cryptosporidium*: the application of advances in detection and diagnosis. *Trends Parasitol* **29**(5):237–251 (doi: 10.1016/j.pt.2013.03.001. Epub Apr 6).

13. Risebro HL, Doria MF, Andersson Y et al. (2007) Fault tree analysis of the causes of waterborne outbreaks. *J Water Health* **5**(Suppl 1):1–18.

14. Smith A, Reacher M, Smerdon W et al. (2006) Outbreaks of waterborne infectious intestinal disease in England and Wales, 1992–2003. *Epidemiol Infect* **134**(6):1141–1149 (Epub 2006 May 11).

15. Bouzid M, Hunter PR, Chalmers RM & Tyler KM (2013) *Cryptosporidium* pathogenicity and virulence. *Clin Microbiol Rev* **26**(1):115–134 (doi: 10.1128/CMR.00076-12).

16. Robinson G, Elwin K & Chalmers RM (2008) Unusual *Cryptosporidium* genotypes in human cases of diarrhoea. *Emerg Infect Dis* **14**(11):1800–1802 (doi: 10.3201/eid1411.080239).

17. Xiao L. (2010) Molecular epidemiology of cryptosporidiosis: an update. *Exp Parasitol* **124**(1):80–89 (doi: 10.1016/j.exppara.2009.03.018. Epub Apr 7).

18. Morse TD, Nichols RA, Grimason AM et al. (2007) Incidence of cryptosporidiosis species in paediatric patients in Malawi. *Epidemiol Infect* **135**(8):1307–1315 (Epub 2007 Jan 15).

19. Chalmers RM & Davies AP (2010) Minireview: clinical cryptosporidiosis. *Exp Parasitol* 124(1):138–146 (doi: 10.1016/j.exppara.2009.02.003. Epub Feb 11).

20. Clayton F, Heller T & Kotler DP (1994) Variation in the enteric distribution of cryptosporidia in acquired immunodeficiency syndrome. *Am J Clin Pathol* **102**(4):420–425.

21. Hunter PR, Hughes S, Woodhouse S et al. (2004) Sporadic cryptosporidiosis case-control study with genotyping. *Emerg Infect Dis* **10**(7):1241–1249.

ANSWERS

MCQ	Feedback
1. *Cryptosporidium parvum* is commonly found in which host or hosts? a. Humans only b. Humans and mammals c. Birds only d. Mammals and birds e. Humans and birds	The major species infecting humans are *C. hominis* and *C. parvum*; they account for >90% of human cases of cryptosporidiosis. *C. parvum* has a human and animal infection cycle, and the source of transmission can be anthroponotic (human to human) or zoonotic (animal to human). It is a common cause of diarrhoea in calves, lambs, and goats that are not yet weaned. *C. hominis* has a human infection cycle and is transmitted anthroponotically. There is no defined animal host, although low density of the parasite has been detected occasionally, and without evidence of onward transmission, in cattle, sheep, and goat faeces.
2. With respect to *Cryptosporidium hominis* and *C. parvum* which statement is incorrect? a. *C. parvum* is zoonotic. b. *C. hominis* can be transmitted via swimming pools. c. Both species are a cause of waterborne outbreaks. d. Both species may be transmitted anthroponotically. e. Both species may be transmitted zoonotically.	For *C. parvum*, the source of transmission can be anthroponotic (human to human) or zoonotic (animal to human). Despite occasional reports in livestock, *C. hominis* appears to be transmitted anthroponotically. In recreational waters, *Cryptosporidium* is the leading microbial cause of outbreaks in both the UK and the US.
3. Which is the most common symptom of *Cryptosporidium* infection in humans? a. Abdominal pain b. Vomiting c. Fever d. Diarrhoea e. Nausea	The most common symptom of *Cryptosporidium* infection is diarrhoea (98%), which is typically watery (81%) and lasts a median of 5–10 days. Other symptoms may include abdominal cramps (60–96%), vomiting (49–65%), fatigue, nausea (35%), loss of appetite, and fever (36–59%). Extraintestinal manifestations may occur, including biliary tract disease (sclerosing cholongitis, acalculous cholecystitis, pancreatitis) and respiratory cryptosporidiosis. Severely immunocompromised persons are most at risk of extraintestinal symptoms.
4. *Cryptosporidium* oocysts in stool samples can be identified by which stain? a. Modified Ziehl–Neelsen b. Iodine c. Crystal violet d. Trichrome e. Acridine orange	Laboratory verification is required to confirm a diagnosis of *Cryptosporidium*. This is usually done by the detection of oocysts in stool samples by a microscopic examination of smears stained with tinctorial stain (usually acid fast, such as modified Ziehl–Neelsen stain, fluorescent stain (such as auramine-phenol), or immunoflourescent stain. Alternatively, an enzyme immunoassay (EIA) or polymerase chain reaction (PCR) may be used.
5. Which is not considered to be a general precaution against *Cryptosporidium* infection? a. Handwashing prior to eating b. Thorough washing of fruit before consumption c. Boiling all drinking water d. Handwashing after contact with animals e. Adequate treatment of nonpotable water	General precautions against *Cryptosporidium* infection include handwashing prior to eating or preparing food and after contact with animals, adequate treatment of nonpotable water, and thorough washing of fruit and vegetables prior to consumption. Suitable handwashing facilities should be provided and used at open farms. In England, the chief medical officer advises that only immunocompromised individuals with T-cell deficiencies should boil all drinking water (including bottled water) to reduce the risk of infection.

GIARDIA OUTBREAKS ON SHIP

Elizabeth Sheridan[1], Allan Johnson[2], and Nabila Mughal[3,4,5]

[1]Poole Hospital NHS Trust, UK.
[2]National Infection Service, Food Water and Environmental Microbiology Laboratory Service, Public Health England, Colindale, London.
[3]Chelsea and Westminster Hospital NHS Foundation Trust.
[4]Imperial College Healthcare NHS Trust.
[5]Imperial College London.

A luxury around-the-world cruise ship is full to capacity, carrying 1200 passengers (average age 66) and 500 crew. It set sail from Southampton three weeks earlier and docked in Spain, the Azores, the Caribbean Islands, and Mexico, before sailing through the Panama Canal and on to California. Passengers explored onshore at all destinations, where most dined at local restaurants and cafés. Water for drinking, personal hygiene, and making ice, as well as for a freshwater swimming pool and four spa pools, is produced on board via reverse osmosis (RO) from seawater taken on only while the ship is sailing in open water or is supplied from water of drinking standard taken on board (bunkered) in port. There is also a seawater pool that is filled daily from the open ocean. Food is a mixture of frozen and refrigerated items from the port of embarkation with some additional fresh food brought on board along the way, sourced from reputable, international suppliers subject to rigorous quality checks. Three passengers consult the ship's doctor with symptoms of diarrhea and abdominal discomfort just before the ship docks in San Francisco.

INVESTIGATION OF THE INCIDENT

As the ship's doctor, you collect stool samples that are sent to a private laboratory in San Francisco. Two days later, just as the ship sets sail again for Hawaii, you receive a call on your satellite phone saying that two of the three samples have tested positive for *Giardia*. That day, four more people visit your office complaining of similar symptoms.

CLINICAL MANAGEMENT

Giardia infections are treated with supportive therapy to prevent dehydration and with antiprotozoal treatment as detailed below. Most exposed people will have an asymptomatic infection. Not all those with symptoms will require treatment because the symptoms are often self-limiting. Those who are immunosuppressed are more likely to have a protracted illness needing treatment. Remember to warn passengers not to drink alcohol while taking metronidazole. You don't have a large supply of drugs on board ship, so you need to arrange to obtain additional supplies at your next stop. If kept well hydrated, people are unlikely to require hospitalization. Any infants, pregnant women, or frail older people need to be monitored more closely to ensure they are adequately hydrated.

PREVENTION OF FURTHER CASES

This situation looks very much like an outbreak: Could the cases be the start of a steep epidemic curve? Your first task is to establish whether *Giardia* is the true and only etiological agent responsible for the diarrhoea. Is it a genuine outbreak of *Giardia*? Is it not an outbreak at all, just isolated cases from exposure in port or detection of asymptomatic carriage? Or is it an outbreak of something else, such as norovirus, with *Giardia* detected incidentally as asymptomatic carriage?

Once you have established that *Giardia* is the responsible agent, you need to establish whether the infection was acquired on board. If it was acquired on board, was it a point exposure to a food source (for example, contaminated salad) or could it result from contamination of the water supply?

Given the relatively long incubation period, you are concerned that you may have many more cases over the coming days. Your working hypothesis must be that this is a real outbreak of *Giardia* and act accordingly.

Finding the source of the infection

A waterborne source could be a problem with the water-storage tanks or distribution system, inadequate disinfection, contaminated water acquired in port, or defective water backflow that has led to contamination of water supplies with sewage. Another possibility is an infected individual shedding cysts to which other people are exposed, for example in swimming pool water or through food handling if he or she happens to be a crew member.

Pipework on the vessel will be complex and difficult to inspect; clean and dirty water can become mixed or water can become contaminated during maintenance work. There may be more than one distribution system and different sections of the ship commonly have separate supplies. There should be adequate prevention of backflow from one part of the system to another wherever potable water serves a nonpotable use. The strength of the method used to prevent backflow will vary according to risk, that is, water serving high-risk systems such as toilets would be fitted with air breaks.

Water for systems such as laundry, recreational water facilities, or boilers comes from dedicated tanks, through a break tank, or through a backflow preventer. Lower-risk systems such as steam ovens and ice machines will have backflow preventers installed. The engineers will have drawings of all systems, but these should be used with caution as they are rarely updated. It is important to look at maintenance records for all parts of the water systems.

Start your investigation by talking to the captain and chief engineer. You need to find out about all potential routes of infection and anything that might have led to a breach in the water supply systems. In this instance, the engineer admitted sheepishly that there had been issues with the RO unit (Figure 7.1).

Clean drinking water had been taken on board in Spain, St Lucia, and again in Mexico, as per usual practice. The ship has 10 tanks for potable water. Normal procedure is to bunker (fill) one tank at a time. The bunkering hose is normally around 75 mm in diameter and should be owned by the ship. The hoses are normally disinfected between use and stored in a dedicated locker with the ends protected by caps or hygienic plastic bags. Water bunkering hoses used to bring water aboard have unique connections, making it

Figure 7.1. **Photo of a ship's reverse-osmosis (RO) plant.**

impossible to connect them to other systems, such as sewage discharge or oil bunkering and discharge.

When water is taken from shore, the most usual practice is to take a shore-side sample of the water and samples from the filled tanks. These samples are tested for *Escherichia coli* and other coliforms with an on-board kit (Colilert, Colisafe, and Colitag are the most common) and the tanks only used if the test results are confirmed as negative. Water produced on board by reverse osmosis would be sent to a separate tank to bunkered water. Transfer of water between tanks should take place only in emergencies to stabilize a ship. You therefore ask to see the records and talk to the crew members who handled the process of taking on the water.

Nothing irregular emerges as far as food supplies are concerned. There are no reports of diarrheal illness in the crew, and your thorough investigation of the catering department reveals no cases in their staff. You take samples of salads and frozen soft fruit.

Reporting the outbreak

At this point the outbreak must be reported to public health authorities. Giardiasis is a reportable disease under EU legislation and elsewhere, including the US. There is a requirement to report any incidence of disease known, or thought, to be of an infectious nature to the appropriate authority at the next port, by submitting a Maritime Declaration of Health certificate. In the US, this authority is the Centers for Disease Control and Prevention (CDC), which has a dedicated team of ship inspectors in their Vessel Sanitation Program. In the US, it is standard practice for the laboratory that conducted testing to report positive results for reportable illnesses directly to the CDC, specifically the Vessel Sanitation Program (VSP) for gastrointestinal illnesses on cruise ships. Once this report is received, the VSP's epidemiologist will contact the cruise line and begin an investigation of this cluster of cases. The investigation will begin by email and telephone and may also include an on-board investigation. You will coordinate with the head office of the cruise line, which may contact Public Health England (PHE) for further advice.

Water sources need to be investigated and, based on investigation results, other vessels warned. In the US, the CDC VSP control the whole process. In Europe, it is a combination of Port State Control (for example, the Maritime Coastguard Agency (MCA) in the UK) and the country's agency for port health, although there are variations between countries. Port State Control and VSP have a global communication network to warn the next port that the vessel is coming and to allow follow-up inspection.

Keep the captain informed of all developments and actions. The cruise line press office may also want to be informed so that it can have a press statement ready in case news reaches the public. In this case, Southampton, as the home port should also be informed as any major remedial work could be organized by way of an extended turnaround time (that is, the time between the disembarkation of one set of guests and the embarkation of the next set of guests) at the end of a voyage. After an outbreak caused by a contagious pathogen, such as norovirus, thorough sanitation procedures are implemented. You need to reduce the likelihood that the illness will spread on this voyage and on the next. The cruise line will also wish to minimize the damage to its reputation. Outbreaks such as this influence the public for years, depending on how the media report the details and it is vital to handle the situation well. Take advice and be frank and open at all times. Never be afraid to ask for help.

You will need to inform passengers and crew at this stage that there is a potential outbreak, let them know what you are doing to stop the outbreak, communicate steps they can take to stay healthy, and keep them updated as the situation develops.

In-depth investigation

You decide to perform an in-depth epidemiological investigation to find and eliminate the source of the infection. You first question the affected individuals about their food and water exposure and then perform a formal case-control study. You develop a working case definition based on the symptoms because microbiological results are not available for all affected people. The cabin numbers of patients are very important because they may receive water from the same contaminated potable water circuit. If that has occurred, then disinfecting the water circuit would be a priority. Help with questionnaires is available from PHE, online, and over the phone. The self-contained environment of the ship facilitates outbreaks but also aids epidemiological investigation. Comprehensive data collection is easier than on land as all affected passengers and crew members should be reporting to you as the central medical officer. People are likely to self-medicate for their symptoms, causing cases to be missed, and you need to encourage reporting. Offer free consultation for people showing symptoms consistent with the outbreak.

Preventing further exposure

To prevent further cases, you need to stop any ongoing exposure to fresh water until it can be ruled out as the source. Ideally all drinking water should be bottled, filtered (using the correct pore size), or boiled, although in practice this may not be possible on board a ship. You should therefore disinfect all filters in the potable system, including the RO unit, and inspect them for damage. The membrane pore size is typically 0.1 μm, which will remove *Giardia* oocysts during normal operation unless the membrane was damaged. Any equipment thought to be sources of infection, such as softeners, should be cleaned and disinfected. Any water still in tanks from a shore source is either treated or dumped. A tank can then be cleaned, disinfected, refilled, and dosed to the higher level of chlorine, for example, 5 ppm. If the membranes of the RO unit are functional and undamaged, then water production can resume as long as the ship is at sea. Production usually begins 200 miles from shore in water that is >200 m deep. Clean water from this tank can then be used for drinking. Ice should not be in contact with food or used in drinks until the ice machine filters have been changed.

The freshwater pool and the spas should be closed for disinfection. The saltwater pool can still be used; this is unlikely to be the primary source of infection because *Giardia* typically is not found in seawater in the open ocean. Saltwater pools are not usually in circulation mode while at sea and therefore, if somebody were to shed the organism, there would be a chance that other bathers would ingest it. Consequently, you instruct the pool to be put in circulation mode to ensure the water is cleaned by circulation through the filters (Figure 7.2) and maintenance of the water chemistry with halogen and pH control.

Water can still be used for personal hygiene, although there is a potential risk of ingesting cysts, as when swallowing shower water and brushing teeth. Because a food-related source has not been ruled out, only cooked or peeled food should be served and items that could act as a vehicle for the infection (for example, salad) should be disposed of. Food storage areas should be cleaned carefully. Note that it is illegal under international shipping regulations to dispose of food overboard near the coast.

Emphasize the importance of hand hygiene to passengers and crew, specifically after using the toilet and before eating or handling food. Isolate symptomatic crew members in their cabins or the medical centre. Until they are free of symptoms, crew members should not serve food and drinks. Avoid self-service buffets.

Sampling to identify the source

Samples of water need to be taken from all storage tanks and many different outlets. Large quantities of water (about 1000 L) need to be studied because cysts are likely to be present only in low numbers and samples have to be collected using a filtration pump. The ship agent portside is charged with identifying a suitable laboratory. Any suspect food and ice should also be sampled prior to disposal. More patient samples should also be tested if any further cases occur.

Environmental measures

Unlike many bacterial pathogens, *Giardia* does not multiply in the environment, but potential sources still need to be disinfected. Storage tanks and pipework will have to be cleaned in stages by emptying and super-chlorinating them before refilling with clean water.

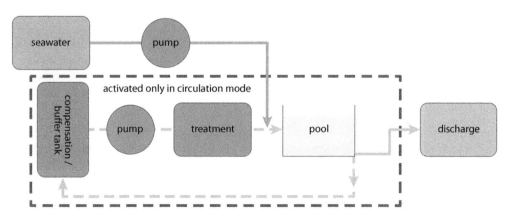

Figure 7.2. **Schematic of backwash system showing circulation mode.**

Pools and spa: After closure, these can be disinfected by dosing with a flocculant that is super chlorinated with 10–40 ppm free chlorine for between 25.5 to 6.5 h, depending on the dose, while being passed continuously through the sand filters. The sand filters are then thoroughly backwashed to sea or bilge before the water is drained to sea (see Figure 7.2). If the facilities seem to be at risk, the sand and filter media can be changed. If cartridge filters were used, they would be changed. The surfaces of the pool and spas and their balance tanks should be cleaned manually and disinfected.

The filters in the RO units are changed in case they were contaminated. Water softeners would also be cleaned and all sand filters subjected to prolonged backwash.

Isolation precautions: Strict isolation is not necessary, although it would be for a more infectious agent such as norovirus. It is not necessary to confine passengers to their cabins but they should be encouraged to stay there and use en-suite facilities if having diarrhoea. The importance of good personal hygiene, in particular hand washing, must be emphasized. Passengers with symptoms must be told specifically not to use the pools and spas once these are reopened. Consider giving pool staff a list of affected passengers, with photos, although passenger confidentiality should be respected.

Containment in the laboratory is not necessary.

Cleaning and waste disposal: No special cleaning is indicated. The housekeeping department should be alerted to enable cleaners to deep clean toilets if necessary and to be equipped with any personal protective equipment (PPE) seen as necessary, that is, when using chlorine in high concentrations. No specific waste disposal or decontamination procedures are needed.

PPE: none needed

What happens next?

The epidemiological analysis does not identify any common food source or any pattern of cabin occupancy associated with any one storage tank. There is a weak association with exposure to spa pools.

Investigation of water intakes shows bunkering of water mains in Vigo and St Lucia. Treated water was taken on in Mexico. Not all the ship's testing records of water for coliforms can be found, however. A few more people report to you with symptoms of diarrhoea over the next 14 days. You continue active surveillance.

Fortunately, the next stop is Honolulu where the ship has an extended stay, enabling a full inspection and disinfection of all the water facilities, including the RO unit. This is clearly much easier to do without passengers aboard, so a choice of activities ashore is organized, free of charge. Any passengers not able or unwilling to leave the ship must be informed that there will be disruption. The crew receives similar information.

The whole system, not just the tanks, is disinfected. Cold and hot water systems should be dosed with chlorine at 20 ppm for 2 h or 50 ppm for 1 h, which is drawn through the entire system. The time starts when the levels are correct at all outlets (and there are thousands).

The system is then drained, refilled, and run until chlorine levels have returned to a concentration of 0.2 ppm (no more than 5.0 ppm).

The hot system should have the temperature raised to as high as possible, at least to 70°C and ideally 80°C, but this depends on the system's materials, as plastic systems become leaky above 70°C. High water temperature should be maintained and hot water circulated for up to 3 days, after which each tap would be flushed at full temperature for 5 min. This water temperature poses clear risk of scalding, so passengers and crew must be warned. It is best done with passengers off the ship, but it has been done with passengers and crew aboard.

Investigations carried out by Port Health after it was notified about the outbreak show that the source was most likely water bunkered in St. Lucia, where a breach of the mains supply was identified. This breach resulted in contamination of the water supply with wastewater and led to an outbreak of *Giardia* on the island, with several large hotels being affected. Other ships that had used the same water source were warned and were able to clean out their water systems with no additional cases being discovered. Unfortunately, there are no records to be found of test results for this water on your ship. *Giardia* is, however, detected in two of the systems on board, one of which supplied the spa pool and a swimming pool.

Prevention of water and food-borne outbreaks on ships

Ships have historically been recognized for their role in transmitting infection. Because they are enclosed environments, it is possible for a range of enteric and respiratory pathogens to cause outbreaks by person-to-person transmission, aerosols, and fomites. Norovirus is a major problem due to the low infective dose, high attack rate, short incubation period, sudden onset of symptoms, long survival time in the environment, and rapid turnover of passengers, and it has plagued cruise companies for many years.

There have been many documented water and food-borne outbreaks associated with passenger, naval, cargo, and fishing vessels. Infection control standards to ensure safe water, food, and sanitation are key to preventing outbreaks, to the health of passengers, and to the success of the shipping industry.

Water has been found to be the cause of many outbreaks on ships and data from 21 documented ship outbreaks worldwide between January 1970 and June 2003 showed that 6400 people were infected. The actual number of infections may be significantly higher than recorded because many cases were mild or asymptomatic, small numbers of passengers or crew members were affected, the duration of trips onboard ferries vs. cruises is short prior to arriving at their destination, crew members may have underreported the number of cases to avoid delays at ports, or ships simply failed to publish information about the outbreak. Travellers returning from short ferry trips, who subsequently become unwell, should be investigated with a view to identifying a potential common source that may involve other travellers.

Waterborne disease outbreaks may occur on passenger ships due to failures in water safety systems.

Water aboard ships is always transported by way of water tanks and complex distribution systems and may additionally be used recreationally in swimming pools and spas. A range of pathogens can cause infections, which usually occur following ingestion of either contaminated water or ice or by contact with water through bathing or swimming. Infections also may occur through inhalation of aerosols. Of these infections, 86% of outbreaks were associated with passenger ships, and, in 57% of those cases, water or ice was confirmed as the source and was additionally suspected in 33% of cases. One-third of outbreaks were attributed to *Enterotoxigenic Escherichia coli* (ETEC), however other species such as *Salmonella* sp., *Shigella* sp., *Cryptosporidium*, norovirus, and *Giardia* were also documented. In 81% of outbreaks, contributing factors were identifiable and the documented outbreak associated with *Giardia* resulted from contaminated water loaded onto a ship and inadequate disinfection.

Cruise ships use huge volumes of fresh water. Ships are able to desalinate seawater by reverse osmosis or distillation at sea, although these processes, especially distillation, are expensive. Cruise ships routinely take fresh water from shore for convenience and necessity when they are in port for longer periods. In addition, desalinated water is unpleasant to drink unless it is passed through a mineralizer and shore water is often taken to refresh the tanks and systems. This process may introduce pathogens into the system, through a contaminated source water or contamination of the filling hose, filling line, or shore side and barge connections. Water will also be chlorinated in the tank at the time of bunkering, normally up to 5 ppm to give a residual of 2 ppm. The chlorine will degrade (blow off) rapidly because of turbulence during bunkering. Further treatment through filtration, more chlorination, UV radiation, silver ionization (Ag), and activated carbon filters can also be used. Note that parasites such as *Giardia* are resistant to chlorination and require additional filtration.

There is commonly a third source of water, which is condensate from the A/C system. This source will produce many tons a day in hot climates, such as the Caribbean. Although it is usually directed to enclosed systems such as laundry, it can pose risks. In waterborne outbreaks, it is normal practice to disinfect the A/C filters and check free drainage.

In addition to contaminated water being bunkered, contamination can also occur during storage. In particular, contamination can arise from defective systems leading to backflow or from cross contamination of potable water with contaminated water, damage to or poor maintenance of pipe works, or inadequate temperature control and decontamination. Biofilms containing pathogens such as *Legionella* and *Pseudomonas* can build up in rarely used outlets or if the system has dead legs (plumbing cul-de-sacs).

The following measures are important in preventing waterborne outbreaks aboard ships:

- Water should be bunkered from safe, reliable sources at port. For water supplied from a recognized water utility, the microbiological and chemical quality are the responsibility of the producer. Technically, the ship should obtain a certificate of analysis from the supplier before bunkering, although they often do not bother. Instead of obtaining the certificate, ships will conduct an on-board microbiological analysis for total coliforms and *E. coli*.
- Water should be loaded properly at port, with particular care to avoid cross connections. Connections should be designed to prevent accidental cross connection.
- Water should be chlorinated in the tank at the time of bunkering, normally up to 5 ppm to give a residual of 2 ppm. The chlorine will degrade (blow off) rapidly because of turbulence during bunkering.
- The water quality should be checked using tests for coliforms and *E.coli*, turbidity, pH and residual chlorine.

- Chlorination and pH levels will be checked and adjusted automatically after leaving the tank and before entering the distribution system.

- Extra treatment of water through filtration, chlorination, UV radiation, silver ionization (Ag), and activated carbon filters can also be used, if the water is suspected of being contaminated. Note that parasites such as *Giardia* are resistant to chlorination and require additional filtration. Residual disinfectants in on-board distribution systems should be monitored routinely.

- Seawater only should be taken on in deep water away from potentially contaminated shallow coastal areas. Seawater should not be taken on during the discharge of sewage or food waste to sea.

- Regular inspection and maintenance of potable water systems, including storage tanks, distribution pipes, hoses, and backflow preventers is needed.

- Adequate training and supervision of crew assigned to these tasks is important to ensure that they are able to identify potential irregularities and prevent outbreaks.

- Detailed records of monitoring, maintenance, training, disinfection, and corrective measures should be kept for at least 12 months. They should be available for inspection and to assist in the management of any infection or outbreak.

A comprehensive approach to water safety on ships can be achieved through the adoption of Water Safety Plans (WSP), in line with the World Health Organization (WHO) and Inland Waterways Association (IWA) guidance, to cover design, construction, operation, routine inspection, and maintenance. WSPs build on the Hazard Analysis Critical Control Point (HACCP) approach, which has gained the approval of the food industry for controlling food quality. A team consisting of crew or other trained personnel responsible for the WSP implementation should be designated. The team should include managers, engineers, water quality controllers, medical staff, facilities managers, and technical crew.

EPIDEMIOLOGY

Giardia lamblia (also called *G. intestinalis* or *G. duodenalis*) is a flagellate protozoan that occurs worldwide, but is more prevalent in resource-poor settings and areas of poor sanitation where waterborne outbreaks cause endemic and epidemic diarrhoea. Children are affected more than adults. Stool positivity can range from 2–5% in industrialized countries to 20–30% in developing countries. Infection in the Northern Hemisphere commonly occurs from July to October among those under 5 years old and adults from ages 25 to 39. It is associated with unfiltered water, swimming in fresh water, or care in day care centres and institutions. Approximately 4000 cases were reported in England and Wales to Public Health England in 2012, and the prevalence of cyst passage may be as high as 20–50% for children in day care centres. In the US, it is a common cause of waterborne diarrhoea.

Reservoir

The gut of infected humans and both wild and domestic animals can act as a reservoir. *Giardia* can be differentiated based on the host of origin, for example, *G. lamblia* in humans, *G. muris* in mice, *G. agilis* in amphibians, and *G. psittaci* in parakeets. Although *G. lamblia* has been found in cats, dogs, cows, sheep, and beavers, many genotypes found in these animals are not found in human isolates, suggestive of a relatively low zoonotic potential. In some areas, however, nearly all domestic cattle may become infected, and there has been a documented waterborne outbreak linked to surface contamination of unprotected water resources by beavers in Canada. Food-borne transmission of *Giardia* in commercial food establishments is also recognized, with the potential to cause outbreaks.

Mode of transmission

Transmission of cysts from faeces often occurs from person to person by hand-to-mouth transference in settings where there is overcrowding, poor sanitation, or where an infected person is involved in food preparation. As few as 10 cysts can cause infection once ingested. Transmission can also be facilitated, as with other gastrointestinal pathogens, through anal intercourse. *G. lamblia* cysts are able to survive well in the environment and even in the cold. Fecally contaminated water supplies with cysts can lead to large outbreaks if the water is not treated prior to distribution. Chlorination levels in swimming pool water take up to 45 min to kill the *G. lamblia* cysts. Streams or lakes may become contaminated with both animal and human faeces. Outbreaks commonly result from waterborne infection. Campers who do not adequately treat stream water are at risk. As cysts are relatively resistant, they may not be inactivated by people's use of commercial kits to disinfect surface water for drinking. Although boiling can eradicate cysts, previously frozen cysts may survive even this process if not boiled long enough.

Hand hygiene is very important to prevent continued transmission in the community, as is the exclusion of symptomatic individuals and food handlers from swimming pools, schools, the workplace, and other institutions where they may pass on infection. Isolation of patients within the hospital setting and in an outbreak and the use of gloves and aprons for standard enteric precautions assist in containment of infection.

Incubation period

Although the median incubation time is 7–10 days, this ranges from 3 to 25 days or longer. Stool examination may be negative early on in disease as the time from ingestion of cysts to detection is often longer than the incubation period. The severity of symptoms and duration of disease are not related to the size of inoculum ingested and up to 60% of patients can be asymptomatic. Asymptomatic infection may be more common in children and in those who have been previously infected.

Period of communicability

Patients continue to act as reservoirs and transmit throughout their period of infection, which can be up to several months, and asymptomatic carriers such as children can continue to act as reservoirs for up to six months. Symptoms can resolve and return later. Shedding of cysts in stool is typically intermittent.

Susceptibility

The infection is common and widespread throughout the developing world, especially in children. In developed countries, there is a higher rate of asymptomatic carriage and disease among men who have sex with men, where transmission may be higher as a result of sexual practices. Travelers to endemic countries, especially if hiking and camping, and those in childcare or residential institutions are at most risk. Recurrent infections can occur.

BIOLOGY

Giardia is a flagellate protozoon first described in 1681 by Van Leeuwenhoek, who observed them while examining his own stools under the microscope. Although initially thought not to cause disease, it is now one of the 10 major parasites affecting humans worldwide. Several species of the genus *Giardia* have been recognized and *G. lamblia* (also called *G. intestinalis* or *G. duodenalis*) is regarded by some as a species complex that is pathogenic in humans. *G. lamblia* genotypes typically associated with human infection are A and B.

Commonly inhabiting the guts of many vertebrates, it is a flagellated unicellular protozoa causing asymptomatic and diarrheal illness in adults, children, and animals. It has distinct nuclei with a nuclear membrane and cytoskeleton.

The two lifecycle stages for *Giardia* are vegetative trophozoites and environmentally resistant cysts. Viable cysts from contaminated food, water, and environments are ingested, causing infection in the susceptible host. Acids in the stomach allow excystation to occur a few hours later in the proximal part of the intestine, releasing two new trophozoites that can infect the intestinal cells. By adhering to the luminal lining, *Giardia* causes diarrhea and may cause malabsorption of nutrients. Trophozoites are able to colonize the upper part of the small intestine until the mid-jejunum, attaching by their concave ventral disk to prevent further intestinal transport and to attain nutrients. They continue to multiply asexually by binary fission, causing disease, and then pass to the terminal region of the intestine, where exposure to biliary fluids and starvation from cholesterol induce them to form new cysts.

Cysts are then released in the stools of the infected host and, although there is variable excretion, they survive well in the environment and are able to subsequently infect new hosts. Cysts contain four nuclei, are 5–7 μm in diameter, and are covered by a wall that is 0.3–0.5 μm thick. The wall is composed of an outer filamentous layer and an inner membranous layer, with two membranes, which allows cysts to be more environmentally stable.

Although no tissue invasion occurs, many trophozoites are found in the duodenal crypts, and lymphocyte infiltration and villous blunting are seen. The mechanism for disease and malabsorption is not fully understood. Although *Giardia* is able to release cytopathic substances such as thiol proteinases and lectins, there is no known toxin, and an immunopathological process with cytokine release and mucosal inflammation, as well as disruption of the brush border, have been postulated.

The importance of the humoral immune response is suggested by increased severity of disease in patients with hypogammaglobulinemia and possibly in those with isolated IgA deficiency. *Giardia* has been shown to exhibit antigenic variation, which may possibly explain why chronicity of disease can occur if it is able to evade the immune system. Genetic differences between isolates may confer virulence. However, an antibody response, although limited because of the intraluminal presence of trophozoites, may provide some protection following evidence of passive transference of antibodies in gerbils and in maternally acquired antibodies from mothers who had previously had infection to their children, as shown in a study in India.

DISEASE

Patients infected with *G. lamblia* can have a spectrum of disease from asymptomatic cyst passage and acute self-limiting diarrhoea to a chronic syndrome of diarrhoea with severe weight loss and malabsorption.

Most patients will develop symptoms of diarrhoea for up to one week, although this can stretch to several weeks, often following a relapsing and remitting course, thus often prompting antimicrobial treatment. An estimated 25–50% of those ingesting *G. lamblia* cysts become acutely symptomatic. In addition, 5–15% become asymptomatic cyst passers and 35–70% have no trace of infection.

Malabsorption with flatulence, cramps, and abdominal pain accompanied by foul-smelling stools with a highfat and mucus content (but no blood) are common symptoms leading to weight loss, bloating, malaise, and anorexia in patients. In keeping with infectious diarrhoea, patients may have a fever at the beginning of disease. In contrast to other bacterial and viral infectious causes, however, symptoms may persist for several weeks to months and result in weight loss. A careful and full exposure and travel history with clinical examination is essential. Diffuse or epigastric abdominal pain may be present on examination, as may extraintestinal manifestations. Although these symptoms are rare and should prompt the clinician to think of other causes, *Giardia* can cause apthous ulceration, polyarthritis, urticaria, and a maculopapular rash.

Patients may experience malabsorption of fats, carbohydrates, sugars, and vitamins. Reduced intestinal disaccharidase activity may persist even after *Giardia* has been treated, the most common being lactose deficiency, which occurs in 20–40% of cases and may present as failure to thrive in children.

PATHOLOGY

Diagnosis

In order to accurately identify the causative agent of symptoms in individuals or in outbreaks, faeces samples should be sent to the microbiology laboratory for ova, cysts, and parasite examination by microscopy. Because of the variable shedding, at least three consecutive stool samples should be sent to increase sensitivity of detection (following BSOP 31 guidelines). A single stool will allow the detection of 60–80% of infections, two stool samples will allow the detection of 80–90%, and three stool samples should allow the detection of over 90%. Despite testing three samples, approximately a third of samples may be found to be negative.

Characteristic cysts of 5–7 μm diameter containing four nuclei can be seen on microscopy.

G. lamblia trophozoites are easily identified by their oval shape, being shaped like a split pear, and by their size, measuring 12–15 μm long and 5–9 μm wide. The cytoskeleton consists of two median bodies (which is unique to *Giardia*), two axonemes, four pairs of flagella, and a ventral disk. Two nuclei are located symmetrically on each side of the midline in the trophozoite, such that they resemble a face with four pairs of flagella from the midline being directed backward. The flagella are important for motility but not attachment to the intestine. The trophozoites are very fragile and prone to osmotic pressure; a wet mount often yields the best results and has a high specificity, although the motile trophozoites can also be visualized by staining with Lugol's iodine.

In patients with suspected giardiasis but negative microscopy on three samples, a diagnosis can also be sought by the use of enzyme immunoassays (EIA) or immunofluorescent antibody microscopy (IFA). EIAs are often suited for processing large numbers of clinical samples and can be done in the microbiology laboratory. IFAs enhance the visualization of *Giardia* by the use of fluorescent-labeled antibodies that react with *Giardia* cysts and the methodology are better suited for use in detecting *Giardia* in food and environmental samples. However, serodiagnosis may still not help to differentiate between previous and current infection.

The examination of invasively obtained samples such as gastric or duodenal aspirates may aid diagnosis in difficult cases. Samples can be obtained by a jejunal aspirate or the use of the string test, in which a string with a capsule end is swallowed and after four hours it is removed and examined for trophozoites. Biopsies, while invasive and a less sensitive method of diagnosis, may be helpful to diagnose *Giardia* and also exclude other more serious causes of weight loss such as inflammatory bowel disease or lymphoma.

Treatment

Giardia can be treated effectively, usually with a single course of treatment. Quinacrine was first introduced as an antimalarial in 1930 and then used for the treatment of giardiasis, but it causes many side effects such as psychosis and haemolysis, especially in those with glucose-6-phosphate dehydrogenase deficiency. Consequently, treatment has been replaced by the use of the nitroimidazoles.

The nitroimidazole class of agents includes metronidazole, tinidazole, ornidazole, and secnidazole. This class, discovered in 1955, was found to be highly effective against several protozoan infections. Patients are treated effectively with metronidazole, which utilizes the anaerobic metabolic pathways present in *Giardia* and enters through the trophozoite. However, trophozoites within cysts may be less affected by nitroimidazoles, possibly because of poor penetration of the drug through the cyst wall, and resistance to metronidazole has been induced *in vitro*.

Metronidazole is well absorbed and has good penetration into tissue and secretions. It is metabolized in the liver and excreted in the urine. The main side effects include nausea and a metallic taste can sometimes result in noncompliance. Patients should also be made aware of avoiding concurrent alcohol and metronidazole intake in order to prevent the disulfiram-like reaction characterized by severe vomiting, flushing, headache, and gastrointestinal pain. Other side effects include a transient neutropenia, pancreatitis, and central nervous system toxicity.

Effective treatment regimens consist of a 5-to-10-day course of metronidazole with median efficacy of 92% in adults and children and is well tolerated as compared with treatment with a single high dose of metronidazole, where efficacy is 36–60%, rising to 67–80% if given for two days.

Tinidazole, another nitroimidazole with a longer half-life, can also be used in a single effective dose and is widely used throughout the world for treatment of giardiasis. Adverse effects include a bitter taste, gastrointestinal upset, and vertigo.

Furazolidone is a nitrofuran compound that was first used for the treatment of *G. lamblia* in the 1950s. Also available in a liquid preparation, it can be used in children and undergoes reductive activation in the trophozoite. It is readily absorbed from the gut and metabolized in the tissues, resulting in low concentrations in the urine and serum. Although its efficacy has been considered to be slightly lower than metronidazole and quinacrine, curative rates between 80 and 96% have been reported for 7-to-10-day courses. Adverse effects include nausea and diarrhoea as well as a brown discoloration of the urine and haemolysis in G6PDH-deficient patients. The drug has a monoamine oxidase (MAO) inhibitory effect and should never be given concurrently to individuals already taking MAO inhibitors. It can also cause a disulfiram-like reaction when taken with alcohol. It is contraindicated in infants younger than one month of age, who could potentially develop haemolytic anaemia because of their normally unstable glutathione.

Albendazole and mebendazole have been used to treat *G. lamblia* infection, and they are thought to be useful agents because of the secondary effect on helminth infections in many of the endemic countries and their good side-effect profile. They work by binding to the *G. lamblia* α-tubulin cytoskeleton, which causes both inhibition of cytoskeleton polymerization and impaired glucose uptake. They are absorbed poorly from the gastrointestinal system and there is negligible excretion by the kidneys. Short-term side effects include anorexia and constipation and long-term side effects include reversible neutropenia and elevated liver enzymes. Use in pregnancy is contraindicated because of concerns about teratogenicity.

Pregnancy and lactation

Symptomatic management can prove challenging. For asymptomatic women and those with mild disease, treatment should be avoided, especially in their first trimester. However, for those women in their first trimester with severe symptoms, treatment may be necessary. Metronidazole is one option. It has been shown to be carcinogenic and mutagenic in rats and, although it enters the fetal circulation from the mother, it has been used extensively in pregnancy for the treatment of trichomoniasis and no carcinogenicity has been demonstrated in humans. Some studies show a possible increase in fetal malformations with metronidazole use, whereas others do not. Hence metronidazole should be used with caution in the first trimester. High-dose, short-course regimens are not recommended. Metronidazole also is excreted in breast milk at concentrations similar to plasma, but it is thought to be relatively safe at low concentrations.

Paromomycin, which is a nonabsorbable aminoglycoside, has been used. It is thought to have a lower risk of teratogenicity when compared with other agents. It inhibits *G. lamblia* protein synthesis by interfering with the 50S and 30S ribosome units and is poorly absorbed from the intestinal lumen, achieving only minimal concentrations in the blood and urine of patients, even when large doses are given. Although not as effective as metronidazole, it is safe in pregnancy and excreted almost 100% unchanged in the faeces. It is effective and should be considered if treatment is required in the first trimester and also is safe during breast-feeding. However, caution is still advised in patients with renal impairment and, if absorbed systemically, ototoxicity and nephrotoxicity may occur.

Asymptomatic infections

Several factors should be considered before initiating treatment, because many people found to be carrying *Giardia* cysts are completely asymptomatic. In developed countries, asymptomatic carriers may not always require treatment outside of an outbreak. However, treatment of children may prevent onward transmission in households and prevent outbreaks, reducing community prevalence of *Giardia*. In cases of recurrent diarrhoea attributed to *Giardia* in a day-care setting or in a household with a returning traveller, screening and treatment of asymptomatic carriers may help to control an outbreak and prevent reinfection. In addition, settings in which onward transmission are important, such as in a food-handler, it is imperative to treat asymptomatic carriers to prevent food-borne outbreaks.

In areas of endemicity, the treatment of children who are likely to become rapidly re-infected can be problematic and not particularly cost-effective. However, in cases of failure to thrive or co-infection with helminths, treatment may allow the child to catch up with growth. The use of albendazole, if well tolerated, in such co-infection may be useful.

Relapse and recurrence

It may be difficult to differentiate true resistance with treatment failure, from cure followed by re-infection or from post-*Giardia* related lactose intolerance.

Although resistance in *Giardia* can occur to all of the commonly used drugs, clinicians should initially resend samples to the laboratory to check for ova, cysts, and parasites or send a *Giardia* antigen to document true and persistent infection. In cases where samples test positive, a careful history to elucidate re-infection accompanied with advice about hand hygiene and prevention measures should be conducted, especially in settings with poor feco–oral hygiene or endemicity.

In vitro sensitivity testing does not necessarily correlate with resistance *in vivo* or clinical success. Repeat courses of treatment with longer duration or higher doses may be effective as may switching to a different class of therapy. However, often combination courses of treatment with two agents for at least two weeks are needed in refractory treatment. The possibility of an underlying immunodeficiency should also be considered if symptoms persist, as in patients with hypogammaglobulinemia, which can be difficult to treat and may require prolonged courses of therapy.

If stools test negative for *Giardia* and other causes have been excluded, post-*Giardia* lactose intolerance should be considered for ongoing diarrhoea and can occur in 20–40% of patients. Although it may take several weeks to resolve, a trial of excluding and avoiding lactose-containing foods may be beneficial.

Prevention

The most common sources of documented outbreaks and disease with *Giardia* are water and food-borne contamination, and preventative measures should be primarily focused on these. Person-to-person transmission is also likely to be significant, especially for sporadic cases. Hand hygiene remains an important control measure for all patients but especially in children, who are prone to infection through feco–oral transmission.

Hand hygiene is very important to prevent onward transmission in the community, as is the exclusion of symptomatic individuals and food-handlers from schools, the workplace, and other institutions where they may pass on infection. Isolation of patients within the hospital setting and in an outbreak with the use of gloves and aprons for standard enteric precautions assist in containment of infection, along with thorough cleaning of all equipment and patient facilities.

Additionally, in the community or for travelers, *Giardia* cysts can be removed by water filtration and rendered non-infectious by boiling. Low-level chlorination and iodine are unreliable at killing *Giardia* cysts and should not be recommended as the only water treatment. Filters of 1μ or smaller can remove cysts, but users should be careful not to contaminate filters.

Public and private water supplies and swimming pools should be treated to remove enteric pathogens, including *Giardia*. A range of approaches is used, including source water protection from agricultural run-off or sewage, sedimentation, and filtration to remove organisms and cysts, and a number of disinfection processes, including chlorination, ozone, and UV light, may be used.

QUESTIONS

1. Which is the most common clinical manifestation of *Giardia* infection worldwide?
   a. Malabsorption syndrome
   b. Bloody diarrhoea
   c. Asymptomatic carriage
   d. Prolonged watery diarrhoea
   e. Acute diarrhoea and vomiting

2. Which of the following can be sources of *Giardia* infection?
   a. Washed salad
   b. Undercooked meat
   c. Drinking water from mountain streams
   d. Water mains contaminated with sewage
   e. Swimming pools

3. What parts of the digestive tract can *Giardia* trophozoites colonize?
   a. Stomach
   b. Duodenum
   c. Jejunum
   d. Terminal ileum
   e. Colon

4. Which of the following methods can be used by travellers to remove *Giardia* cysts from drinking water?
   a. Boiling
   b. High level chlorination
   c. Iodine
   d. Filter (pore size 2 μ)
   e. Freezing

5. Which of the following measures can be reliably used to remove enteric pathogens from a ship's water systems?
   a. Chlorination at 20 ppm for 30 min
   b. Heat treatment at 60°C
   c. UV sterilizer lamps
   d. Point of use filters with a pore size of 0.1 μm
   e. Sand filtration

GUIDELINES AND REFERENCES

1. Heymann DL (ed) (2014), Control of Communicable Diseases Manual, 19th ed., American Public Health Assn.

2. Public Health England (2000–2001), Giardia lamblia Laboratory reports: all identifications reported to the Health Protection Agency England and Wales.

3. Mandell et al., Principles and Practice of Infectious Diseases, 7th ed.

4. Investigations of parasites other than blood, National Standards Methods, BSOP31.

5. Gardner TB et al., Treatment of giardiasis, *Clin Microbiol Rev* Jan. 2001, **14**:114–128.

6. World Health Organization (1993). Guidelines for Drinking-Water Quality, 2nd ed. WHO. Geneva.

7. RM Roon et al. (2004) Public Health Reports, A Review of Outbreaks of Waterborne Disease Associated with Ships: Evidence for Risk Management.

8. Yund J (1998). Giardiasis aboard an amphibious vessel. *Naval Medical Surveillance Reports*, April–June:9.

9. European Manual for Hygiene Standards and Communicable Diseases Surveillance on Passenger Ships, 2016.

ANSWERS

MCQ	Feedback
1. Which is the most common clinical manifestation of *Giardia* infection worldwide? a. Malabsorption syndrome b. Bloody diarrhoea c. Asymptomatic carriage d. Prolonged watery diarrhoea e. Acute diarrhoea and vomiting	Most *Giardia* infections are probably asymptomatic, although these individuals are still infectious. Malabsorption syndrome is classically associated with *Giardia*, but is only one of many manifestations. Bloody diarrhoea is more likely to be caused by *Entamoeba histolytica* or bacterial pathogens such as *Shigella*. *Giardia* can cause prolonged diarrhoea, which may be watery due to malabsorption. Acute infections tend to be characterized more by diarrhoea than vomiting. Vomiting should make you think of a viral infection or food poisoning with preformed toxin.
2. Which of the following can be sources of *Giardia* infection? a. Washed salad b. Undercooked meat c. Drinking water from mountain streams d. Water mains contaminated with sewage e. Swimming pools	Even if salad has been washed, this is not a reliable means of removing cysts, given the low infectious dose. Additionally using contaminated water to wash it can make the problem worse. Meat is more likely to be a source of zoonotic infection (for example, *Salmonella*, where the meat has become contaminated during butchering, or *Campylobacter* found within deep tissues. *Giardia* arises predominantly from fecally contaminated water, especially where the mains supply has been affected. Mountain streams may look safe and pure, but remember the organism survives well in the environment in its cyst form and there is always the possibility that seemingly clean water has been contaminated upstream, as from campers defecating too close to the water source. Swimming pools are particularly risky if a person shedding cysts has been in the water, although the number of cysts will be reduced over time by water treatment. *Cryptosporidium* is more commonly associated with swimming pool outbreaks.
3. What parts of the digestive tract can *Giardia* trophozoites colonize? a. Stomach b. Duodenum c. Jejunum d. Terminal ileum e. Colon	In contrast to most other enteric pathogens, *Giardia* only colonizes the proximal small intestine. It is only found further down in cyst form.
4. Which of the following methods can be used by travellers to remove *Giardia* cysts from drinking water? a. Boiling b. High level chlorination c. Iodine d. Filter (pore size 2 μ) e. Freezing	Boiling is a good way of making water safe. Chlorination would work but would make the water undrinkable. Commercial chlorine tablets should be effective but this depends on how badly contaminated the water is and whether it has been left long enough for the chlorine to act, so it is not ideal. Iodine is not reliable and is rarely sold nowadays, also it is not recommended to consume large amounts of iodine because there is a risk of thyroid disorders and controversy over what is a safe dose. Filters are good, but only if the pore size is one micron or less and as long as the filtered water is not re-contaminated or the filter damaged. *Giardia* cysts survive freezing temperatures very well for long periods.
5. Which of the following measures can be reliably used to remove enteric pathogens from a ship's water systems? a. Chlorination at 20 ppm for 30 min b. Heat treatment at 60°C c. UV sterilizer lamps d. Point of use filters with a pore size of 0.1 μm e. Sand filtration	Chlorination needs to be of a sufficient concentration and duration of exposure to eradicate all cysts. The higher the concentration and temperature, the lower the concentration needed. This concentration would need at least 2 h to work in a cold water system; heat needs to be at least 70°C. Temperatures of 70°C and over are necessary to eradicate cysts, although vegetative bacteria would be killed at lower temperatures. Generally a combination of measures is used to ensure complete eradication. In contrast to chemical treatment, UV sterilizer lamps only work at the point of disinfection, and would not work if infection is introduced downstream of this, as there is no residual effect. Sand filtration will only work if a flocculant is added to the water, and even then it is not guaranteed to be fully effective. Point-of-use filters are an option if the system cannot be decontaminated reliably and are used in various settings, including healthcare.

HIV

Sam Douthwaite and Chris Ward

Directorate of Infection, Guys and St Thomas' NHS Foundation Trust, London, UK

A 30-year-old female surgical doctor presented to a hospital emergency department at midnight, having sustained an injury while assisting in a complex vascular operation 30 minutes previously during her evening shift at the same hospital. She had been wearing two pairs of gloves at the time, but the scalpel, which had been in the patient's wound and was visibly bloodstained, still pierced her skin. She immediately washed the injury with cold water while attempting to make it bleed and then presented for a risk assessment.

INVESTIGATION OF THE CASE

A risk assessment was undertaken. The source patient for the injury was a known intravenous drug user (IVDU), whose vascular complications were the result of a previous self-inflicted arterial puncture while attempting to inject heroin. He had moved to the area one year ago and had tested negative for HIV, hepatitis B, and hepatitis C on a previous visit at the same hospital nine months ago. At the point of assessment of the surgical doctor, the patient was still recovering from his anaesthetic and did not have capacity to consent to further serological testing.

CINICAL MANAGEMENT

The surgical doctor had received a full course of hepatitis B vaccination and knew that she had a protective level of immunity; in line with UK guidelines, she was therefore offered a booster dose of hepatitis B vaccination. She was informed of the need for surveillance for hepatitis C infection over the next three months. After some discussion, she decided to take HIV post-exposure prophylaxis (PEP) until the source patient could give his consent for testing. Her pregnancy test was negative, she was not on any other medication, and she was given a combination of emtricitabine and tenofovir as well as lopinovir and ritonavir (the UK PEP guidelines at the time of incident). She was advised to contact her occupational health department the next morning to arrange further follow-up.

The next day the source patient consented to testing for bloodborne viruses and was found to be negative for hepatitis C and hepatitis B, but positive for HIV infection. The surgical doctor continued her PEP for one month and, other than some initial nausea and vomiting, did not suffer from any side effects.

As is the standard UK occupational health process, she was advised that there were no restrictions on her medical practice during the period she was taking PEP or subsequent follow-up. She was also counselled on standard infection control precautions and the need to use barrier contraception until she had completed her follow-up and was advised to report any acute illness to her occupational health physician. Twelve weeks after completing her PEP her HIV test remained negative and she was discharged.

EPIDEMIOLOGY

Acquired immune deficiency syndrome (AIDS) was identified as a clinical entity in 1981 and subsequently linked to the virus that eventually came to be known as human immunodeficiency virus type 1 (HIV-1). There are four lineages of HIV-1, termed groups M, O, N, and P, each of which represents an independent transmission from chimpanzees in West Africa at some point in the twentieth century. Group M is the predominant virus responsible for the global pandemic, and the other three groups are confined to much smaller populations in West Africa. It is possible to genetically separate group M viruses further into multiple subtypes, circulating recombinant forms, and unique recombinant forms. The predominant subtype in sub-Saharan Africa is C, and for the US and Western Europe subtype B is most common.

A separate lentivirus with a similar clinical picture, called human immunodeficiency virus type 2 (HIV-2), crossed from sooty mangabeys to humans on eight separate occasions in West Africa. This virus tends to follow a milder clinical course and largely remains restricted to West Africa. Further discussion of this virus is beyond the scope of this case.

According to the Global Burden of Disease, Injuries, and Risk Factors Study in 2015 there were an estimated 38.8 million people living with HIV/AIDS and 1.2 million deaths. The burden of disease remains highest in sub-Saharan Africa (29 million people living with HIV/AIDS and 800,000 deaths in 2015), however, the disease has a global distribution. The nature of the epidemic varies across regions, with heterosexual spread in the general population being the main mode of transmission in sub-Saharan Africa and specific behaviours (male–male sex, injection drug use, and sex work) being the predominant risk factors elsewhere. However, even within specific risk groups in an individual country, the prevalence of HIV can vary significantly from region to region. Where implemented appropriately, healthcare interventions are able to reduce transmission and influence the course of the epidemic. The UK adopted needle-exchange programs as a public health intervention following an influential report in 1986 that showed an HIV prevalence in IVDUs in Edinburgh of 51% (prevalence in equivalent IVDU populations: Ukraine 2006, 41%; Russia 2007, 37%; US 2002, 15%). Following the widespread adoption of needle-exchange programs and other harm reduction interventions, the current HIV prevalence in UK IVDUs is 1.6%.

Transmission

HIV-1 is a bloodborne virus and transmission is by contact with the body fluids of an infected individual, although not all fluids have the same risk. Blood, blood products, semen, vaginal secretions, donor organs and tissue, and breast milk have all been implicated in transmission events. Body fluids considered to be high risk for transmission are:

- cerebrospinal fluid (CSF)
- peritoneal fluid

- pleural fluid
- pericardial fluid
- synovial fluid
- amniotic fluid
- semen
- vaginal secretion
- breast milk

Other fluids may be considered as high risk if contaminated with blood, for example, saliva during dental procedures.

There are many other factors that affect transmission rates, including mode of inoculation (percutaneous, mucous membranes, or intact skin) and the viral load of the infected individual. Accurate, evidence based, estimates of risk of transmission following specific types of exposure are made difficult because the data are observational and variables cannot fully be accounted for. Estimated risks of transmission are listed in Table 8.1.

Risk of transmission increases 2.5-fold for each log increase in viral load, and transmission from a source with an undetectable viral load is rare. Transmission from male to a female sexual partner is more efficient than female to male partner. Male circumcision reduces transmission from female to male sexual partner, but not from male to female partner.

Healthcare worker HIV exposure

Since 1984, there have been five documented cases of HIV transmission in health care workers following occupational exposure and a further 47 probable cases in the UK (probable, because no baseline negative HIV test was obtained). Several factors are thought to increase the risk of transmission following percutaneous injury, including deep injury, a device visibly contaminated with blood, a procedure involving the needle being placed in the source patient's vein or artery, and advanced HIV disease in the source patient.

Table 8.1. **Estimated risk for transmission for various bodily fluids**

Mode of Transmission	Risk of HIV acquisition per 1000 exposure events
Blood transfusion	900
Perinatal exposure	130 to 450
Needle sharing by injecting drug users	6.7
Receptive anal intercourse	5 to 32
Percutaneous needle-stick injury	3
Receptive penile–vaginal intercourse	1 to 3
Mucous membrane exposure in a healthcare context	0.9
Insertive anal intercourse	0.65
Insertive penile–vaginal intercourse	0.3 to 0.9
Receptive oral intercourse	0.1
Insertive oral intercourse	0.05

General prevention of sharps injuries for health care workers

In the UK employers are legally obligated, through the Health and Safety at Work Act (HSWA) and Control of Substances Hazardous to Health (COSHH) regulations, to protect employees from hazards in the workplace. For sharps injuries in a healthcare setting, this means having appropriate procedures and training in place to minimize the risk of exposure. Employees are required to use appropriate personal protective equipment for procedures (for example, gloves, gown, mask, eye visor). Other considerations apply to the equipment used to carry out procedures (for example, syringes with advanceable needle guards, blunt suture needles, safety needle holders) and practices to enhance safety (for example, not resheathing needles, use of sharps bins, only one person handles a sharp instrument at any one time). Employees should make themselves aware of local protocols for the prevention of bloodborne virus transmission. Although the clinical environment is the most obvious area of risk, there are other areas that should be considered. These include laboratory staff working with human samples, appropriate waste disposal, and correct laundering of linen.

The same legislation obliges healthcare institutions to protect patients from bloodborne virus exposure. These protocols will include screening of new staff, hepatitis B vaccination, and restrictions on procedures that are prone to exposure. Other protocols require disinfection or sterilization of reusable equipment, including surgical instruments, endoscopes, and dialysis machines. The guidelines at the end of this case provide further details on these practices.

Prevention of bloodborne virus transmission in the IVDU community

The sharps injury described in the opening case involved an IVDU and therefore an individual from a group associated with higher prevalence of HIV. Estimating prevalence of HIV infection in IVDU populations is difficult, but best estimates suggest wide variation both between countries and areas within countries. Evidence for effectiveness of interventions designed to reduce transmission in this group is difficult to gather for similar reasons. However, there are data to show that provision of sterile injecting equipment, community outreach to IVDUs, and implementation of interventions when seroprevalence is still low limits the spread of HIV in such a population.

Post-exposure prophylaxis

There is no high quality clinical trial evidence on the use of antiretroviral therapy (ART) as post-exposure prophylaxis (PEP) to prevent HIV infection for percutaneous exposures. There is, however, a significant amount of data from animal models and observational studies to suggest potential benefit of ART for those exposed to HIV in the healthcare setting, from injecting drug use, as post-exposure prophylaxis following sexual exposure (PEPSE), or for infants at risk of mother-to-child transmission (MTCT). The efficacy of PEP/PEPSE declines with time following exposure, and guidelines usually recommend receiving the first dose at least within 72 hours and ideally within one-hour post-exposure. The choice of which drugs to use is largely based on tolerability, however, information regarding the virus (for example, drug resistance mutations) from the source patient and the exposed patient's other medical conditions will also play a role in the decision. Further details of drug choice and management protocols are beyond the scope of this book. This information can readily be found in the guidelines listed at the end of this case.

BIOLOGY

HIV-1 is a single-stranded RNA virus belonging to the Retroviridae family. The two defining features of its life cycle are reverse transcription (conversion of RNA genome to double-stranded DNA intermediate) and integration (permanent insertion of the double-stranded DNA intermediate into the host cell genome). Reverse transcription is carried out by the reverse transcriptase (RT) enzyme that is packaged into viral particles before they are released from the previously infected cell. RT has no editing ability and therefore is not able to correct mutations in the DNA transcripts that it produces. The inability to correct mutations, combined with the high viral load and rapid rate of replication, means that the virus quickly mutates and diversifies within an individual. The resulting diversity allows for evasion of the host's humoral and cell-mediated immune response by selection of mutations in epitopes to which the immune system is responding.

Integration is carried out by the viral enzyme integrase (INT) and ensures that the viral genome is present within the cell until it dies. Cell death is frequently the consequence of HIV-1 infection. In a small subset of cells that evade this fate, however, the virus can remain dormant and is able to reactivate at a later time. This is the reason that combined antiretroviral therapy (cART) is unable to cure patients of HIV-1 infection, as it does not eradicate these integrated viral genomes.

HIV-1 is an enveloped virus, which means its genome and protein core are surrounded by a lipid bilayer. The viral envelope protein (Env) is a glycoprotein present in the envelope that mediates binding to receptors (CD4) and co-receptors (CCR5 or CXCR4) on target cells within the host. Once bound to receptor and co-receptor, the virus is able to fuse its envelope with the cell's plasma membrane and release the viral content into the cell. Specificity for CD4 and CCR5 or CXCR4 is a large part of what determines which cells the virus is able to infect. Individuals with mutations that result in a lack of co-receptor expression have been shown to be relatively resistant to HIV infection. *In vivo,* the most significant cells to be infected are activated CD4 positive T-helper cells, however, macrophages and dendritic cells are also infected. It is the infection of and loss of CD4 positive T-cells that defines a large part of HIV-1 disease and its progression to AIDS.

PATHOLOGY

The exact mechanisms by which HIV-1 establishes infection in a host and causes the well-documented pathological consequences remain to be fully elucidated at a cellular level. The most frequent route of transmission is via the anogenital mucosa during sexual intercourse. It is thought that the first cells the virus is likely to infect are dendritic cells within the mucosal tissues. These cells then convey the virus to draining lymph nodes and spread the infection to helper T cells. From this point the virus disseminates to the rest of the body, and there is a rapid rise in the number of viral genome copies (viral load) detectable in the blood. This rise in viral load is closely followed by a rapid fall in the concentration of CD4 positive T-cells both in plasma and other tissues.

As the adaptive immune system, particularly CD8 positive cytotoxic T-lymphocytes (CTL), begins to respond to infection, the viral load falls and the CD4 positive T cell concentration in the blood rises. However, by this time severe depletion of CD4+ cells in lymphoid tissues has taken place. The viral load eventually plateaus at a viral set point and this, along with CD4 T cell count, remains relatively stable for a prolonged period. Eventually, after a period of months to years, the CD4 T cell count begins to fall and, in the absence of effective interventions, the patient will progress to AIDS. During this final phase of the illness, the patient's viral load again rises.

The viral and host factors that determine both the viral set point and the rate at which a patient progresses to AIDS are still being discovered. With respect to viral factors, the viral load of the person from whom infection was acquired may play a role in determining viral set point. Lower diversity in the viral envelope protein and a virus with lower replicative capacity have both been shown lead to better patient control of the virus. Finally, infection with a virus that harbours mutations for drug resistance may limit an individual's treatment options.

The study of two groups of patients, long-term nonprogressors (patients who maintain a good CD4 count for many years in the face of active viral replication) and elite controllers (patients who naturally reduce viral replication to undetectable levels without cART), has been used to inform our understanding of host factors affecting disease progression. These studies have suggested various immune factors including human leukocyte antigen (HLA) subtype, HIV specificity of CD4 and CD8 T-cells, humoral responses and antibody dependent cellular cytotoxicity, mutations in co-receptors, and improved natural killer (NK) cell function, may all play a role in this process.

The causes of the final decline in CD4 T cell count that leads to AIDS are not clear. In certain primates, infection with simian immunodeficiency virus (SIV) never leads to a condition comparable to AIDS. Similarly, AIDS is far less frequent in patients infected with HIV-2 compared with those infected with HIV-1. Direct killing of infected CD4 cells, either by virus or immune system, combined with failure of the body to replace them contributes only a small portion of the total CD4 T cell population that is infected with virus. It is thought that exhaustion of the immune system from a chronic inflammatory response plays some role. There is also evidence of bystander killing, with an increased rate of loss of non-infected CD4 T cells through an unknown mechanism.

The introduction of cART has drastically changed disease progression and improved the prognosis of HIV-1 infection. Therapy is able to reduce the viral load to undetectable levels and, once initiated, CD4 cell counts rise considerably. There is some debate as to whether cART completely inhibits viral replication or maintains it below the level of detection. However, it is clear that the immune response seen with uncontrolled viral replication abates once a patient is successfully established on treatment. Equally, while the viral load remains undetectable on cART, there is no development of resistance to the antiviral therapy used. Current treatment is unable to eradicate integrated provirus and therefore cessation of therapy in established infection almost invariably results in a rapid rebound in viremia.

DISEASE

The clinical features of HIV-1 infection can be separated into early and late and into those caused by the virus versus those resulting from the associated immunosuppression. After acquiring HIV-1 infection, 40–90% of patients will experience a seroconversion illness within 2–4 weeks. This is thought to correspond to the high viremia seen early in infection prior to immune control of viral replication. Symptoms are usually those of a nonspecific viral illness or glandular fever, with fever, fatigue, myalgia, sore throat, maculopapular rash, and lymphadenopathy. Less frequent symptoms can include mucosal ulcers, gastrointestinal upset, and neurological symptoms (ranging from headache, through cranial nerve palsies to encephalitis and Guillain-Barré syndrome). In addition, some patients may contract opportunistic infections during the brief period when their CD4 cell count is suppressed. In most cases, although duration and severity vary, symptoms are self-resolving. Patients may then remain asymptomatic for many years, until their CD4 count begins to decline.

The common complications arising from immunodeficiency in the latter stages of the disease are listed in Table 8.2. Malignancies associated with HIV infection are listed in Table 8.3. It should be noted that, although many organisms will predominantly affect one organ system, many infections may be disseminated in the severely immunosuppressed. In addition, there are multiple possible complications arising from the effects of the virus itself. These include HIV-associated nephropathy (HIVAN) (a form of focal sclerosing glomerulosclerosis) and HIV-associated neurocognitive disorder (HAND). The final group of complications are multifactorial in origin and include an increased risk of cardiovascular disease, decreased bone mineral density, and dyslipidemia.

Diagnosis of adults with HIV continues to be largely based on serological testing. Newer tests that incorporate identification of viral antigen as well as host antibodies decrease the window period (time after infection in which diagnostic tests remain negative) to four weeks. Patients who may have been exposed to HIV within this window period and test negative are recommended to have a repeat test three months later. Best practice is to confirm an initial positive test with repeat testing on a separate sample.

Table 8.2. **Common complications of late HIV infection**

System infected	Opportunistic infection
Respiratory infections	Bacterial pneumonia
	Tuberculosis
	Pneumocystis jiroveci pneumonia
	Atypical mycobacterial infection
	Cytomegalovirus (CMV) pneumonitis
Neurological infections	Toxoplasmosis
	Cryptococcal meningitis
	Tuberculous meningitis or tuberculoma
	Progressive multifocal leukoencephalopathy
	CMV retinitis
Gastrointestinal infections	Parasitic infections (*Cryptosporidium* sp., microsporidiosis)
	Esophageal candidiasis
	CMV colitis
Skin infections	Molluscum contagiosum
	Herpes zoster, including disseminated disease
Other	Leishmaniasis (with appropriate travel history)

Table 8.3 **Malignancies associated with HIV infection**

Category of malignancy	HIV-associated malignancy
AIDS-defining malignancies	Kaposi's sarcoma
	Non-Hodgkin's lymphoma
	Invasive cervical carcinoma
Non-AIDS defining malignancies	Hodgkin's lymphoma
	Testicular germ-cell tumour
	Non-small-cell lung cancer

Current UK guidelines recommend people with HIV start cART regardless of CD4 count or other HIV-related factors. These guidelines also recommend which therapy to initiate, but individual patient factors (comorbidities, drug interactions, patient preference, side effects, and viral resistance markers) should also be taken into account. Once initiated on therapy patients should be monitored for viral response, side effects, and adherence. Patients need to be at least 95% compliant with therapy in order to avoid the risk of generating resistance in the virus. In addition to monitoring viral load, CD4 cell count and antiretroviral medications, there are a number of other factors to consider in providing full holistic care for the patient. These include vaccination status, risk factors for cardiovascular and bone disease, sexual health, and the patient's mental health and well-being.

QUESTIONS

1. What is the standard quoted risk of acquiring HIV-1 infection form a needle-stick injury when the source patient is known to be HIV positive?
   a. 30%
   b. 3%
   c. 0.3%
   d. 1%
   e. 0.1%

2. Which of the following is not a bodily fluid that is high risk for HIV transmission?
   a. Saliva during a dental procedure
   b. Semen
   c. CSF
   d. Faeces
   e. Breast milk

3. Which of the following situations has the highest risk for HIV transmission from an HIV-positive patient with an uncontrolled viral load?
   a. Blood splash to eye during surgical procedure
   b. Scratch from patient that breaks the skin of healthcare worker
   c. Needle-stick injury that breaches the skin following venesection
   d. Blood splash to mouth during surgical procedure
   e. Bite from patient that breaks the skin of healthcare worker

4. Which of the following presentations warrants consideration of a seroconversion illness and an HIV test?
   a. Fever malaise and diffusely enlarged lymph nodes
   b. Fever with an unexplained rash
   c. Diarrhoea and weight loss
   d. Progressive ascending sensory loss and motor weakness
   e. All of the above

5. Which of the following is not an AIDS-defining illness?
   a. Kaposi's sarcoma
   b. Hodgkin's lymphoma
   c. Oesophageal candidiasis
   d. Disseminated shingles
   e. *Pneumocystis jiroveci* pneumonia

GUIDELINES

HIV diagnosis and management:

1. www.bhiva.org/clinicalguidelines.aspx
2. www.bashh.org/BASHH/Guidelines/Guidelines/BASHH/Guidelines/Guidelines.aspx
3. www.eacsociety.org/Guidelines.aspx
4. Aberg JA, Gallant JE, Ghanem KG et al. (2013). Primary care guidelines for the management of persons infected with HIV: 2013 update. HIV Medicine Association of the Infectious Diseases Society of America. *Clin Infect Dis* (doi: 10.1093).

Guidelines related to sharps injuries:

5. Guidance for clinical health care workers: protection against infection with blood-borne viruses. London: Department of Health; 1998.
6. HIV Post-Exposure Prophylaxis: Guidance from the UK Chief Medical Officer's Expert Advisory Group on AIDS. London: Department of Health, September 2008.
7. Health services information sheet 7—Health and safety (Sharp Instruments in Healthcare) regulations 2013. London: Department of Health, 2013.
8. Public Health England 2014: The management of infected healthcare workers who perform exposure prone procedures: update guidance, January 2014.
9. www.who.int/hiv/topics/prophylaxis/ en/

REFERENCES

1. Kuhar DT, Henderson DK, Struble KA et al. (2013). Updated US public health service guidelines for the management of occupational exposures to human immunodeficiency virus and recommendations for postexposure prophylaxis. *Infect Control Hosp Epidemiol* **34**: 875–892 (doi:10.1086/672271).
2. Sharp PM & Hahn BH (2011) Origins of HIV and the AIDS pandemic. *Cold Spring Harb Perspect Med* **1**:a006841.
3. www.who.int/gho/hiv/en/
4. Buonaguro L, Tornesello ML & Buonaguro FM (2007) Human immunodeficiency virus type 1 subtype distribution in the worldwide epidemic: pathogenetic and therapeutic implications. *J Virol* **81**(19):10209–10219.

5. GBD 2015 HIV Collaborators (2016) Estimates of global, regional, and national incidence, prevalence, and mortality of HIV, 1980–2015: the global burden of disease study 2015. *Lancet HIV* **3**:e361–387.

6. Mathers BM, Degenhardt L, Phillips B et al. (2008) Global epidemiology of injecting drug use and HIV among people who inject drugs: a systematic review. *Lancet* **372**:1733–1745.

7. Powers KA, Poole C, Pettifor AE & Cohen M (2008) Rethinking the heterosexual infectivity of HIV-1: a systematic review and meta-analysis. *Lancet infect Dis* **8**:553–563.

8. Donnell D, Baeten JM, Kiarie J et al. (2010) Heterosexual HIV-1 transmission after initiation of antiretroviral therapy: a prospective cohort analysis. *Lancet* **375**:2092–2098.

9. Huang Y-F, Yang J-Y, Nelson KE et al. (2014) Changes in HIV incidence among people who inject drugs in Taiwan following introduction of a harm reduction program: a study of two cohorts. *PLoS Med* **11**(4):e1001625.

10. Nicolosi A, Leite MLC, Mussico M et al. (1994) The efficiency of male-to-female and female-to-male sexual transmission of the human immunodeficiency virus: a study of 730 stable couples. *Epidemiol* **5**(6):570–575.

11. Cardo DM, Culver DH, Ciesielski CA et al. (1997) A case control study of HIV seroconversion in health care workers after percutaneous exposure. *N Engl J Med* **337**(21):1485–1490.

12. Attia S, Egger M, Muller M et al. (2009) Sexual transmission of HIV according to viral load and antiretroviral therapy: systematic review and meta-analysis. *AIDS* **23**:1397–1404.

13. Tokars JI, Marcus R, Culver DH CI et al. (1993) Surveillance of HIV infection and zidovudine use among health care workers after occupational exposure to HIV infected blood: the CDC cooperative needlestick surveillance group. *Ann Intern Med* **118**:913–919.

14. Health Protection Agency, Health Protection Services; Public Health Wales; Public Health Agency Northern Ireland; Health Protection Scotland. Eye of the Needle. United Kingdom Surveillance of Significant Occupational Exposure to Bloodborne Viruses in Healthcare Workers. London: Health Protection Agency. December 2012.

15. Woode Owusu M, Wellington E, Rice B et al. (2014) Eye of the Needle, United Kingdom Surveillance of Significant Occupational Exposures to Bloodborne Viruses in Healthcare Workers: data to end 2013. Public Health England, London.

ANSWERS

MCQ	Feedback
1. What is the standard quoted risk of acquiring HIV-1 infection from a needle-stick injury when the source patient is known to be HIV positive? a. 30% b. 3% c. 0.3% d. 1% e. 0.1%	The value of 0.3% or 3 per 1000 exposure events is for a needle-stick injury from a hollow bore needle which has just been used in a patient known to be HIV positive and is based on what little data are available. Remember that other factors could possibly modify this risk, for example, the viral load of the patient.
2. Which of the following is not a bodily fluid that is high risk for HIV transmission? a. Saliva during a dental procedure b. Semen c. CSF d. Faeces e. Breast milk	Options a, b, c, and e are all high-risk fluids. Note that, in general, saliva is not considered a high-risk fluid, however, it becomes high risk if it is stained with blood.
3. Which of the following situations has the highest risk for HIV transmission from an HIV-positive patient with an uncontrolled viral load? a. Blood splash to eye during surgical procedure b. Scratch from patient that breaks the skin of healthcare worker c. Needle-stick injury that breaches the skin following venesection d. Blood splash to mouth during surgical procedure e. Bite from patient that breaks the skin of healthcare worker	Percutaneous injuries involving blood carry the highest risk of HIV transmission. Exposure of intact skin, even to a high-risk body fluid, does not carry a risk of transmission. Exposure of mucous membranes to high-risk fluids does carry a risk of HIV transmission, but not as high as inoculation with a sharp instrument. Note that for an exposure to have occurred, a high-risk bodily fluid needs to be involved. For example, option b is not an exposure, as no high-risk bodily fluid was involved.
4. Which of the following presentations warrants consideration of a seroconversion illness and an HIV test? a. Fever malaise and diffusely enlarged lymph nodes b. Fever with an unexplained rash c. Diarrhoea and weight loss d. Progressive ascending sensory loss and motor weakness e. All of the above	Identifying patients infected with HIV and promptly initiating appropriate therapy is extremely important. Many patients living with HIV are not diagnosed with the illness until late in the disease when they develop immune deficiency. Seroconversion illness can present with almost any symptoms and it is therefore important to have a low threshold for testing patients for HIV.
5. Which of the following is not an AIDS-defining illness? a. Kaposi's sarcoma b. Hodgkin's lymphoma c. Oesophageal candidiasis d. Disseminated shingles e. *Pneumocystis jiroveci* pneumonia	A person living with HIV is said to have AIDS if his or her CD4 T cell count falls below 200 cells/mm³ or the person develops one of a set of conditions (mostly opportunistic infections) that are characteristically linked with HIV and only occur when the immune system is significantly compromised. These conditions are known as AIDS-defining illnesses. Although Hodgkin's lymphoma can be associated with HIV, it is not an AIDS-defining illness.

HUMAN PAPILLOMAVIRUS (HPV)

Nigel Field and Richard Gilson

Institute for Global Health, University College London, UK.

THE PROBLEM OF PREVENTING HPV

Although human papillomavirus (HPV) transmission is often clinically silent and infection does not present an immediate threat to health, HPV causes a significant burden of ill health. Not only are genital warts the most common viral infection triggering individuals to seek sexual health care, but HPV types associated with cancer are necessary (although not sufficient) for almost all cervical cancers and a range of other squamous cell malignancies.

In 2007, Australia rolled out the first nationally funded HPV vaccination program in the world, targeting girls aged 12 years and including a catch-up program for those aged 13–26. Within just four years, in agreement with modelling predictions, a decrease was observed in the incidence of genital warts diagnosed in young women, with warts all but eliminated in women younger than 21 by 2011. Modelling predictions had not been as confident about the effect in men, but a parallel highly significant decrease in wart diagnoses was observed in young heterosexual men, pointing to a form of herd protection afforded to the unvaccinated men. Although too early to detect the effect of HPV vaccine on cervical and other cancers, these data indicated the remarkable potency of HPV vaccination in making substantial improvements to public health at a population level.

In England, about 3000 cases of cervical cancer are diagnosed each year with around 800 deaths. In the wake of these findings, it was necessary for the UK to decide whether to implement a national immunization program to improve control of HPV.

INVESTIGATION

In the UK, the Joint Committee on Vaccination and Immunization (JCVI) is the statutory expert committee that provides impartial advice to UK health departments on immunization. The JCVI first considered HPV vaccines in 2005 when the data available (mostly unpublished) provided good evidence of vaccine efficacy from randomized controlled trials (RCT). However, evidence of vaccine effectiveness when introduced at population scale in real-world settings was lacking. Decisions were further complicated because two competing vaccines were available: Cervarix®, a bivalent vaccine protecting against the two HPV types responsible for most cervical cancer, and Gardasil®, a quadrivalent vaccine providing additional protection against the two HPV types responsible for most genital warts.

From 2005, the JCVI was convened 10 times over two years to deliberate a proposed HPV vaccine program. The JCVI considered several key areas of work, including both published and unpublished research, with a specific remit to identify gaps and limitations in the available evidence. These key areas are:

- Vaccine efficacy studies
  Trials of both candidate vaccines showed good safety and high efficacy in protecting against cervical cancer precursors, with evidence of prolonged immunity and some cross-protection to nonvaccine types.

- Whether the program would be cost-effective
  Making evidence-based assumptions about the natural history of infection, burden of infection, and expected health benefits, and by building a model of sexual transmission, a cost-effectiveness analysis suggested that routine vaccination of girls aged 12 to 14 years could be expected to be cost-effective at 80% vaccine coverage.

- Attitudinal work
  Qualitative in-depth interviews explored parents' responses to HPV vaccination being added to the immunization program and found most parents to be very positive about a vaccine to prevent cervical cancer.

In providing evidence-based but practical recommendations, the JCVI had to consider some of the following public health issues:

- Which vaccine to use?
  The JCVI advised that the choice should depend primarily on cost-effectiveness. Although the quadrivalent vaccine was preferable because it offered combined protection from warts as well as cancer, if the bivalent vaccine were cheaper, the differential in price could compensate for the lack of protection against warts.

- Whom to vaccinate?
  The JCVI recommended vaccinating girls aged 12 to 13 years, that is, prior to initiating sexual activity for most girls. The available evidence suggested that vaccinating older girls was unlikely to be cost-effective because the vaccine would be "wasted" in those already exposed. Nevertheless, JCVI recognized that some older girls during the first phase of vaccine introduction would benefit from immunization and a time-limited catch-up program of girls aged 13 to 18 years was recommended.

- Whether to vaccinate boys?
  The decision not to vaccinate boys was based on the estimated cost-effectiveness at the time, which included the assumption that most boys would be protected through herd immunity, if high coverage was achieved in girls. This decision has being revisited, in part because of increasing evidence about the role of HPV in anal, penile, head, neck, and other cancers, and in part because of the lack of herd protection afforded to men who have sex with men (MSM). MSM are an important group because they suffer higher rates of HPV-associated disease, particularly anogenital warts and anal cancer, but they are difficult to target because they may not identify themselves before their sexual debut and later vaccination may not be cost-effective. They are also unlikely to be protected by herd immunity through vaccination of girls and may be discriminated against if not protected. The JCVI considered this further in 2014–2015 and recommended that a targeted program of vaccination for MSM attending sexual health clinics should be introduced to address this issue. A pilot program was started in England in 2016, and a full roll-out is expected from 2018; it has already been introduced in other parts of the UK.

- How to deliver the vaccine?

  The JCVI recommended delivering the HPV vaccine in three doses to girls in school year 8, using a school-based system co-administered with the tetanus, diphtheria, and polio vaccine (Td/IPV) teenage 3-in-1 booster. This approach was felt to be practicable and likely to be acceptable, and it was anticipated that it would achieve high vaccination coverage.

- How to monitor and evaluate the vaccination program (surveillance)?

  Surveillance of vaccine coverage was argued to be an essential and integral component of the vaccination program. A comprehensive plan was therefore drawn up to monitor and evaluate the vaccine introduction, with the expectation that this would inform any further public health initiatives to control cervical cancer in the UK.

PUBLIC HEALTH INTERVENTIONS

From 2008 in the UK, girls aged 12–13 years and a catch-up group of those up to 18 years were offered three doses of a bivalent HPV vaccine, Cervarix®, which effectively prevents infection with the two high-risk HPV (HR-HPV) types (16 and 18) responsible for most cervical cancer. From September 2012, the program switched to using the quadrivalent vaccine, Gardasil®, which also protects against low-risk HPV (LR-HPV) types (6 and 11) that cause most genital warts.

The vaccines contain nonviable virus-like particles (VLP), which do not contain DNA and cannot cause HPV-associated diseases. VLPs are synthesized as recombinant proteins, using either yeast systems or baculovirus-infected insect cells. Both vaccine types are extremely effective at preventing premalignant cell changes associated with HPV types 16 and 18, and both have an extremely good safety profile. Although there are very few groups who cannot receive the HPV vaccine, the advice is that pregnant women should avoid HPV vaccination. This advice is given because there is a lack of evidence rather than because the vaccines are known or expected to lead to any negative health consequences for mother or baby.

To date, the UK HPV vaccination program is deemed to have been successful. National vaccine coverage has been high and above the level required for herd immunity. For example, 86% of girls in the target group in England received three doses from 2012 to 2013. Vaccinating girls in this way, before first sex, is predicted to prevent at least 60% of cervical cancer over the next decades. Although there is also some evidence that the initial HPV-16/18 vaccination program has some cross-protection and prevented genital warts, the introduction of the quadrivalent vaccine means that much larger reductions in genital warts are now expected. The vaccine schedule used for the 12- to 13-year-old girls was changed in 2014 from three to two doses, as the immunogenity data suggest that two doses are equally effective in this age group.

Analyses of routine surveillance data on vaccine coverage found no evidence of a relationship between coverage at age 12 and areas of social deprivation, but there was evidence that deprivation might be associated with lower coverage for older girls eligible for the catch-up program. Further work is required to ascertain what individual factors might be associated with vaccine coverage in order to better understand the characteristics of those remaining at risk of persistent HPV infection and to explore what effect this might have on HPV prevention in subgroups of the population (Figure 9.1).

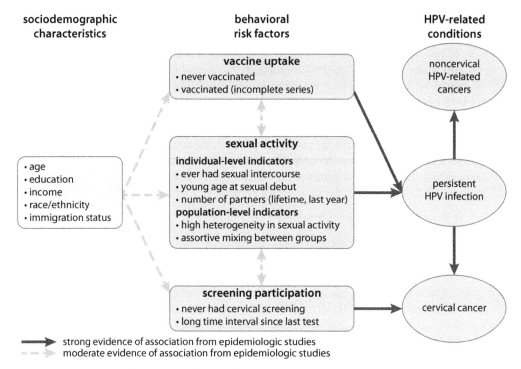

Figure 9.1. **Conceptual framework of the different pathways linking sociodemographic characteristics, behavioral risk factors such as vaccine uptake, and HPV-related cancers.** (From Brisson M, Drolet M & Malagón T [2013] *J Natl Cancer Inst* **105**:750–751. With permission from Oxford University Press.)

Future developments in HPV immunization include a nonavalent vaccine, which has been licensed recently, and will increase the proportion of cervical cancers that can be prevented. Whether to offer HPV vaccine to boys has been reviewed recently, but it is unlikely that this will be cost-effective, given the very high coverage in girls.

In England, the cervical cancer-screening program aims to reduce cervical cancer incidence and the number of women who die from this disease. All women aged 25 to 64 years are routinely offered cervical screening tests using an automated invitation generated through general practitioner (family doctor) patient lists. Screening is offered every three years for those aged 25 to 49 and every five years from 50 to 64 years; those with human immunodeficiency virus (HIV) are offered screening annually. The test, called liquid-based cytology (LBC), involves taking a cervical brush sample (necessitating a speculum examination), with the cells suspended in a liquid buffer. Samples are processed onto a slide, which is examined to detect evidence of cervical intraepithelial neoplasia (CIN), using a well-established grading system. More recently, the screening program clinical management algorithms have been revised so that samples with borderline or mild dyskaryosis on LBC are tested for HR-HPV, which is used to decide whether women are offered colposcopy. This has been shown to improve the sensitivity of screening for CIN and reduce the number of women requiring interventions, which may prove to be unnecessary.

In the UK, uptake of cervical screening is high (around 80%). However, both incidence and mortality of cervical cancer are associated with deprived areas. Of concern is that the incidence of cervical cancer has been increasing in young women, who are also less likely to access screening.

EPIDEMIOLOGY

Anogenital HPV is typically acquired through sexual contact in adolescence or early adulthood, but there is also evidence of vertical and horizontal HPV transmission in young children and there may be a bimodal age distribution. A review of eight studies found HPV DNA was detected in 9 to 56% of children ranging in age from 0 to 12 years, and concordance has been demonstrated between maternal and child HPV types. Although there is little evidence of transmission via fomites, infection may also be transmitted between hands (HPV type-2) and genitals in children. Nevertheless, sexual transmission remains the primary mode of transmission, and more than half of all sexually active women will acquire one or more HPV type during their lifetime.

Most HPV acquisition occurs soon after sexual debut, through sexual contact with an infected partner, and the risk of infection is strongly associated with the number of sexual partners. Risk of infection is therefore much higher after introduction of a new sexual partner, although the level of risk will be determined by the sexual history of the new partner. Note that the use of condoms substantially reduces the risk of infection but does not eliminate transmission because HPV may be present on the skin not covered by the condom.

Although adult genital HPV infection is often asymptomatic and undiagnosed, genital warts remain the most frequently diagnosed viral sexually transmitted infection (STI) in UK sexual health clinics. In part, this results from recurrence; although usually readily treatable, recurrent warts make up around 40% of all genital warts diagnoses. In the UK, among those aged 10–49 years, the estimated prevalence of LR-HPV type 6 is 8% in men and 16% in women, and for type 11 is 2% in men and 6% in women. Genital wart diagnoses in Genitourinary Medicine (GUM) clinics in the UK had been increasing each year until 2009, with over 91,000 diagnoses in 2009. However, for the first time in more than a decade, a small drop in the number of diagnoses occurred between 2008 and 2009. Against the trend of other STIs, it is particularly noticeable that wart diagnoses have declined in the past few years among adolescents. Whether or how this relates to the introduction of a national HPV vaccination program is uncertain, given that the vaccine in use from 2008 to 2012 was the bivalent vaccine, which does not protect against HPV 6 and 11. Nevertheless, in England, GUM clinic diagnoses among girls aged 15 to 19 years, who correspond to the first English girls to received HPV vaccination, fell by 22% between 2009 and 2012.

The prevalence of HR-HPV types is estimated to be 16% in women and 8% in men aged 16 to 44 years, but high-risk groups, such as HIV-positive MSM, may have much higher prevalence.

BIOLOGY

The family of human papillomaviruses includes nearly 200 types, with more than 30 types that are sexually transmitted. This section focuses on anogenital HPV.

HPV is a double-stranded DNA virus, which characteristically integrates into the host cell genome. HPV preferentially infects squamous epithelial cells, including skin and anogenital and respiratory mucosa. Most HPV infections are cleared rapidly; the median duration of a new infection is estimated to be around eight months, with 70% of infections cleared within one year and 90% within two years. Latent infections are not well defined, but an estimated 10% are transforming type infections—probably where the viral episome is maintained in the basal epithelial layer. The process of transformation is driven, at least in part, by two viral oncogenes, E6 and E7. These genes encode products that subvert the host cellular machinery to the advantage of the virus and its replication strategy, preventing apoptosis (programmed cell death) and promoting cell-cycle progression. Cell-cycle deregulation is thought to involve many other steps, and in some cases may be associated with the integration position of the viral episome (viral genetic material) in the host genome, which can affect transcription of host and viral gene products.

HPV is usually transmitted through direct skin-to-skin contact, and it has been shown that infection of the basal epithelium depends on micro wounding and repair of the epithelium. The incubation period for HPV ranges widely from a few weeks to months or years. Although HPV causes no or minimal viremia, most infected persons seem to respond with an effective cell-mediated immune response, such that viral DNA can no longer be detected and clinical remission is sustained. This effective immune response is typically targeted to early phase nonstructural viral proteins. Circulating antibody may be detectable but seroconversion rates are low and vary by site of infection.

DISEASE PATHOLOGY

HPV and cancer

Cervical cancer is the second most common cause of cancer in women in the world, and it is estimated that 500,000 cases and 270,000 deaths occur each year. At least 13 HR-HPV types encode oncogenic proteins that disrupt the cell cycle in skin and mucosal tissues and can drive transformation of infected cells. Although most HR-HPV infections are thought to be transient, with no long-term consequences, a minority of infections persist and lead to premalignant and malignant changes in infected cells. Persistent infection with HR-HPV types, particularly types 16, 18, 31, and 45, is necessary for development of cervical cancer (HR-HPV infection is found in over 99% of cases), and infection is also associated with a subgroup of oropharyngeal and esophageal carcinomas. In Europe, HPV type 16 is associated with 50% and HPV type 18 with 15% of cervical cancers. In England, where a systematic routine cervical screening program has been in place since 1988, about 3000 cases of cervical cancer are diagnosed each year, and it is estimated that the program prevents around 4500 deaths per year. Nevertheless, there are still around 800 cervical cancer deaths each year in England.

HPV and warts

LR-HPV types cause both anogenital and nongenital warts, as well as juvenile and adult recurrent respiratory papillomatosis. Warts are benign lesions, although some lesions may contain co-infection with HR-HPV types, and around 90% of anogenital warts are caused by HPV types 6 or 11. The life-time risk of acquiring warts in high-income countries is estimated to be 10%. Although benign, these infections represent a substantial disease burden and a significant financial cost to healthcare budgets.

QUESTIONS

1. With respect to human papilloma virus (HPV), which of the following statements are true and which are false?
   a. Infection with HPV type 16 or type 18 in the cervical squamous epithelium is sufficient to cause cervical invasive neoplasia (CIN).
   b. HPV types 6 and 12 cause most anogenital warts.
   c. HPV is an RNA virus that integrates into the host genome.
   d. Although the virus life cycle does not include a viremic phase, most infected persons develop an effective cell-mediated immune response.
   e. HPV is not thought to be effectively transmitted through fomites.

2. Which **one** of the following statements about HPV vaccination is incorrect?
   a. Girls older than 21 are less likely to benefit from HPV vaccination.
   b. The vaccine is made from synthesised proteins and is therefore deemed to be safe for pregnant women.
   c. Over the coming two decades, the vaccination program might reduce cervical cancer by as much as 60%.
   d. Vaccine coverage in excess of 80% in girls is predicted to give herd immunity, including protection to heterosexual boys.
   e. HPV vaccines have an excellent safety record to date.

3. Which **one** of the following statements about HPV transmission is incorrect?
   a. HPV may be transmitted from mother to child at birth.
   b. HPV acquisition is strongly associated with reported lifetime number of sexual partners.
   c. HPV infection is more common in people who smoke.
   d. HPV transmission is prevented by using condoms.
   e. HPV incubation can be long, from weeks up to several months or even years.

4. In giving advice about cervical cancer screening, which of the following statements are true and which are false?
   a. Patients with HIV should be offered cervical cancer screening every two years.
   b. Women with a family history of cervical cancer should be offered cervical screening from the age of 16 years.
   c. Women aged 50–64 years are offered cervical cancer screening every five years.
   d. Cervical cancer screening using liquid-based cytology requires a speculum examination.
   e. HPV-testing for high-risk types can improve the sensitivity of cervical cancer screening.

5. The public health benefits of HPV vaccination can be increased. Which of the following are true and which are false?

   a. The introduction of the nonavalent vaccine will increase the proportion of genital wart cases that can be prevented.

   b. Vaccination of boys will result in early benefits in cervical cancer prevention in the UK.

   c. The UK has a high uptake of vaccine in girls, by international standards.

   d. Oropharyngeal cancer is being increasingly recognized as an adverse health outcome of HPV infection.

   e. The current policy of vaccinating young girls will indirectly lead to a reduction in anal cancer incidence in MSM.

GUIDELINES

1. Greenbook, Chapter 18a https://www.gov.uk/government/uploads/system/uploads/attachment_data/file/317821/Green_Book_Chapter_18a.pdf

2. Joint Committee on Vaccination and Immunization (JCVI) – HPV Subgroup minutes and statements https://www.gov.uk/government/uploads/system/uploads/attachment_data/file/294845/Minutes_HPV_Subcommittee_meeting_Jan_2014_final.pdf

3. http://webarchive.nationalarchives.gov.uk/20130107105354/http://www.dh.gov.uk/prod_consum_dh/groups/dh_digitalassets/@dh/@ab/documents/digitalasset/dh_094739.pdf

4. ECDC HPV vaccination policy documents. http://ecdc.europa.eu/en/publications/publications/20120905_gui_hpv_vaccine_update.pdf

5. http://www.ecdc.europa.eu/en/aboutus/organisation/Director%20Speeches/Marc-Sprenger-overview-HPV-vaccination-Doha-December-2013.pdf

4. CDC HPV vaccine information http://www.cdc.gov/hpv/vaccine.html

REFERENCES

1. Read TRH, Hocking JS, Chen MY et al. (2011) The near disappearance of genital warts in young women 4 years after commencing a national human papillomavirus (HPV) vaccination programme. *Sex Transm Infect* **87**:544–547 (doi:10.1136/sextrans-2011-050234).

2. Ali H, Donovan B, Wand H et al. (2013) Genital warts in young Australians five years into national human papillomavirus vaccination programme: national surveillance data. *BMJ* **346**:f2032–f2032 (doi:10.1136/bmj.f2032).

3. Joint Committee on Vaccination and Immunisation - Groups - GOV.UK. https://www.gov.uk/government/groups/joint-committee-on-vaccination-and-immunisation

4. JCVI statement on Human papillomavirus vaccines to protect against cervical cancer.

5. Jit M, Vyse A, Borrow R et al. (2007) Prevalence of human papillomavirus antibodies in young female subjects in England. *Br J Cancer* **97**:989–991 (doi:10.1038/sj.bjc.6603955).

6. Johnson AM, Mercer CH, Beddows S et al. (2012) Epidemiology of, and behavioural risk factors for, sexually transmitted human papillomavirus infection in men and women in Britain. *Sex Transm Infect* **88**:212–217 (doi:10.1136/sextrans-2011-050306).

7. Stanley M, O'Mahony C & Barton S (2014) HPV vaccination. *BMJ* **349**:g4783–g4783 (doi:10.1136/bmj.g4783).

8. Schmitz M, Driesch C, Jansen L et al. (2012) Non-Random Integration of the HPV Genome in Cervical Cancer. *PLoS ONE* **7**:e39632 (doi:10.1371/journal.pone.0039632).

9. Ho GY, Bierman R, Beardsley L et al. (1998) Natural history of cervicovaginal papillomavirus infection in young women. *N Engl J Med* **338**:423–428 (doi:10.1056/NEJM1998021233807).

10. Sapp M & Day PM (2009) Structure, attachment and entry of polyoma- and papillomaviruses. *Virology* **384**:400–409 (doi:10.1016/j.virol.2008.12.022).

11. Stanley M. (2010) HPV-immune response to infection and vaccination. *Infect Agent Cancer* **5**:19 (doi:10.1186/1750-9378-5-19).

12. Peto J, Gilham C, Deacon J et al. (2004) Cervical HPV infection and neoplasia in a large population-based prospective study: the Manchester cohort. *Br J Cancer* **91**:942–953 (doi:10.1038/sj.bjc.6602049).

13. Muñoz N, Bosch FX, de Sanjosé S et al. (2003) Epidemiologic classification of human papillomavirus types associated with cervical cancer. *N Engl J Med* **348**:518–527 (doi:10.1056/NEJMoa021641).

14. Winer RL, Feng Q, Hughes JP et al. (2008) Risk of female human papillomavirus acquisition associated with first male sex partner. *J Infect Dis* **197**:279–282 (doi:10.1086/524875).

15. Peto J, Gilham C, Fletcher O et al. (2004) The cervical cancer epidemic that screening has prevented in the UK. *Lancet* **364**:249–256 (doi:10.1016/S0140-6736(04)16674-9).

16. Gilson R. (2015) UK National Guidelines on the Management of Anogenital Warts.

17. Antonsson A, Karanfilovska S, Lindqvist PG et al. (2003) General acquisition of human papillomavirus infections of skin occurs in early infancy. *J Clin Microbiol* **41**:2509–2514.

18. Rombaldi RL, Serafini EP, Mandelli J et al. (2009) Perinatal transmission of human papillomavirus DNA. *Virol J* **6**:83 (doi:10.1186/1743-422X-6-83).

19. Sinal SH & Woods CR (2005) Human papillomavirus infections of the genital and respiratory tracts in young children. *Semin Pediatr Infect Dis* **16**:306–316 (doi:10.1053/j.spid.2005.06.010).

20. STD Facts—Human papillomavirus (HPV). http://www.cdc.gov/std/HPV/STDFact-HPV.htm

21. Koutsky L (1997) Epidemiology of genital human papillomavirus infection. *Am J Med* **102**:3–8.

22. Sonnenberg P, Clifton S, Beddows S et al. (2013) Prevalence, risk factors, and uptake of interventions for sexually transmitted infections in Britain: findings from the National Surveys of Sexual Attitudes and Lifestyles (Natsal). *Lancet* **382**:1795–1806 (doi:10.1016/S0140-6736(13)61947-9).

23. STI Annual Data Tables. https://www.gov.uk/government/statistics/sexually-transmitted-infections-stis-annual-data-tables

24. Desai S, Chapman R, Jit M et al. (2011) Prevalence of human papillomavirus antibodies in males and females in England. *Sex Transm Dis* **38**:622–629 (doi:10.1097/OLQ.0b013e31820bc880).

25. Public Health England (2014) Declines in genital warts since start of the HPV immunisation programme. **8**.

26. Machalek DA, Poynten M, Jin F et al. (2012) Anal human papillomavirus infection and associated neoplastic lesions in men who have sex with men: a systematic review and meta-analysis. *Lancet Oncol* **13**:487–500 (doi:10.1016/S1470-2045(12)70080-3).

27. Public Health England. Human papillomavirus (HPV), *Green Book Chapter 18a* https://www.gov.uk/government/uploads/system/uploads/attachment_data/file/317821/Green_Book_Chapter_18a.pdf.

28. Jit M, Chapman R, Hughes O et al. (2011) Comparing bivalent and quadrivalent human papillomavirus vaccines: economic evaluation based on transmission model. *BMJ* **343**:d5775.

29. Szarewski A, Skinner SR, Garland SM et al. (2013) Efficacy of the HPV-16/18 AS04-adjuvanted vaccine against low-risk HPV types (PATRICIA randomized trial): an unexpected observation. *J Infect Dis* **208**:1391–6. doi:10.1093/infdis/jit360.

30. Howell-Jones R, Soldan K, Wetten S et al. (2013) Declining genital warts in young women in England associated with HPV 16/18 vaccination: an ecological study. *J Infect Dis* **208**:1397–1403 (doi:10.1093/infdis/jit361).

31. Hughes A, Mesher D, White J et al. (2014) Coverage of the English national human papilloma-virus (HPV) immunisation programme among 12 to 17 year-old females by area-level depri-vation score, England, 2008 to 2011. *Euro Surveill Bull Eur Sur Mal Transm Eur Commun Dis Bull* **19**.

32. NHS Cervical Screening Programme (NHSCSP). http://www.cancerscreening.nhs.uk/cervical

33. HPV triage and test of cure in the cervical screening programme in England. http://www.cancerscreening.nhs.uk/cervical/hpv-triage-test-of-cure.html

34. Montz FJ (2000) Management of high-grade cervical intraepithelial neoplasia and low-grade squamous intraepithelial lesion and potential complications. *Clin Obstet Gynecol* **43**:394–409.

35. Health and Social Care, Information Centre, Screening and Immunisations team. Cervical Screening Programme, England 2011-12. The Health and Social Care Information Centre 2012. http://www.cancerscreening.nhs.uk/cervical/cervical-statistics-bulletin-2011-12.pdf

36. Shack L, Jordan C, Thomson CS et al. (2008) Variation in incidence of breast, lung and cervical cancer and malignant melanoma of skin by socioeconomic group in England. *BMC Cancer* **8**:271 (doi:10.1186/1471-2407-8-271).

37. Foley G, Alston R, Geraci M et al. (2011) Increasing rates of cervical cancer in young women in England: an analysis of national data 1982–2006. *Br J Cancer* **105**:177–84 (doi:10.1038/bjc.2011.196).

38. Lancuck L, Patnick J & Vessey M (2008) A cohort effect in cervical screening coverage? *J Med Screen* **15**:27–29 (doi:10.1258/jms.2008.007068)

ANSWERS

MCQ	Feedback
1. With respect to HPV, which of the following statements are true and which are false? a. Infection with HPV type 16 or type 18 in the cervical squamous epithelium is sufficient to cause cervical invasive neoplasia (CIN). b. HPV types 6 and 12 cause most anogenital warts. c. HPV is an RNA virus that integrates into the host genome. d. Although the virus life cycle does not include a viraemic phase, most infected persons develop an effective cell-mediated immune response. e. HPV is not thought to be effectively transmitted through fomites.	a. False. Although highly oncogenic, infection with HPV type 16 or type 18 (or another oncogenic type) is required but is not sufficient to trigger cancer in cervical epithelial cells. b. False. HPV 6 and 11 cause most anogenital warts. c. False. HPV is a DNA Virus. d. True. it not well understood why the host produces such an effective cell-mediated response to HPV infection without a viraemic phase e. True.
2. Which one of the following statements about HPV vaccination is incorrect? a. Girls older than 21 are less likely to benefit from HPV vaccination. b. The vaccine is made from synthesised proteins and is therefore deemed to be safe for pregnant women. c. Over the coming two decades, the vaccination program might reduce cervical cancer by as much as 60%. d. Vaccine coverage in excess of 80% in girls is predicted to give herd immunity, including protection to heterosexual boys. e. HPV vaccines have an excellent safety record to date.	Although there is no known risk associated with giving vaccines like HPV, which are synthetic proteins, and the vaccines are unable to cause HPV-associated disease, a precautionary approach is recommended. The UK Green Book (Immunisation Against Infectious Disease) advises completing the pregnancy before finishing any HPV vaccination course if started before or inadvertently during pregnancy.
3. Which one of the following statements about HPV transmission is incorrect? a. HPV may be transmitted from mother to child at birth. b. HPV acquisition is strongly associated with reported lifetime number of sexual partners. c. HPV infection is more common in people who smoke. d. HPV transmission is prevented by using condoms. e. HPV incubation can be long, from weeks up to several months or even years.	Although condoms can reduce the likelihood of HPV transmission, they are not an effective method to prevent transmission, in part because of the presence of HPV particles on skin not covered by the condom.
4. In giving advice about cervical cancer screening, which of the following statements are true and which are false? a. Patients with HIV should be offered cervical cancer screening every two years. b. Women with a family history of cervical cancer should be offered cervical screening from the age of 16 years. c. Women aged 50–64 years are offered cervical cancer screening every five years d. Cervical cancer screening using liquid-based cytology requires a speculum examination. e. HPV-testing for high-risk types can improve the sensitivity of cervical cancer screening.	a. False: Patients with HIV should be offered screening annually. b. False: There is no evidence for initiating screening at a younger age in women with a family history of cervical cancer. c. True d. True e. True
5. The public health benefits of HPV vaccination can be increased. Which of the following are true and which are false? a. The introduction of the nonvalent vaccine will increase the proportion of genital wart cases that can be prevented. b. Vaccination of boys will result in early benefits in cervical cancer prevention in the UK. c. The UK has a high uptake of vaccine in girls, by international standards. d. Oropharyngeal cancer is being increasingly recognized as an adverse health outcome of HPV infection. e. The current policy of vaccinating young girls will indirectly lead to a reduction in anal cancer incidence in MSM.	a. False: Additional genotypes covered are all HR-HPV types. b. False: High coverage rates in girls mean that early benefits will be limited. Long-term benefits expected include a reduction in risk for the residual population of unvaccinated women and for disease in MSM. c. True d. True e. False: Herd immunity will have little impact on disease in MSM.

A LABORATORY INCIDENT LINKED TO EXPOSURE TO BOTULINUM TOXIN

CASE 10

John Holton

Department of Natural Sciences, School of Science & Technology, University of Middlesex, UK.
National Mycobacterial Reference Service - South, Public Health England, UK.

A 42-year-old male patient was admitted to the hospital with nausea, weakness, difficulty in speaking, and double vision. He also complained that his legs felt tired. The patient had been abroad and had complained of a diarrheal episode some 2–3 weeks previously. He had recently eaten some olives, a salad, tinned tuna, a pepperoni pizza, honey yogurt, and ice cream. His full blood count (FBC), urea and electrolytes (U&E), liver function tests (LFT), and C-reactive protein (CRP) were all normal. The magnetic resonance imaging (MRI) of the head was normal and his cerebrospinal fluid (CSF) was unremarkable. As part of the clinical investigation as to the cause of the symptoms, specimens of olives, honey yogurt, and tuna were received at a laboratory for investigation of the presence of botulinum toxin.

During the processing of one of the specimens, the bag containing the olives, which were being macerated in a Stomacher® (a blender used in food safety testing) burst, distributing the contents over the machine, the bench top, and onto the skin of the technician.

INVESTIGATION OF THE INCIDENT

An incident occurs with accidental or deliberate release of a biological agent into the environment, thereby exposing any individuals to that biological agent.

A laboratory incident has three major components. The immediate response is decontamination of any individuals, the second component is a risk assessment of the incident to determine the circumstances and severity of the incident, and the final component is management of contaminated individuals.

The initial response to the incident should be dealt with by the appropriate environmental release (spill) policy, for example, a liquid spill, an aerosol release, or a broken centrifuge tube. If a spill has occurred, it should be diluted or neutralized. A contaminated individual should remove any contaminated laboratory wear or, if the spill is extensive, may need to shower. If an aerosol has been released, there is a potential for a greater number of individuals being exposed and it is handled by evacuation of the area. A biological agent must be inactivated by gaseous means.

The second part of the investigation is to complete an incident form detailing the circumstances surrounding the incident. A generic incident form is shown in Figure 10.1. Frequently, an investigation of an incident will demonstrate the reasons for the incident, which may be mechanical failure, lapse of standard operating procedure (SOP), or an

Details of person completing this report
Full name ....
Staff/Student/Visitor/Contractor (circle relevant category)
School/Department/Course - if staff or student ....
Address & postcode if visitor or contractor...
Telephone contact number ....
Signature... Date...

Details of accident/incident
What happened. Give cause (how and why) if known ....
....
....
....
....
When it happened: date .... time ....
Where it happened ....

Details of any persons injured
Full name ....
Staff/Student/Visitor/Contractor (circle relevant category)
School/Department/Course if staff or student ....
Address & postcode if visitor or contractor ....
....
Telephone contact number..
School/Department/Course ....
Nature of injury ....
....
....
Treatment given..
Treatment given by ....
Taken to hospital - yes/no
If yes, which hospital and how taken..
Off work as a result of accident/incident - yes/no   If yes - number of days............................

For completion by the Health and Safety Officer
Accident/Incident investigated - yes/no
Written investigation report necessary - yes/no
Written investigation report completed - yes/no
RIDDOR reportable - yes/no. If yes, date reported ....
EA/DEFRA reportable - yes/no. If yes, date reported ....
Charity Commission, serious incident reportable - yes/no   If yes, date reported ....

Witness details, statements, etc - continue report overleaf if necessary.

Investigator's guide
In order to determine the cause of the accident or incident, it may be appropriate to interview parties who were involved. First think about the questions you ultimately want to answer, for example:
  a) Was a safe work procedure used?
  b) If there were safe working procedures were they up to date?
  c) If there were safe working procedures and instructions, were they realistic, accurate and adequate?
  d) If there were safe working procedures, etc were they readily available to those carrying out the work?
  e) If there were safe working procedures, were they enforced/monitored/supervised?
  f) Were training needs for the activity identified?
  g) Was any required Personal Protective Equipment available for use and used?
  h) Had conditions changed to make the normal procedure unsafe?
  i) Were the appropriate tools and materials available?
  j) Were safety devices working properly?
  k) Were the correct materials/substances being used?

Figure 10.1. **A generic incident form.**

unforeseen problem. Such investigations will lead to a change in procedures or in the SOP to prevent future incidents. The risk assessment will also indicate the degree of exposure of any individual and will indicate whether or not exposed individuals require prophylaxis.

The third responsibility after a laboratory incident is management of the individuals who have been exposed. This will in part occur during initial decontamination of contaminated individuals, but the risk assessment of the incident will determine if prophylaxis is required. Exposure could involve skin contact, inhalation, ingestion, or inoculation. It is important to have knowledge of the hazard group of the organism, its usual epidemiology, diseases, and treatment. Dealing with exposed individuals usually involves either a watch-and-wait approach, after ensuring the individual understands the signs and symptoms to watch out for, collection of a baseline blood sample, or prophylactic antimicrobial agents. Factors that suggest an increased susceptibility to infection by laboratory workers, for example, pregnancy or immunosuppression, will be likely to affect the management of the person as they may be more likely to receive prophylaxis

In this particular incident, the liquid was spilled onto the Stomacher®, the bench, and the skin of the individual. Subsequent testing of the macerated food confirmed it contained botulinum toxin in a high concentration.

An incident form was completed and it revealed the bag had burst because of a retained olive stone. In this incident, it was also revealed that the individual did not wear gloves when handling the specimen, which was contrary to the standard operating procedure. The spilled material was inactivated by dilution with peracetic acid. The SOP for dealing with specimens possibly containing botulinum toxin was reviewed for all staff members.

CLINICAL MANAGEMENT

The individual was not given botulinum antitoxin because the toxin is not absorbed through intact skin. He was, however, reminded of the protocol for working with toxins. Where absorption of botulinum toxin has occurred, the treatment is to give equine antitoxin, after a skin test for allergy, followed by periodic assessment of respiratory function and signs of anaphylaxis. An intravenous injection of the trivalent antitoxin (A, B, E) has a half-life of about 5–8 days. The antitoxin neutralizes only unbound toxin, and the earlier it is given, the better the outcome.

PREVENTION OF FURTHER INCIDENTS

There are currently strict protocols and procedures available to prevent laboratory-acquired infections. First, there is a risk-based analysis of biosafety from levels 1–4. The criteria are infectivity of the organism, severity of the illness, transmissibility of the organism, and efficacy of treatment. Each level has its own protocols. Biosecurity level (BSL) 1 refers to organisms not known to cause disease in healthy individuals. Organisms belonging to BSL2 can cause human disease by ingestion or contact, but are not generally serious. They are unlikely to be transmitted to another individual and an effective treatment exists for them. Work with BSL1–2 organisms can be carried out on an open bench. Organisms belonging to BSL3 can be transmitted by aerosols, and onward transmission of the organism is likely. They may cause serious or lethal disease, but an effective treatment exists. Organisms belonging to BSL4 are high-risk organisms that are life-threatening, transmission can occur by aerosols, onward transmission of the organism is likely, and no treatment is available. BSL3 requires a separate laboratory room with laminar-flow safety cabinets and restricted access. BSL4 requires a specifically designed laboratory, often built away from large conurbations, with Class 3 laminar-flow cabinets and personal protective equipment (PPE) that may include a full body suit with oxygen supply. Small countries usually have only one specially designated BSL4 laboratory. Organisms are categorized according to the hazard group (HG) associated with the biosafety level and handled accordingly. For example, *Staphylococcus epidermidis* belongs to HG1, toxigenic *Staphylococcus aureus* belongs to HG2, *Mycobacterium tuberculosis* belongs to HG3, and Marburg virus belongs to HG4. Overall in HG3 there are 38 bacteria, 10 fungi, 3 helminths, 6 protozoa, and 70 viruses. All HG4 organisms are viruses, of which there are 21. All prions belong to HG3.

Biohazard protection involves both personal and environmental factors. PPE at a basic level includes gloves and a laboratory coat. In some cases, glasses (goggles or a face visor) may be required to prevent splashes onto mucous membranes or the conjunctiva. PPE can also include face masks. There are three categories of face mask. FFP1 has a filter efficiency of 80% and an inward leak of ambient air <22%, FFP2 has a filter efficiency of 94% and an inward leak of <8%, and FFP3 has a filter efficiency of 99% and an inward leak of <2%. A higher grade of PPE is provided by total nuclear chemical biological (NCB) suits for work in BSL4 laboratories (Figure 10.2).

Analysis of aerosol production demonstrates the proportion of a volume, V, that is converted to an aerosol depends on the height at which a liquid falls before impacting a surface (usually the floor). A fall of 1 m compared to 10 cm will convert significantly more of the liquid to an aerosol. Liquids only produce aerosols on impact or if expressed from a container under pressure, for example a blocked syringe, whereas powders will produce aerosols as they fall through the air. The laboratory incidents that produce the

Figure 10.2. **Nuclear chemical biological suits worn by two microbiologists working in a BSL4 laboratory.** (Courtesy of CDC/Scott Smith.)

largest concentration of aerosols are dropping a fungal plate, dropping a volume 0.75 m, a centrifuge incident, and a blocked syringe. A spray factor (SF) can be determined from the ratio of the concentration in colony forming units (cfu/mL) of organisms in the aerosol (A) divided by the concentration in the liquid (L). $SF = A/L$. If the SF is known for different types of incidents, the concentration A can be calculated from this formula.

Environmental protection includes a correctly designed laboratory and is governed in part by the following UK Regulations: Health and Safety at Work Act, Management of Health and Safety at Work Regulations, Control of Substances Hazardous to Health Regulation, and the Construction, Design and Management Regulations. Microbiology laboratories should have a flexible design to allow for changes of work patterns, for example the provision of general and local exhaust facilities, temperature and humidity controls, and a minimum working space of 11 m^3 per individual.

There are certain mandated requirements in the UK for microbiology laboratories. These include the following:

- access only for authorized persons
- safe storage of HG2 and HG3 organisms
- bench tops that are resistant to organic solvents, acids, and alkalis, and impervious to water
- BSL3 laboratory floors must have the same characteristics as bench tops
- specified disinfection procedures
- aerosol generating procedures are carried out in a biological safety cabinet
- BSL3 laboratories are isolated from other activities in the building and have a viewing port in order to see the occupants
- BSL3 laboratories are sealable to allow fumigation and the containment of its own equipment
- BSL3 laboratories are maintained at negative atmospheric pressure
- efficient extraction of air from BSL3 laboratories using a high-efficiency particulate (HEPA) filters

The minimum recommended pressure differential for a BSL3 laboratory to achieve negative pressure is -30 to -50 Pa. It is also recommended that BSL3 laboratories have a lobby that increases the containment by a factor of 100 in case of a biological spill, as well as making the laboratory more secure against unauthorized individuals.

Air quality is also an important aspect of microbiology laboratories. Proper air quality is achieved by filtration of air into a laboratory (input or outlet) or to a biological safety cabinet. There are various grades of filters. In Europe, HEPA filter grades range from E10 to U17. The grade recommended for BSL3 laboratories and biological safety cabinets is H14, retaining >99.995% of particles ≥ 0.3 μ. In the US HEPA filters remove 99.97% of particles ≥ 0.3 μ.

There are three main types of biological safety cabinets. A Class 1 biological safety cabinet protects the operator but does not protect the work from extraneous contamination. A Class 2 biological safety cabinet not only protects the operator but removes contaminated air from the cabinet by a HEPA-filtered laminar flow of air and therefore also protects the culture from contamination. In a Class 3 biological safety cabinet, there is a physical barrier between the operator and the specimen and work is conducted through glove ports. The inlet and outlet air are both HEPA filtered. A Class 3 safety cabinet is the only one suitable for work with biosafety level 4 pathogens.

EPIDEMIOLOGY

Laboratory-acquired infections have been recorded for over a century. Three of the first recorded infections were typhoid in 1885, cholera in 1894, and diphtheria in 1898, the latter two associated with oral pipetting. Between 1893 and 1968, there were 1917 recorded laboratory-acquired infections of which 299 were associated with oral pipetting. Other recorded routes of infections were a case of syphilis following accidental inoculation of a finger (needle stick) and 15 cases of Q fever in a laboratory building, probably from aerosol dissemination. Between 1969 and 1976 in Europe and the US, there were 3921 laboratory-acquired infections caused by 16 different microorganisms. Of these infections, laboratory staff in the UK accounted for 95 infections that were caused by seven of the organisms. Over time there has been a decrease in the prevalence of laboratory-acquired infections. In the UK, between 1988 and 1999, the prevalence was 82.7/100,000 but in 1994 to 1995, it had reduced to 16.2/100,000. In the US, between 2002 and 2003, 33% of laboratories reported at least one infection.

Not all laboratories are the same—some are clinical diagnostic laboratories, either state run or private, some are veterinary diagnostic laboratories, and others are research or reference laboratories. Each deals with separate groups of organisms and the laboratory staff is therefore susceptible to different infections. Differences also exist depending on the country of location. Worldwide, the most common reported laboratory-acquired infections are brucellosis, Q fever, hepatitis (B and C), typhoid, tularemia, tuberculosis, dermatomycoses, Venezuelan equine encephalitis, psittacosis, and coccidioidomycosis, with the majority of deaths attributed to typhoid. In clinical diagnostic laboratories the common bacteria causing infections are *Shigella* spp., *Brucella* sp., *Mycobacterium tuberculosis*, *Salmonella* spp., *Staphylococcus aureus*, and *Neisseria meningitidis*. The prevalence of infections in laboratory staff for *Brucella* is 64.1/100,000 compared to 0.08/100,000 in the general population, and for *Neisseria meningitidis*, the figures are 25.3/100,000 compared to 0.62/100,000. The most common fungal laboratory-acquired infections are caused by the dimorphic fungi *Coccidioides immitis*, *Histoplasma capsulatum*, and *Blastomyces dermatidis*, as a result of aerial dissemination of spores. The most common parasitic laboratory-acquired infections are caused by *Toxoplasma gondii*, *Trypanosoma cruzi*, and *Plasmodium* spp., the latter typically resulting from sharps injuries when making blood films.

Despite the fall in prevalence of laboratory-acquired infections, there are incidents of acquired infection that often make newspaper headlines. In 2000 in the US, 12 laboratory workers acquired *Brucella abortis* infections following an accidental break of a centrifuge tube. In 2009, a research worker in the US developed plague and subsequently died. In 2011 at the same institution in the US another research worker developed a serious skin infection with *Bacillus cereus* following contamination of an open wound, which was not correctly covered. Also in the US, 12 laboratory staff became infected with *Francisella tularensis* from clinical specimens. A Gram stain indicated it was a Gram-negative coccobacillus, but it did not grow on primary isolation media and was initially identified as a *Haemophilus* spp. The clinician had not informed the laboratory of the patient's diagnosis. This situation emphasizes the importance of clinicians informing the laboratories of high-risk specimens. Following the deliberate release of *Bacillus anthracis* in the US, an individual developed cutaneous anthrax after handling a vial of *B. anthracis* and contaminating a shaving cut. The laboratory worker was not wearing gloves.

BIOLOGY

Botulinum toxin is produced principally by *Clostridium botulinum*, but may also be produced in rare instances by *C. butyricum*, *C. baratti*, or *C.argentinense*. *Clostridia* are anaerobic, spore-bearing Gram-positive bacilli. *C. botulinum* was first identified by Emile van Ermengem. *C. botulinum* can produce seven serotypes (A, B, C, D, E, F, G) of botulinum neurotoxin that target the neuromuscular junction. Toxins A, B, E, and F cause human disease, whereas toxin E is frequently associated with contaminated fish. A strain of *C. botulinum* usually produces only one serotype of toxin, but sometimes may produce more than one serotype.

Botulinum toxin is produced during outgrowth from spores and is inactivated by temperatures of 90°C for 5 min, 0.1M sodium hydroxide for 20 min, 3500 ppm peracetic acid for 10 min, and 10,000 ppm available chlorine for 30 min. The toxin may be absorbed from the intestinal tract (food and infant botulism) or from broken skin (wound botulism). The toxin is not absorbed through intact skin but pre-formed toxin can be absorbed through the respiratory route (by deliberate release or laboratory accident as occurred in Europe in 1962).

Botulism can be acquired by the oral route (food-borne and infant botulism), through the skin (wound botulism including intravenous drug use), and the aerial route (deliberate or accidental release).

Food botulism is associated with eating food that has been incorrectly processed or stored and that contains *C. botulinum* toxin. Home preparation of canned vegetables is a common source. Disease may also be acquired from eating contaminated fish products. Infant botulism is acquired from giving honey (contaminated with *C. botulinum*) to the neonate, but the organism may be acquired from the environment.

Wound botulism is acquired after trauma in which the wound is contaminated with *C. botulinum* and is caused by the toxin it produces. Wound botulism occurs most often in relation to the injection of illegal drugs.

Aerial botulism would most likely result from deliberate release as a bioterrorist weapon.

Rarely, botulism can be acquired in an adult with no obvious food-related cause and the organism is probably derived from the environment. The individual also may have a dysbiosis related to immunosuppression. Finally, botulinum toxin is used in cosmetic surgery, rectal surgery to treat anal fissures, and the treatment of cervical dystonia, strabismus, axillary hyperhidrosis, and migraine.

In the UK between 1978 and 2009, there have been 33 cases of food-borne botulism, with one large outbreak from hazelnut yogurt, and 13 cases of infant botulism. Between 2000 and 2010, there have been 144 cases of wound botulism associated with IVDUs.

In the US between 2000 and 2011, there have been 1518 cases of botulism of which 197 were food-borne, 1045 were infant botulism, and 276 were wound botulism associated mainly with IVDUs.

DISEASE

Botulism can present with acute onset or a more gradual onset over several days, depending on the route and the dose of toxin absorbed. Patients may complain of dysphagia, dysarthria, diplopia, ptosis, dry mouth, and muscle weakness affecting the head and neck, with a symmetric descending paralysis of respiratory muscles of the chest, arms, and legs.

The autonomic nervous system may also be affected, and the patient may have dilated pupils, constipation, urinary retention, and orthostatic hypotension.

On examination, the patient may have reduced tendon reflexes and incoordination, but a preserved sensory system and normal cognition.

A full blood count and urea and electrolytes are normal. The patient usually does not have a temperature. Specimens of faeces, serum, gastric aspirate, vomitus, and suspect food should be sent for investigation. Radiological investigation is used only to exclude other neurological diseases. Characteristic electromyography changes may be present.

PATHOLOGY

The molecular weight of the toxin is 150 kDa, and it is composed of an A and B subunit. The A subunit (50 kDa) is a zinc-endopeptidase. It hydrolyses SNARE proteins, such as vesical-associated membrane proteins (VAMP), present on presynaptic vesicles carrying acetylcholine (Ach) and thereby prevents the release of the vesicle contents and thus signal transmission at the neuromuscular junction. The B subunit (100 kDa) targets the toxin to presynaptic receptors and facilitates entry of the active subunit into the nerve.

QUESTIONS

1. The following are important parameters for determining the management of a laboratory incident. True or false?
   a. The Advisory Committee on Dangerous Pathogens (ACDP) Hazard Group to which the organism belongs
   b. The temperature at which the organism was cultured
   c. The production of aerosols
   d. The height at which a liquid falls
   e. The health of laboratory personnel contaminated

2. With regard to the epidemiology of laboratory incidents, are the following true or false?
   a. They have only recently been recognized as a problem.
   b. *Brucella* infections in laboratory workers are about 100 times more common than in the general population.
   c. *Aspergillus flavus* is the most common fungal pathogen associated with laboratory-acquired infections.
   d. The prevalence of an organism causing laboratory-acquired infection depends on geography.
   e. Failure of SOPs is frequently linked to laboratory-acquired infection.

3. With regard to *Clostridium botulinum* and its toxin, are the following statements true or false?
   a. It is the only organism to produce botulinum toxin.
   b. The toxin is produced from the spores.
   c. The toxin produced by *C. botulinum* is heat resistant.
   d. *C. botulinum* produces three serotypes of botulinum toxin with different clinical effects.
   e. *C. botulinum* toxin acts upon spinal reflexes to produce its symptoms.

4. With regard to *Clostridium botulinum*, are the following statements true or false?
   a. Laboratory staff must wear an FFP2 mask.
   b. It belongs to HG4.
   c. It causes descending paralysis.
   d. The half-life of antitoxin in the serum is 58 days.
   e. Toxin E has been associated with eating fish.

5. Are the following statements true or false?
   a. In a microbiology laboratory, the maximum allowable work space per individual is 6 m^3.
   b. Lassa fever is an HG3 virus.
   c. Botulinum toxin is absorbed through intact skin.
   d. A CL3 laboratory must be maintained at negative pressure.
   e. In a Class 2 safety cabinet there is an H14 filter.

GUIDELINES

1. Miller et al. (2012) Guidelines for safe working practices in human and animal medical diagnostic laboratories. *MMWR* **61**:1-101.

2. Advisory Committee on Dangerous Pathogens (2001) The Management, Design and Operation of Microbiological Containment Laboratories. HMSO, Norwich.

3. Heptonstall J & Gent N (2006) CBRN Incidents: Clinical Management and Health Protection, Health Protection Agency.

4. MRC health and safety policy notes: Risk Management (guidance Note 1); A Guide to Risk Assessment (guidance Note 2) (extra.mrc.ac.uk/hss/pdfs/risk_assessment_guidance.pdf).

5. University of Edinburgh: Laboratory Self Inspection Checklist (www.safety.ed.ac.uk/safenet/self_inspec/wordlab_pdf.pdf).

6. ILO, International Hazard Datasheets on Occupation: 'Laboratory worker' (last update 2000) http://www.ilo.org/public/english/protection/safework/cis/products/hdo/htm/ wrkr_labor.htm#hazard

7. Centers for Disease Control and Prevention (CDC): 'The 1, 2, 3's of Biosafety Levels' http://www.cdc.gov/od/ohs/symp5/jyrtext.htm

8. Homer LC et al. Guidelines for Biosafety Training Programs for workers assigned to BSL-3 Research laboratories. *Biosecur Bioterror* 2013 11v 10-19

REFERENCES

1. Addressing the Continuing Threat of Laboratory-Acquired Infections https://wwwn.cdc.gov/cliac/pdf/Addenda/cliac0908/Addendum%20E.pdf

2. Bennett A & Parks S (2006) Microbial aerosol generation during laboratory accidents and subsequent risk assessment. *J Appl Microbiol* **100**:658–663.

3. University of Michigan, US Occupational Safety & Environmental Health (www.oseh.umich.edu/research/lab-accidents.shtml).

4. Owens B (2014) Anthrax and smallpox errors highlight gaps in US biosafety. *Lancet* **384**:294.

5. Kumar P et al. (2013) Heat fixed but unstained slide smears are infectious to laboratory staff. *Indian J Tuberc* **60**:142–146.

6. Traxler RM et al. (2013) A literature review of laboratory acquired brucellosis. *J Clin Microbiol* **51**:3055–3062.

7. Rake BW (1978) Influence of crossdrafts on the performance of a biological safety cabinet. *App Envir Microbiol* **36**:278–283.

8. Kenny MT & Sabel EL (1968) Particle size distribution of *Serratia marcescens* aerosols created during common laboratory procedures and simulated laboratory accidents. *Appl Microbiol* **16**:1146–1150.

9. Collins CH (1984) Safety in microbiology: a review. *Biotechnol Genet Eng Rev* **1**:141–165.

10. Parker SL & Holliman RE (1992)Toxoplasmosis and laboratory workers: a case control assessment of risk. *Med Lab Sci* **49**:103-106

ANSWERS

MCQ	Feedback

1. The following are important parameters for determining the management of a laboratory incident. True or false?
   a. The Advisory Committee on Dangerous Pathogens (ACDP) Hazard Group to which the organism belongs
   b. The temperature at which the organism was cultured
   c. The production of aerosols
   d. The height from which a liquid falls
   e. The health of laboratory personnel contaminated

a. True

b. False

c. True
d. True
e. True
The severity of an incident is related to the hazard group of the organism: a Group 4 organism is very likely to cause infection for which there is no good treatment, whereas a Group 2 organism is a low-virulence organism that would not affect healthy individuals. The height from which a liquid falls is related to the production of aerosols, which itself is related to the spread of the organism and thus the severity of the incident. Laboratory workers who are in some way compromised are more likely to become infected, and this relates to how an incident should be managed, for example, giving prophylaxis.

2. With regard to the epidemiology of laboratory incidents, are the following statements true or false?
   a. They have only recently been recognized as a problem.
   b. *Brucella* infections in laboratory workers are about 100 times more common than in the general population.
   c. *Aspergillus flavus* is the most common fungal pathogen associated with laboratory-acquired infections.
   d. The prevalence of an organism causing laboratory-acquired infection depends on geography.
   e. Failure of SOPs is frequently linked to laboratory-acquired infection.

a. False

b. True

c. False

d. True

e. True
Laboratory acquired infections were noted over a century ago. *Brucella* infections are much more common in laboratory workers than the general population because it is a relatively uncommon infection. In the laboratory it is present in high concentrations during the diagnostic procedure. The most common fungal laboratory-acquired infections are the dimorphic fungi *Coccidioides*, *Histoplasma*, and *Blastomyces*. Because microorganisms are geographically varied, so are the types of laboratory-acquired infections. Mechanical failure or failure to adhere to SOPs are common causes of laboratory infections.

3. With regard to *Clostridium botulinum* and its toxin, are the following true or false?
   a. It is the only organism to produce botulinum toxin.
   b. The toxin is produced from the spores.
   c. The toxin produced by *C. botulinum* is heat resistant
   d. *C. botulinum* produces three serotypes of botulinum toxin with different clinical effects.
   e. *C. botulinum* toxin acts upon spinal reflexes to produce its symptoms.

a. False

b. False
c. True

d. True

e. False
Botulinum toxin can also be produced by other clostridial species, for example, *C. butyricum*. The toxin is produced during the process of sporulation, not actually from the spores, and it is heat resistant. There are seven serotypes of botulinum toxin that act on the neuromuscular junction (not the spinal synapses) with the same clinical effect.

MCQ	Feedback

4. With regard to *Clostridium botulinum,* are the following true or false?
  a.   Laboratory staff must wear an FFP2 mask.
  b.   It belongs to HG4.
  c.   It causes descending paralysis.
  d.   The half-life of antitoxin in the serum is 5-8 days.
  e.   Toxin E has been associated with eating fish.
	a. False
	b. False
	c. True
	d. True
	e. True
	Handling *C. botulinum* does not require a face mask, nor a BSL3 laboratory. Clinically, it may present with a symmetric descending paralysis. The half-life of the antitoxin is about 5–8 days, and toxin E has been associated with fish.

5. Are the following are true or false?
  a.   In a microbiology laboratory, the maximum allowable work space per individual is 6 m³.
  b.   Lassa fever is an HG3 virus.
  c.   Botulinum toxin is absorbed through intact skin.
  d.   A CL3 laboratory must be maintained at negative pressure.
  e.   In a Class 2 safety cabinet there is an H14 filter.
	a. False
	
	b. False
	c. False
	d. True
	
	e. True
	The minimum work space per person in a microbiology laboratory is 11m³. Lassa virus is an HG4 organism, as are other haemorrhagic fever viruses. Botulinum toxin is not absorbed through intact skin but may enter through cuts in the skin. BSL3 laboratories are at negative pressure to prevent the inadvertent release of pathogenic organisms. A Class 2 safety cabinet protects both the work and the worker and is fitted with an H14 filter.

LEGIONELLA PNEUMOPHILA INFECTION

CASE 11

Peter Wilson

Department of Microbiology and Virology, University College London Hospitals, UK.

A 55-year-old man had been receiving outpatient treatment for non-Hodgkin's lymphoma but was admitted from the clinic with fever, diarrhoea, headache, and vomiting. He had a dry cough and then a productive cough for two days. He was a smoker. He had not travelled abroad in the last year and had no recent restaurant meals. He had no contacts with anyone with pneumonia or serious infection, but his partner had a viral illness two weeks earlier. On examination, he was febrile 38.5°C, tachypneic, and intermittently confused. There were widespread crackles in the chest with dullness at the left base. A chest X-ray showed patchy shadowing in both lungs and an area of collapse consolidation in the left base.

INVESTIGATION OF THE CASE

The patient's blood gases showed mild hypoxemia of (P_aO_2 10 kPa.) The white cell count was 16×109/L with a neutrophilia. The C-reactive protein (CRP) was 90 mg/L. Urinary antigen for *Legionella* was positive. Blood and urine cultures were negative at 48 h. Sputum was cultured for *Legionella*, and an isolate sent to the reference laboratory was later confirmed as *Legionella pneumophila* serogroup 1. Liver function tests showed a mild transaminitis (alanine transaminase [ALT] 63 U/L, aspartate transaminase [AST] 52 U/L) and hyperbilirubinemia (1.5 mg/dL). Serology for atypical pneumonia was negative.

Public Health authorities were aware of three cases of Legionnaires' disease in the area in the preceding two months, but they were not linked and not related to the hospital. There were no other hospital cases. The patient gave no history of recent foreign travel or staying in hotels. The patient had regularly attended the Haematology Day Care Unit. The unit was housed in an old hospital building. Within that unit, examination of the plumbing diagram showed there were dead legs (that is, plumbing cul de sacs) and temperature records at outflow taps showed the hot water supply temperatures were between 31° and 43°C, representing a significant risk of *Legionella* proliferation. Additionally, there had been prior reports of the cold water supply being lukewarm, also acting as a source for infection. Every other weekend the unit was unused, so the first use of water on Monday morning would have higher risk of causing infection because the water flow had been static over two days. There were spray fittings on the outlets and on one shower, which may produce aerosols.

On investigation in the ward, one hot tap showed a different strain of *Legionella* at a low level. Culture of water from a shower showed *Pseudomonas aeruginosa* only. Remedial action was taken promptly and chlorine dioxide was used as a biocide. There was a report of four other staff on the unit being sick in those months, including three with flu-like symptoms, but there was no evidence supplied that this resulted from Legionnaires' disease or Pontiac Fever. Urinary antigen tests on three staff were negative. Nevertheless, the isolation of another strain of the organism suggested that conditions were appropriate for growth and that there was a risk of infection. In the absence of a travel history or other likely source of exposure, infection in the hospital was likely. The water supply at the patient's home was free of the organism, but it is possible he could have been infected from another unknown environmental source.

CLINICAL MANAGEMENT

After consultation with the microbiologist, the patient was treated with ciprofloxacin. Amoxicillin and clarithromycin were also given. His partner tested negative for legionella antigen. By the second day after admission, fever and breathlessness were improving. He was afebrile by the third day and CRP settled. Amoxicillin was stopped when the reference laboratory report was made. Ciprofloxacin was continued for another two weeks. He improved over the next month and was able to resume treatment for his lymphoma.

PREVENTION OF FURTHER CASES

Nosocomial cases should prompt a retrospective review of nosocomial pneumonia over 3–6 months plus prospective laboratory testing. Plumbing controls involve elimination of dead legs, correct temperature control in both hot and cold water systems by keeping hot water above 50°C and cold water below 20°C, and reducing stagnation in the water systems. To avoid scalding in hospitals, thermostatic mixing valves have to be installed. Environmental control depends on adherence to national guidance for risk assessment and maintenance. Monitoring of biocide and thermal control is essential. Culture of potable water is performed in hospitals where patients may be immunocompromised. Remedial action is required if positive cultures are obtained.

An appropriate maintenance protocol for legionella control, which includes planned monitoring of the hot and cold water temperature, frequency of outlet use, a chlorine dioxide system, and removal of dead legs if possible, is important for preventing further cases. A protocol and action plan in the event of a hospital-acquired case needs to be put in place and followed.

The legionella testing records of the building were examined. Although the microbiologist was unaware, three routine water samples had shown *L. pneumophilia* serogroup 1 at low level in the last six months. Each time the outlet was descaled and disinfected and upon retesting no *Legionella* had been detected. One of the positive samples was from a handwashing basin in the open area of the unit. On ward inspection, the taps were clean and the outflow did not go straight into the drain. However, descaling had already taken place as part of a planned preventative program. Temperature monitoring was manual as no building management system was available. Thermal control showed that temperatures at some points in the system were repeatedly recorded below 40°C.

All outlets in the building were then tested for *Legionella* spp. *L. pneumophilia* serogroup 1 was reported in 5% of outlets and in some cases at a level of 300 colony forming units (cfu)/L. Removal of dead legs was not practical so temporary filters were applied to outlets in patient areas.

The source of the patient's *Legionella* could not be confirmed because no contemporaneous culture of a similar strain was available. However, the presence of *Legionella* in the hospital building was a possible causative link. Outbreaks related to potable water are usually in a single building and cases occur over many months. Infected water systems should be disinfected, usually by hyperchlorination.

EPIDEMIOLOGY

Legionella spp. are found commonly in fresh water and soil. The number of bacteria in the source is a determining factor in the development of Legionnaires' disease. Hospital-acquired outbreaks of Legionnaires' disease have been reported repeatedly and are associated with potable water supplies rather than cooling towers. Many cases are not diagnosed because the signs and symptoms are nonspecific and many hospitals do not routinely test patients with hospital-acquired pneumonia. Detection of *Legionella* spp. in the water supply on routine testing increases the suspicion of Legionnaires' disease and makes diagnosis more likely. Colonization of water supplies in large buildings is common, but showering is not a common means of transmission in hospitals. Most cases result from aspiration of oropharyngeal secretions and, hence, patients with chronic lung disease or recent surgery are most at risk.

The organism proliferates at temperatures of 20–45°C. When disseminated as an aerosol, inhaled organisms can infect alveolar macrophages and epithelium causing severe pneumonia (Legionnaires' disease) or a flu-like illness (Pontiac fever). The reported prevalence is 0.4–0.6 per 100,000 population, but this probably represents less than 1 in 20 of the true number. Outbreaks are often related to drinking water systems, but most cases are sporadic. Infection rates in healthcare settings are higher than in the community. Nosocomial cases have been associated with aerosols from contaminated showerheads and hot water taps. Up to 60% of samples have been reported to be positive at these sites, with as many as 1000 cfu/mL. Lower water temperature and stagnation in these distal areas can allow proliferation of large numbers of organisms. Hydrotherapy pools, whirlpool baths, and spa pools are sources of *Legionella* spp. and have to be carefully maintained and disinfected. Other sources of infection are water sprays in supermarkets and on golf courses and even windshield wiper sprays. For example, long-distance truck drivers have a higher seroprevalence compared to the general population.

The type of *Legionella* spp., the means of transmission, and the susceptibility of the host are important factors in the development of disease. *L. pneumophila* accounts for 90% of cases and serogroup 1 is the most virulent. In the immune-suppressed patient, up to 20% of cases are caused by other species that are not detected by the urinary antigen test. Hospital water systems are colonized with various species. Inhalation of a contaminated aerosol transmits the infection when the bacteria are in an aerosol of droplet of 1-5 µm, which can be generated by nebulizers and showers. Water hitting a handwash basin, shower, or bath generates aerosols, and the concentration of bacteria in the water and the amount of water dispersed are critical to the risk of causing disease.

Sources such as cooling towers provide much longer exposure times than baths or showers. Smaller particles (<5 µm) remain airborne for long periods and penetrate the lungs, but larger ones contain more bacteria. Smoking, male sex, alcoholism, chronic obstructive pulmonary disease, cancer, diabetes, immune suppression, surgery, and organ transplantation all increase the risk of the disease. Several studies have failed to show a correlation between the numbers of *Legionellae* present in the water and cases of legionellosis. Nevertheless, it is necessary to suppress the organism when it is found. In some cases, a different source such as a cooling tower has been implicated in the absence of positive isolates from the water system under investigation. The risk of a single high-level exposure may be similar to that of multiple low-level exposures. In nosocomial outbreaks, the potable water supply is usually implicated.

In the UK in 2014 there were 331 confirmed cases of Legionnaires' disease, of which 69% were in males. Of 4402 cases of Legionnaires' disease between 1980 and 2002, 264 were hospital-acquired, but the diagnosis is often missed in complicated cases or when sero-conversion is delayed.

Guidelines on the control of *Legionella* spp. in water systems form the basis for preventative maintenance to limit the level of colonization and prevent outbreaks. Methods of control include heat, ultraviolet light, sonication, compressed air, use of chemicals to prevent scale, biocides and charcoal filters, elimination of dead legs, and regular flushing of outlets. Colonization can be prevented by maintaining the temperature of the water below 20°C or above 55°C. In an outbreak, heating and flushing with hyperchlorination can be used, but it only reduces levels for a few weeks. Close control of temperature in the entire system is difficult, so hyperchlorination or a thermal shock followed by continuous chlorination may be more effective, but prolonged hyperchlorination is expensive, corrosive, and potentially toxic.

Survival of *Legionella* spp. in water depends on temperature, nutrients, and the presence of protozoa. The growth range of 20–45°C is likely where the water flow is slow, or in a dead leg or mixing valve. Bacteria adhere to metal or plastic by producing slime, and together with fungi, algae, protozoa, and debris form a biofilm. A suitable temperature and presence of nutrients such as sediment, sludge, scale, or rust, are needed for biofilm to develop. *Legionella* is predominately (98%) localized in biofilm on the surface of the pipe. Biofilms facilitate interaction between different microorganisms and *Legionella* spp. are frequently located within amoebae as facultative intracellular parasites, providing protection from many disinfection processes. Colonization of biofilm in hot water systems also renders *Legionella* spp. resistant to halogenation and ionization. Thermal treatment will disinfect biofilm but can also increase its formation.

The prevention of *Legionella* outbreaks requires identification of the sources of risk, preparation of adequate risk assessments, and action plans that are executed. Treatment regimens must be properly monitored and maintained, and staff applying control measures must be trained in the risks the organisms pose. Engineering issues such as dead legs in the systems have to be addressed. The risk of infections is reduced by the correct application of the chosen water treatment method by all levels of the healthcare organization. The goal is only to reduce numbers of bacteria as elimination is very difficult except in small-scale carefully controlled systems.

BIOLOGY

Legionella spp. require oxygen for growth. Soluble iron is needed for maximum growth and buffered charcoal containing yeast extract (BYCE) agar is the preferred medium. Isolation takes 2–5 days at 35°C. *L. pneumophila* includes over 16 serogroups and serogroup 1 can be subdivided into 10 subgroups. Serogroup 1 causes 70–90% of cases of Legionnaires' disease. *Legionella* spp. are found in standing water between 5 and 50°C. The highest concentrations are found at 25–40°C.

Amoebae support intracellular growth of the bacteria and promote survival in unfavourable environmental conditions. The organisms pass in and out of biofilm and disruption of biofilm can result in a heavy release, which if aerosolized can be inhaled to cause disease. Air conditioning, cooling towers, water baths, and warm water plumbing are common sources. In old pipes with low water flow, growth in biofilm greatly increases bacterial concentration in the water, particularly if warm. Between 5 and 30% of hot water plumbing systems contain the organism.

DISEASE

Legionnaires' disease is an acute pulmonary infection caused by Gram-negative *Legionella* spp., usually *L. pneumophila*. Although *Legionella* causes only 1–5% of cases of pneumonia, it can be fatal.

The incubation period is between 2 and 10 days, usually 4 to 6 days. Pontiac fever develops 4 h to 3 days after exposure with a 70% attack rate. Approximately 70% of cases are not linked to outbreaks and the diagnosis is often missed. The attack rate in hospital-acquired outbreaks is around 1%. Around half the cases are in travellers. Other cases are related to living near a cooling tower, decorative fountains, nebulizers, humidifiers, recent plumbing work, and older plumbing. Mortality is between 10 and 15%. Men are more susceptible, particularly if smokers or if they have chronic lung disease, renal failure, immune suppression, recent surgery, or are elderly.

The symptoms include fever, dry cough, diarrhoea, confusion, hyponatremia, and abnormal liver function. The prodrome lasts several days with headache, fever, muscular pain, and anorexia, sometimes with diarrhoea. Cough may or may not be productive. Pneumonia with patchy infiltrates and sometimes pleuritic chest pain is present. Headache can be severe and neurological complications can occur. Pontiac fever is a self-limited short febrile illness with headache and muscle pains but no pneumonia. Liver dysfunction, hyponatremia, and haematuria are common features. Diagnosis is principally by detection of antigen in the urine, although this is not effective for all serogroups of *L. pneumophila* or other *Legionella* spp. Detection of urinary antigen has a sensitivity of 65–90% (most sensitive for *L. pneumophila* serogroup 1) and high specificity. Urinary antigen persists in urine after the start of treatment. Respiratory specimens can be cultured on BCYE medium, although the results of culture are more variable than serology.

Treatment with erythromycin, tetracycline, or ciprofloxacin is most effective. Ciprofloxacin or erythromycin plus rifampicin for 14 days is preferred in the critically ill. Erythromycin or tetracycline is given for 10–14 days or ciprofloxacin for 7–10 days in mild pneumonia. Response is usual in 12–24 hours.

In the UK, all cases must be reported to the local Public Health Unit. Pontiac fever is a systemic illness without pneumonia associated with the same bacterium. The urinary antigen is rarely positive and patients generally recover without therapy.

PATHOLOGY

Infection usually occurs by inhalation of aerosol of organisms originating in water. Humidity is an important factor in transmission. *L. pneumophila* has a number of virulence factors important for enhancing growth, adhesion, and the establishment of infection. These factors include proteins related to iron acquisition: the major outer membrane protein (MOMP) and pili, both involved in adhesion to host cells, and the Dot/Icm Type IV secretion system, delivering several hundred different effectors into host cells, principally the proteins SdhA, SidJ, and AnkB. Components of the secretion system, IcmX, are important for pore formation in macrophages and the formation of *Legionella*-containing vacuoles, and DotD DotC proteins are responsible for intracellular survival of *Legionella*. One of the components of the Type II secretion system, TatB, is important for intracellular replication *of Legionella*. The strongly hydrophobic lipopolysaccharide (LPS) of *Legionella* activates Toll-like Receptor 2 (TLR2) and Toll-like Receptor 4 (TLR4). It is also shed as vesicles into the interstitium and arrests phagolysosome development as well as acting as a vehicle for export of a variety of soluble virulence factors. The organisms multiply intracellularly within the lung causing pneumonia and systemic infection. Pontiac fever may be caused in patients by LPS inhalation.

QUESTIONS

1. Are the following statements true or false?
   a. *Legionella* serogroups 2–14 are common causes of human Legionnaires' disease.
   b. Hospital-acquired outbreaks are usually associated with ward ventilation.
   c. Patients having surgery are at particular risk of nosocomial infections.
   d. Pontiac fever is not associated with pneumonia.
   e. Chlorine dioxide may not be effective biocide in showers.

2. Are the following statements true or false? The following are risk factors for Legionnaires' disease:
   a. Smoking
   b. Male sex
   c. Immune suppression
   d. Alcoholism
   e. Diabetes

3. With regard to transmission, are the following true or false?
   a. Person-to-person spread is likely.
   b. The optimum droplet size for transmission is <1 μm.
   c. Correlation between number of *Legionellae* in water and patient cases is often not demonstrated.
   d. Thermal disinfection can increase biofilm formation.
   e. Elimination of *Legionella* from water systems is very difficult to achieve.

4. Regarding *Legionellae*, are the following true or false?
   a. *Legionellae* can be grown on routine respiratory agar plates.
   b. *Legionellae* can grow inside amoebae.
   c. *Legionellae* can grow at 40–50°C.
   d. Disruption of biofilm can increase the risk of transmission.
   e. Urinary antigen tests do not detect serogroups 2–14.

5. Regarding Legionnaires' disease, are the following true or false?
    a. Incubation period is <2 days.
    b. Outbreaks account for only 30% of cases.
    c. The cough is often dry.
    d. Headache is a symptom of Pontiac fever.
    e. Humidity is a major risk factor in transmission.

GUIDELINES

1. Department of Health (October 2006). HTM 04-01: The control of Legionella, hygiene, "safe" hot water, cold water and drinking water systems, Part A - Design, installation and testing.

REFERENCES

1. Edelstein PH & Cianciotto NP (2010) Legionella. In Principles and Practice of Infectious Diseases (Mandell GL, Bennett JE & Dolin R eds.) Elsevier, pp. 2969–2984.
2. Sabria M & Yu VL (2002) Hospital-acquired legionellosis: solutions for a preventable infection. Lancet Infect Dis **2**:368–373.
3. O'Neill E & Humphreys H (2005) Surveillance of hospital water and primary prevention of nosocomial legionellosis: what is the evidence? *J Hosp Infect* **59**:273–279.
4. Molloy SL, Ives R, Hoyt A et al. (2008) The use of copper and silver in carbon point-of-use filters for the suppression of Legionella throughput in domestic water systems. *J Appl Microbiol* **104**:998–1007.
5. van der Kooij D, Veenendaal HR & Scheffer WJ (2005) Biofilm formation and multiplication of *Legionella* in a model warm water system with pipes of copper, stainless steel and cross-linked polyethylene. Water Res **39**:2789–2798.
6. Saby S, Vidal A & Suty H (2005) Resistance of *Legionella* to disinfection in hot water distribution systems. *Water Sci Technol* **52**:15–28.
7. Thomas V, Bouchez T, Nicolas V et al. (2004) Amoebae in domestic water systems: resistance to disinfection treatments and implication in *Legionella* persistence. *J Appl Microbiol* **97**:950–963.
8. Isenman HL et al. (2006) Legionnaires' disease caused by L. longbeachae: clinical features and outcomes of 107 cases from an endemic area. *Respirology* (doi: 10.1111/resp.12808).
9. Borges V et al. (2016) Legionnella *pneumophila* strain associated with the first evidence of person to person transmission of Legionnaires disease: a unique mosaic genetic backbone. Sci Rep (doi: 10.1038/srep26261).
10. Kusk D et al. (2016) Fast label free detection of *Legionnella* spp. in biofilms by applying immunomagnetic beads and Raman spectroscopy. *Syst Appl Microbiol* **39**: 32–140.

ANSWERS

MCQ	Feedback
1. Are the following statements true or false?	
a. *Legionella* serogroups 2–14 are common causes of human Legionnaires' disease.	a. False. Serogroup 1 is most likely linked to disease.
b. Hospital-acquired outbreaks are usually associated with ward ventilation.	b. False. Hospital outbreaks relate to potable water supplies.
c. Patients having surgery are at particular risk of nosocomial infections.	c. True. This results from aspiration associated with endotracheal tubes.
d. Pontiac fever is not associated with pneumonia.	d. True. It presents with a febrile illness and myalgia.
e. Chlorine dioxide may not be effective biocide in showers.	e. True. The concentration may be inadequate to prevent proliferation in stagnant water.
2. Are the following statements true or false? The following are risk factors for Legionnaires' disease:	
a. Smoking	a. True
b. Male sex	b. True
c. Immune suppression	c. True
d. Alcoholism	d. True
e. Diabetes	e. False
3. With regard to transmission, are the following true or false?	
a. Person-to-person spread is likely.	a. False. Spread is mainly through environmental aerosols.
b. The optimum droplet size for transmission is <1 µm	b. False. The optimal size is 1–5 µm.
c. Correlation between number of *Legionellae* in water and patient cases is often not demonstrated .	c. True. Host susceptibility is more important.
d. Thermal disinfection can increase biofilm formation .	d. True. Legionella is protected in biofilm.
e. Elimination of *Legionella* from water systems is very difficult to achieve.	e. True. Most control measures just reduce bacterial numbers.
4. Regarding *Legionella*, are the following true or false?	
a. Legionellae can be grown on routine respiratory agar plates.	a. False. Charcoal containing yeast extract agar is needed.
b. *Legionellae* can grow inside amoebae.	b. True. Intracellular growth allows survival in poor environmental conditions.
c. *Legionellae* can grow at 40–50°C.	c. True. Hot water near the tap is often in this range.
d. Disruption of biofilm can increase the risk of transmission.	d. True. Bacterial concentration in water can rise sharply.
e. Urinary antigen tests detect serogroups 2–14.	e. True. They detect serogroup 1.
5. Regarding Legionnaires' disease, are the following true or false?	
a. Incubation period is <2 days.	a. False. The incubation period is 2–10 days.
b. Outbreaks account for only 30% of cases.	b. True. Most cases are not linked to outbreaks.
c. The cough is often dry.	c. True. Cough is usually nonproductive and dry.
d. Headache is a symptom of Pontiac fever.	d. True. Headache is common in both forms of the disease.
e. Humidity is a major risk factor in transmission.	e. True. Infectious aerosol travels further in high humidity.

MULTIDRUG-RESISTANT TUBERCULOSIS (MDR-TB)

CASE 12

Helen McAuslane, Dominik Zenner, Public Health England, UK.

The authors wish to acknowledge the great support from Dr. Sam Perkins and Dr. Anita Roche, Public Health England, in helping to put together the information about the case.

A 17-year-old college student presented with a two months' history of cough and was diagnosed with sputum smear-positive pulmonary tuberculosis (TB) in late July. Her chest X-ray indicated a consolidation in the left lung and shadowing in the right lung, but no cavitation.

CLINICAL MANAGEMENT

The patient was started on standard first-line TB treatment (isoniazid, rifampicin, pyrazinamide, and ethambutol). However, she failed to improve clinically, and culture results in mid-August showed growth of *Mycobacterium tuberculosis* (later confirmed as Beijing strain) and resistance to all first-line drugs (isoniazid, rifampicin, pyrazinamide, ethambutol, and streptomycin) on direct sensitivity testing (DST). Second-line DST was initiated and later results showed additional resistance to some injectable agents (Amikacin and Kanamycin) and to Ethionamide. She was commenced on a World Health Organization (WHO)-recommended multidrug-resistant tuberculosis (MDR-TB) drug regimen. However, her treatment had to be changed several times because of side effects and in keeping with emerging evidence from DST. For most of her treatment she was treated with moxifloxacin, linezolid, PAS, cycloserine, and prothionamide.

Although still smear positive in the beginning of September, she was discharged from the hospital with a patient contract, which included self-isolation and directly observed therapy (DOT). She remained symptomatic and was readmitted in October. In mid-November her condition had improved enough to be discharged again. However, treatment compliance remained a significant concern, and, despite DOT, her sputum remained smear and culture positive and her blood levels for cycloserine, moxifloxacin, and linezolid were undetectable even in February of the following year. She was investigated for gastrointestinal malabsorption, and the help of a psychologist was enrolled. She started gradually to improve clinically and radiologically over the next few months and she became smear negative from April, culture negative from May, and stopped coughing in June. She successfully completed treatment about 18 months later, 30 months after initial diagnosis.

INVESTIGATION OF THE CASE

Careful contact tracing took place within the household and among close contacts within the college. No secondary cases were found, and latent TB screening did not show any evidence of further transmission. Her isolates were strain typed and her profile was found indistinguishable to isolates of a number of other individuals by 24-loci mycobacterial interspersed repetitive units–variable number tandem repeats (MIRU–VNTR), but no epidemiological links could be identified. Her demographic profile made it unlikely that she had acquired TB from or passed it on to other individuals in this molecular cluster.

PREVENTION OF FURTHER CASES

It was planned that during her infectious period she would be fully isolated. In the hospital, she would stay in a negative-pressure room, and personal protective equipment (PPE), which included disposable masks, gowns, and gloves, should be used by the attending health care staff. Outside the hospital she was to self-isolate in a single-occupancy room. She would be allowed to socialize with her friends, though this was restricted to outside venues and with the use of disposable face masks. She was not allowed to attend college or to go inside public venues. However, in practice, outpatient arrangements had been difficult to police and, at least in the initial phase, it is possible that they had not been adhered to at all times.

When she finally started to improve in spring and summer, she was keen to re-enter college. She then had three smear- and culture-negative sputum results from May and two smear- and culture-negative results from June, but she had scanty acid-fast bacilli in her sputum in early September, just before she planned to return. Although it is possible that these were dead organisms, it is also not unusual for MDR-TB to become smear- or culture-positive again after initial culture conversion. This was a particular concern in view of the resistance profile of her TB and her variable treatment compliance. It was decided to either await culture results of the early September sample or at least confirm smear negativity by serial sputum samples and wait for this culture result before allowing her back to college.

TRANSMISSION AND INFECTION CONTROL

There are three main transmission routes of TB: direct, oral ingestion, and respiratory. In most settings, including the UK, the first two of these are only of historical interest. TB is usually transmitted by inhalation of airborne droplet nuclei containing *M. tuberculosis* bacilli, produced through coughing, sneezing, singing, or other aerosol-generating procedures from an individual with pulmonary TB. Droplets can remain suspended in the air for several hours, facilitating transmission to susceptible individuals. *M. tuberculosis* has a low infective dose with a median infectious dose (ID_{50}) of <10 bacilli. The incubation period for active disease is around three to eight weeks, and cases are infectious for as long as viable bacilli are present in the sputum. The risk of transmission depends on the proximity and duration of contact, the number of viable bacilli in the sputum, and the susceptibility of the contact. Those at highest infection risk are often close contacts, such as those living in the same household with the case or those in frequent prolonged contact, such as caregivers.

The transmission of TB and progression to active disease is multifactorial and influenced by socioeconomic risk factors, such as poor quality or overcrowded housing conditions, poor air quality, and barriers to accessing to health services. Populations living and working in congregate settings, such as prisons, refugee camps and homeless shelters have a particularly high risk of transmission. Staff and patients in healthcare facilities are also at increased risk of infection. Individuals with social risk factors may be more likely to develop drug resistance as a result of difficulties in accessing services or adhering to treatment.

The duration of infectivity depends on the number of viable *Mycobacteria* in the sputum, which in turn depends on success of the treatment. Treatment with isoniazid provides early bactericidal activity to rapidly reduce the number of viable mycobacteria. Rifampicin has a sterilizing activity, which will continue to kill mycobacteria beyond the end of treatment. These two first-line drugs are the most effective in reducing the bacillary load and thus infectivity of the patient. Without isoniazid and rifampicin treatment for MDR-TB is substantially less effective and the infectious period is prolonged.

Assessing infectiousness

There is no evidence that drug-resistant TB is more transmissible than drug-sensitive TB. Therefore the principles of infection control are similar. These include an assessment of the length and degree of infectiousness of the affected patient, an assessment of the environment and appropriate methods of preventing further transmission, and an assessment of the vulnerability of potential contacts. Infection control in MDR-TB is complicated by the difficulties in assessing infectiousness and infectious periods and the potentially more serious consequences of transmission events.

It is important to remember, however, that the global effort to prevent drug-resistant TB also includes the rapid detection, appropriate surveillance, and treatment of new cases at the country and provincial level and implementation of comprehensive infection-control measures to interrupt the transmission cycle, particularly in institutional settings. Careful and appropriate antibiotic use and supporting patients to adhere to treatment are also essential for the prevention of MDR-TB transmission. Stringent infection-control measures are particularly important for halting transmission in high-burden TB countries with coexisting HIV epidemics.

The length and degree of infectiousness of the patient varies with the capacity to produce aerosol with viable *Mycobacteria*, which in turn depends on the extent and location of the disease, other host immune factors, and the duration and effectiveness of treatment. Assessing the infectiousness was a challenge in the case study, not least because the case had lost the effectiveness of very important bactericidal and sterilizing agents, which allow a rapid elimination of viable *Mycobacteria* from the patient's sputum in drug-sensitive TB. In addition, treatment adherence was suboptimal, further extending the infectious period. There was no evidence of further transmission from the patient to her close contacts, but some studies have demonstrated that Beijing strains have increased infectiousness. The duration and degree of infectiousness determines the length of required isolation measures for the patient. The variability of drug resistance patterns and the lack of robust data on minimum treatment duration to achieve sputum sterilization for second-line treatment regimens can make it difficult to assess infectiousness in practice. The WHO recommends at least five negative cultures to confirm treatment success, and more than one negative culture from high-quality sputum samples taken on separate occasions is required to deisolate.

Treatment adherence is key to preventing the spread of MDR-TB. Treatment for MDR-TB takes at least 20 months and many of the second-line antituberculosis drugs are toxic with unpleasant adverse effects including fever, rashes, gastrointestinal disturbances, hearing disturbances, and psychiatric symptoms, which make adherence more difficult.

Measures of treatment adherence include treatment diaries, markers of clinical or radiological improvement, or even measuring of drug levels in the blood. Assessing treatment adherence is important for determining the degree of infectiousness.

In many countries, legislation allows for nonadherent patients to be compulsorily detained in a hospital or treatment facility. Although this is a last resort, it may be necessary on some occasions to safeguard the public health when all other measures to promote voluntary cooperation have failed.

Assessing potential contacts

A key factor to inform the risk assessment of an infectious patient is an assessment of the potential contacts, the length, proximity, and number of these potential contacts and their vulnerability. Strict isolation of infectious TB patients is particularly important in environments where contact with patients who are immunosuppressed (for example, HIV patients) is possible, and decisions for deisolation will be influenced by these risk assessments.

Infection control in healthcare and institutional settings

The principle of TB infection control is to ensure that the exposure to infective TB patients is limited in time, place, and person. The US Centers for Disease Control and Prevention (CDC) suggests three hierarchical levels of infection control at an institutional level. The CDC framework emphasises the importance of administrative controls as the foundation of effective infection control, supported by environmental controls and the use of PPE, which serve to further protect both staff and patients.

Administrative controls include robust infection-control policies and plans and guidance documents, which should outline clear instructions for staff. Limiting exposure could include restricting access to TB patients, through the use of specifically assigned wards, and instituting specific staff training requirements. Standardized risk assessments should be carried out for facilities, procedures, and staff.

The TB risk of healthcare workers can be 1.9 to 5.7 times higher than the general population, depending on the setting. Therefore, staff at healthcare facilities and other institutions can be at particular risk of acquiring or even transmitting TB. Because of the higher risk, it is important to have a robust policy covering occupational health, screening for new employees, as well as follow-up checks at regular intervals (for example, every six months or yearly, depending on risk of exposure). An occupational health screening for new healthcare staff may include an assessment of relevant medical history and HIV status, Bacillus Calmette-Guérin (BCG) vaccination history, and screening for active or latent TB infection.

Interrupting transmission

For all infectious TB patients, robust measures need to be taken to avoid transmission. Broadly these can be categorized into measures to avoid the generation of infective aerosols, environmental controls, and measures to protect exposure of healthy contacts. Special considerations for environmental controls and the use of personal protective equipment are aerosol-producing procedures (for example, sputum induction or collection, bronchoscopy, or gastric lavage) or the transport of infectious cases.

If the patients' clinical condition and community circumstances allow, inpatient treatment for patients with TB and drug-resistant TB should be minimized and outpatient care encouraged to prevent transmission in institutional settings. All infectious TB patients who require admission to hospital must be appropriately isolated. MDR-TB patients should be isolated in a negative-pressure room.

MDR-TB patients might require prolonged isolation, and good, early patient communication is key to ensuring adherence to treatment and infection-control requirements. Engagement with the patient should include training and awareness about the importance of reducing infected aerosols (for example, by adhering to cough etiquette or wearing surgical masks) and may include patient incentives for adherence. Multidisciplinary working including an early referral to other services may be helpful if appropriate. For example, a referral to the social worker proved effective in improving adherence for the young college student described earlier. A patient contract may also be helpful.

Environmental-control measures aim to reduce the concentration of airborne infectious particles in the environment through diluting contaminated air, removing infectious particles, and controlling airflow in patient areas.

An essential component for preventing transmission is adequate ventilation, defined as a minimum of 12 air changes per hour (ACH), equivalent to an air flow of 80 L per second per patient in a room measuring 24 m^3. For infectious MDR-TB patients, this should be a minimum requirement for high-risk areas such as TB clinics or ward areas. Common examples of environmental controls include negative-pressure rooms and mechanical or natural ventilation systems. The direction of airflow in negative-pressure rooms is inbound (from corridors into the isolation room), and this helps to avoid contaminated airflow to other areas. This air can then be directly removed to the outside of the building, where transmission is less likely (Figure 12.1).

Specific technological measures, which can help to reduce or eliminate infectious aerosols from the ambient air, are high-efficiency particulate air (HEPA) filters and ultraviolet germicidal irradiation (UVGI). The latter works through UV light sterilization of airborne droplet nuclei, but its use is limited by cost and health and safety considerations around the use of UV light.

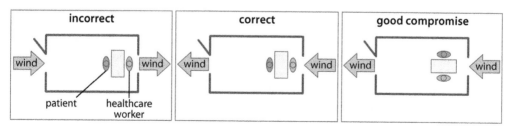

Figure 12.1. **Example of appropriate furniture arrangement in a naturally ventilated consulting room.** (From Implementing the WHO Policy on TB Infection Control in Health-Care Facilities, Congregate Settings and Households, 2009, (http://www.stoptb.org/wg/tb_hiv/assets/documents/TBICImplementationFramework1288971813.pdf).

Another key intervention to reduce transmission is the use of PPE for all healthcare staff and other contacts of infectious patients. PPE includes gloves, aprons, and masks. Within institutional settings and when caring for MDR-TB patients, special protective face masks with a capacity to filter 1 μ particles (for example, N95 or FFP3 respirator mask) should be worn. These masks need to be individually fitted to ensure they provide the required protection.

N95 or FFP3 particulate filter masks are effective for preventing transmission to non-infected individuals but are not recommended for use by the patient. Infectious patients and TB suspects should be given a surgical mask to wear to minimize ambient droplet nuclei.

Infection control in the community

The WHO recommends ambulatory care for MDR-TB patients because it is cost-effective, it reduces nosocomial transmission, it minimizes the social and psychological consequences of lengthy hospital admissions. If a patient is to receive ambulatory care, a risk assessment of infectivity, home environment, and potential contacts should be made and robust arrangements should be put in place to reduce transmission in the community prior to discharge. There should be a full assessment of household contacts, the patient's living accommodation, and education provided to all family members on signs and symptoms of TB, infection control, and cough hygiene.

As illustrated in the opening case, it is vital to ensure community support as well as patient adherence. The arrangements for ambulatory care should be organized and discussed with the patient and relevant community healthcare professionals as soon as possible.

MDR-TB patients should be on DOT and arrangements should be made for supervision of tuberculosis treatment with both patients and caregivers to promote treatment adherence. Preliminary studies have indicated that the use of virtually observed therapy (using smartphones or videophones to film taking the medication) appears to have potential as an alternative to DOT.

Particular consideration should be given to vulnerable household contacts, such as children or immunocompromised family members, who should avoid providing direct care where possible or wear particulate filter masks if this is unavoidable. Children under 5 years of age should have minimal or no contact with the infectious patient and have regular screening for TB symptoms. Unlike drug-sensitive TB, there is currently insufficient evidence on chemoprophylaxis regimens for close contacts of MDR-TB patients to recommend these.

The restriction of work or school, and engaging in other outside activities should be based on a comprehensive risk assessment, which includes the evaluation of infectivity of the patient, vulnerability of contacts, and the environment. Exclusion criteria may be more stringent in patients who teach small children or if their work or study involves lengthy contact with others. Patients who work with vulnerable people as in a healthcare setting, or with young children, or who are assessed as otherwise at risk of transmission to others should have good evidence that they are no longer infectious before they can return to their jobs.

In conclusion, a whole systems approach is required for MDR-TB infection prevention and control. This includes early detection and treatment of cases, robust institutional policies and guidelines, environmental controls, as well as supportive treatment and education for patients and their contacts. Adequate infection control does not need to be resource intensive, but it is vital to ensure that adequate facilities, appropriately trained staff, and resources are available to maintain appropriate infection control, particularly in institutional settings.

EPIDEMIOLOGY OF TB AND MDR-TB

The WHO estimated that about 9.6 million incident cases of TB occurred globally in 2014 and that approximately one-third of these cases had not been notified to national surveillance programs. Approximately 480,000 (about 5%) of all annual incident cases are estimated to be affected by MDR-TB, defined as *M. tuberculosis* strains resistant to both isoniazid and rifampicin (Table 12.1). MDR-TB is more common among previously treated patients, although the increased incidence in new patients, suggesting direct person-to-person transmission of drug-resistant strains, is a growing concern. Worldwide in 2014 approximately 3.3% of newly diagnosed TB cases and 20% of previously treated TB cases had MDR-TB. Some of the highest rates of MDR-TB are found in central Asia and Eastern Europe, where several countries report MDR-TB in more than 20% of new cases and up to 60% of previously treated cases. More than half of all MDR-TB cases notified in 2014 were from the Russian Federation, China, and India, although it has been suggested that MDR-TB incidence rates may be highest in sub-Saharan Africa, where many cases go undetected and the true scale of the MDR-TB epidemic remains unknown.

It is estimated that fewer than 25% of MDR cases are detected. Many high-burden countries have limited capacity for diagnosis, laboratory processing (including drug sensitivity testing), and national surveillance, which contribute to the high rate of under-detection.

Approximately 9.6% of people with MDR-TB have an extensively drug-resistant (XDR) strain that has additional resistance against flouroquinolones and at least one injectable antituberculous drug (see Table 12.1). The first cases of XDR-TB were reported in South Africa in 2006 and by the end of 2015, a total of 105 countries had reported at least one case of XDR-TB. The highest incidence of XDR-TB is reported in Eastern Europe and the former Soviet Union, but, as with MDR-TB, limited capacity for diagnosis and surveillance makes estimating rates difficult.

Table 12.1 **Drug-resistant TB definitions**

Latent tuberculosis infection (LTBI)	An asymptomatic and non-infectious stage of TB disease, where viable *Mycobacteria* are controlled by the immune system.
Multidrug-resistant tuberculosis (MDR-TB)	The patient is resistant to at least isoniazid and rifampicin.
Extensively drug-resistant tuberculosis (XDR-TB)	The patient is resistant to isoniazid and rifampicin, any fluoroquinolone and one of the second-line injectable drugs.
Pre-XDR-TB	The patient is resistant to isoniazid and rifampicin, plus any fluoroquinolone or one of the second-line injectable drugs.
Resistance in new patients	The patient had no history of TB treatment, or a previous treatment that lasted for less than one month. This situation suggests transmission of a resistant strain in the population.
Resistance in previously treated patients	The patient had a history of TB treatment of more than one month. This situation suggests resistance that was acquired during treatment or possible re-infection with a resistant strain.

In England in 2014, approximately 7.4% of all TB cases had drug resistance against any first-line antibiotic and 1.3% had MDR-TB, a rate that has decreased since 2011. In 2014, 11 pre-XDR-TB and three XDR-TB cases were notified in England. The highest number of MDR-TB cases in England was reported among patients between 25 and 34 years old, and almost 90% were non-UK born. Nearly 17% had at least one social risk factor.

The global case fatality rate from MDR is approximately 26% (range 16–58%), representing 150,000 deaths annually (range 53,000–270,000). MDR-TB mortality is significantly affected by comorbidities (for example, HIV), and access to diagnosis and effective treatment. Estimates of mortality depend on having robust surveillance systems in place, although mortality is lower in Organisation for Economic Co-ordination and Development (OECD) countries; in England 4% of those notified in 2012 had died by their last recorded outcome. The case fatality rate of XDR-TB can be much higher.

BIOLOGY

TB infections are caused by *Mycobacterium tuberculosis* complex, which includes the species *M. tuberculosis sensu stricto* an Asian variety, *M. africanum* (biotypes I and II), *M. bovis, M. bovis BCG, M. microti, M. caprae, M. canetti,* and *M. pinnipedi. Mycobacterium* spp. are nonmotile and rod shaped, and they do not stain well with Gram stain. They are obligate aerobes that grow best in tissues with a high-oxygen content, such as the lungs. They are classified as acid-fast bacilli because of their ability to retain dyes when treated with acids, as in the Ziehl-Neelson stain method.

M. tuberculosis bacilli are facultative intracellular parasites and slow growing organisms with a generation time of around 12 to 18 hours. The cell wall of *M. tuberculosis* bacilli is composed of over 60% lipid, containing mycolic acid, cord factor, and wax. Mycolic acid is hydrophobic, forming a lipid shell and affecting permeability, and it is this feature that is thought to be a significant factor in the virulence of *M. tuberculosis.* This lipid cell wall contributes to the survival of the organism inside macrophages, and impermeability makes it difficult to target with drugs, and contributes to the development of drug resistance. *Mycobacteria* are also often found in organs or body parts where the delivery of antibiotics may be more difficult, including pulmonary cavities and solid caseous material.

Drug resistance of *Mycobacteria* either occurs as a result of a selection pressure advantage of spontaneously occurring viable resistance mutations in the bacterial population and incomplete or interrupted treatment with first line drugs can facilitate this process. Drug-resistant strains have been increasingly reported and since the mid-1990s MDR-TB has become global. The development of drug-resistant TB strains is often associated with a history of failed or inappropriate treatment, as a result of poor adherence, limited access to health services, or interruption of drug supply, or any combination of these. Most MDR-TB cases are diagnosed in adults, however it is possible that there is under-ascertainment of paediatric MDR-TB, because of the difficulty of diagnosis or obtaining specimens.

Immunocompromised individuals (for example, persons affected by HIV) are more susceptible to TB and often have a rapid disease progression, but there is no robust evidence to demonstrate an association between HIV and drug-resistant TB. Higher MDR-TB rates in countries with a high burden of both HIV and TB could be explained by difficulties in accessing appropriate treatment in resource-poor settings that are often prevalent in these countries.

Clinical and molecular studies suggest that drug-resistant strains of tuberculosis are probably similarly infectious compared with sensitive strains. Measures of infection control for MDR-TB will follow similar principles to drug-sensitive TB, although they may need to be adhered to more rigorously because of the serious consequences of transmission.

DISEASE

M. tuberculosis can affect any part of the body or be disseminated (miliary TB), and clinical symptoms vary with the site of disease. In most countries, TB in adults is usually pulmonary, often affecting well-ventilated (upper) areas of the lungs.

Symptoms of active pulmonary TB include a persistent cough, fatigue, night sweats, weight loss, and blood in the sputum. Pulmonary TB is associated with necrosis, cavitation, and bleeding of lung tissue. Patients with extrapulmonary or disseminated disease may have unspecific symptoms, depending on the site of the infection.

Diagnosis of TB is complicated because disease onset is often insidious and symptoms are nonspecific. In low-incidence countries, onset of TB is often a result of reactivation of LTBI, an asymptomatic and non-infectious stage of the disease, in which viable *Mycobacteria* are controlled by the human immune system. The WHO estimates that about one-third of the world's population has LTBI. A small proportion of individuals with LTBI will develop TB disease and estimates of progression rates vary between 3–13% within two years, with a 5–10% lifetime risk. In people co-infected with HIV, the risk of developing active disease increases to 10% per year.

MDR-TB is a laboratory diagnosis, usually based on microbiological culture and drug susceptibility testing. Detailed clinical, contact, and travel history, as well as epidemiological knowledge of local drug-resistance patterns may be important to inform appropriate infection control, treatment, and rapid genotypic resistance testing before diagnostic sensitivity testing (DST) becomes available. Drug resistance should be suspected in any patient who fails to respond appropriately to TB treatment and who remains positive on sputum-smear microscopy or culture, despite good adherence to effective treatment.

Sputum is the most common sample used for smear and culture, though many other fluids and tissues can also be cultured. Liquid or solid culture is considered the gold standard for diagnosing TB and performing DST. The test properties of cultures are affected by the quality of the sample and the number of viable *Mycobacteria*; paucibacillary samples (for example, from children or HIV positive patients) may lead to false negative results. At a minimum, three high-quality early morning sputum samples are required. Solid cultures take about six weeks to grow, although liquid cultures can reduce this timescale to around 10–14 days. In resource-limited, high-burden country settings there can be delays in obtaining laboratory results, which in turn delays the start of an appropriate treatment regimen. Delays in diagnosing MDR-TB increase the risk of continued transmission and sub-optimal treatment, leading to increased resistance profiles.

Nucleic acid amplification tests (NAAT) such as GeneXpert MTB/RIF can help to rapidly identify *M. tuberculosis* bacilli in sputum, leading to earlier diagnosis and treatment initiation. GeneXpert MTB/ RIF (Cepheid, US) is a self-contained, automated, real-time molecular test for *M. tuberculosis* and for rifampicin resistance. Detection of rifampicin resistance is mainly based on detection of *rpoB* gene, and because rifampicin monoresistance is uncommon, a positive test result is usually taken as proxy for MDR. GeneXpert MTB/RIF has been shown to have very good test properties compared with culture, with an estimated sensitivity of 90.4% (95% CI; 89.2–91.4%) and specificity of 98.4% (95% CI; 98.0–98.7%) for pulmonary TB (lower for extrapulmonary TB).

GeneXpert MTB/RIF has been recommended as the initial diagnostic test for suspected MDR-TB and HIV-associated TB. The WHO has used this test since 2010 with cerebrospinal fluid (CSF) specimens for patients with suspected TB meningitis, and preferential pricing is available for many developing countries. Although GeneXpert enables early diagnosis and treatment, it does not replace the need for culture and DST.

Another molecular test for MDR-TB is the line probe assay (LPA), which provides rapid diagnosis of tuberculosis, identification of individual members of the *Mycobacterium tuberculosis* complex (MTBC), and identification of genetic mutations for resistance to rifampicin, isoniazid, fluoroquinolones, aminoglycosides, and ethambutol.

PATHOLOGY

When aerosolized *M. tuberculosis* bacilli enter the alveoli of the lung, they can cause active disease shortly after infection (primary disease). More commonly, they are controlled by the immune system and cause a latent infection, which can reactivate at a later stage. Upon infection, *Mycobacteria* are usually phagocytized by alveolar macrophages. The macrophages of the immune system can help prevent the reproduction of bacilli through formation of granulomas.

Latent infection occurs when bacteria inside the granuloma become dormant. In active disease, the host immune response is not sufficient to contain the bacteria and they multiply and destroy the surrounding cells and cause the typical TB disease symptoms. Bacteria can be disseminated by the lymphatic circulation to lymph nodes in the lung, forming the primary complex. Bacteria also circulate around the body via the bloodstream.

QUESTIONS

1. MDR tuberculosis is defined as which of the following?
    a. Resistant to prothionamide
    b. Resistant to rifampicin
    c. Resistant to moxifloxacin and rifampicin
    d. Resistant to rifampicin and isoniazid
    e. Resistant to rifampicin, isoniazid, and moxifloxacin

2. First-line treatment for tuberculosis is:
    a. Rifampicin and isoniazid
    b. Rifampicin, isoniazid, and ethambutol
    c. Rifampicin, isoniazid, ethambutol, and pyrazinamide
    d. Rifampicin, moxifloxacin, ethambutol, and pyrazinamide
    e. Cycloserine, isoniazid, pyrazinamide, and moxifloxacin

3. True or false. Which of the following is important in prevention of further MDR-TB cases?
   a. Positive-pressure rooms
   b. Negative-pressure rooms
   c. Use of linezolid
   d. Masks
   e. Isolation in a single room

4. Which of the following is not important in spread of MDR-TB?
   a. Vitamin D
   b. Droplet nuclei
   c. Coughing
   d. Overcrowding
   e. Poverty

5. True or False. Which of the following are important in development of drug resistance?
   a. Compliance with treatment
   b. Inappropriate treatment
   c. HIV status
   d. Limited access to health services
   e. Smoking

GUIDELINES

1. NICE (2011) Tuberculosis: Clinical diagnosis and management of tuberculosis, and measures for its prevention and control. Clinical Guideline 117 (www.nice.org.uk/nicemedia/live/13422/53638/53638.pdf).

2. NICE (2012). Prevention and control of healthcare-associated infections in primary and community care. Clinical Guideline 139 (www.nice.org.uk/nicemedia/live/13684/58656/58656.pdf).

3. WHO (2008) Implementing the stop TB strategy: a handbook for national tuberculosis control programmes (www.who.int/tb/publications/2008/who_htm_tb_2008_401_eng.pdf?ua=1).

4. The Tuberculosis Coalition for Technical Assistance (2010) Implementing the WHO policy on TB infection control in health-care facilities, congregate settings and households. (www.stoptb.org/wg/tb_hiv/assets/documents/TBICImplementationFramework1288971813.pdf).

5. CDC Tuberculosis Infection Control and Prevention. Infection Control in Health Care Settings (online) (www.cdc.gov/TB/topic/infectioncontrol/default.htm).

REFERENCES

1. WHO (2010). Multidrug and extensively drug-resistant TB(M/XDR TB), 2010 global report on surveillance and response. WHO, Geneva whqlidoc.who.int/publications/2010/9789241599191_eng.pdf

2. WHO (2008). Implementing the stop TB strategy: a handbook for national tuberculosis control programmes WHO, Geneva.[cited 2014 Jun 3]. www.who.int/tb/publications/2008/who_htm_tb_2008_401_eng.pdf?ua=1

3. Toczek A et al. (2013) Strategies for reducing treatment default in drug resistant tuberculosis: systematic review and meta analysis. *Int J Tuberc Lung Dis* **17**(3):299–307.

4. Caminero JA (2010) Multi-drug resistant tuberculosis: epidemiology, risk factors and case finding. *Int J Tuberc Lung Dis* **14**:382–390.

5. Public Health England (2014). Tuberculosis in the UK: annual report data up to 2013. www.hpa.org.uk/webc/HPAwebFile/HPAweb_C/1317139689583

6. Smith I (2003) *Mycobacterium tuberculosis* pathogenesis and molecular determinants of virulence *Clin Microbiol Rev* **16**:463–496.

7.   Shingadia D et al. (2003) Diagnosis and treatment of tuberculosis in children. *Lancet Infect Dis* **3**:624–632.

8.   Guillerm M et al (2007) Tuberculosis diagnosis and drug sensitivity testing: an overview of current diagnostic pipeline. fieldresearch.msf.org/msf/handle/10144/13532

9.   NICE PH37 (2012) Tuberculosis-hard to reach groups: guidance. nice.org.uk/

10. Diel R et al. (2012) Predictive value of interferon-γ release assays and tuberculin skin testing for progression from latent TB infection to disease state: a meta-analysis. *Chest* **142**:63–75

11. NICE (2012) Prevention and control of healthcare-associated infections in primary and community care, CG139. www.nice.org.uk/nicemedia/live/13684/58656/58656.pdf

12. Marais BJ, Lönnroth K, Lawn SD et al. (2013) Tuberculosis comorbidity with communicable and non-communicable diseases: integrating health services and control efforts. *Lancet Infect Dis* **13**(5):436–448.

13. CDC (2013). Tuberculosis Education Chapter 7: Tuberculosis Infection Control. CDC Tuberculosis Education, pp. 189–226. www.cdc.gov/tb/education/corecurr/pdf/chapter7.pdf

14. The Tuberculosis Coalition for Technical Assistance (2010) Implementing the WHO Policy on TB Infection Control in Health-Care Facilities, Congregate Settings and Households. Stop TB Partnership. http://www.stoptb.org/wg/tb_hiv/assets/documents/TBICImplementationFramework1288971813.pdf

15. WHO (2014). Guidelines for the programmatic management of drug-resistant tuberculosis www.who.int/tb/challenges/mdr/programmatic_guidelines_for_mdrtb/en/

16. The Interdepartmental Working Group on Tuberculosis (1998) The prevention and control of tuberculosis in the United Kingdom. UK guidance on the prevention and control of transmission of 1. HIV-related tuberculosis 2. drug-resistant, including multiple drug-resistant, tuberculosis. webarchive.nationalarchives.gov.uk/20130107105354/http://www.dh.gov.uk/prod_consum_dh/groups/dh_digitalassets/@dh/@en/documents/digitalasset/dh_4115299.pdf

17. Nathanson E, Gupta R, Huamani P et al. (2004) Adverse events in the treatment of multidrug-resistant tuberculosis: results from the DOTS-Plus initiative. *Int J Tuberc Lung Dis.* **8**(11): 1382–1384.

18. Health Protection Agency HP, Part 2A Orders (2010). www.legislation.gov.uk/ukdsi/2010/9780111490976/contents

19. Okello D, Floyd K, Adatu F et al. (2003) Cost and cost-effectiveness of community-based care for tuberculosis patients in rural Uganda. *Int J Tuberc Lung Dis* **7**(9 Suppl 1):S72–S79.

20. WHO (2009) Natural ventilation for infection control in health-care settings. www.who.int/water_sanitation_health/publications/natural_ventilation/en/

21. Chiang C-Y, Centis R, Migliori GB (2010). Drug-resistant tuberculosis: Past, present, future. *Respirology.* **15**(3):413–32.

22. WHO (2009) WHO policy on TB infection control in health-care facilities, congregate settings and households. WHO, Geneva.

23. National Institute for Health and Care Excellence (NICE) (2011) Tuberculosis: clinical diagnosis and management of tuberculosis, and measures for its prevention and control. Clinical Guideline 117. publications.nice.org.uk/tuberculosis-cg117/guidance

24. Anderson L (2014) Genotyping and Its Implications for Transmission Dynamics and Tuberculosis. In Clinical Tuberculosis 5e (Davies DO, Gordon SB, Davies G eds) pp. 55–69. CRC Press.

25. Republic of South Africa. Department of Health. Management of Drug - Resistant Tuberculosis. Policy Guidelines (2011)

26. Abubakar I, Zignol M, Falzon D et al. (2013) Drug-resistant tuberculosis: time for visionary political leadership. *Lancet Infect Dis* **13**(6):529–539.

27. Toczek A, Cox H, du Cros P et al. (2013) Strategies for reducing treatment default in drug-resistant tuberculosis: systematic review and meta-analysis. *Int J Tuberc Lung Dis* **17**(3):299–307.

28. WHO (2013) Global tuberculosis report 2013. www.who.int/tb/publications/global_report/en/

29. Kant S, Maurya AK, Kushwaha RAS et al. (2010) Multi-drug resistant tuberculosis: an iatrogenic problem. *Biosci Trends* **4**(2). https://www.ncbi.nlm.nih.gov/pubmed/20448341

31. Wade VA, Karnon J, Eliott JA & Hiller JE (2012) Home videophones improve direct observation in tuberculosis treatment: a mixed methods evaluation. *PLoS ONE* **7**(11):e50155.

32. Virtual Monitoring Could Aid Adherence to TB Medication (2013). www.infectioncontroltoday.com/news/2013/09/virtual-monitoring-could-aid-adherence-to-tb-medication.aspx

33. Public Health Agency of Canada (2012) *Mycobacterium tuberculosis* complex—pathogen safety data sheets. www.phac-aspc.gc.ca/lab-bio/res/psds-ftss/tuber-eng.php#footnote20

34. Todar, Mycobacterium tuberculosis and tuberculosis. Online Textbook of Bacteriology. textbookofbacteriology.net/tuberculosis.html

35. Gillespie SH (2002) Evolution of drug resistance in *Mycobacterium tuberculosis*: clinical and molecular perspective. *Antimicrob Agents Chemother* **46**(2):267–274.

36. Cole ST, Brosch R, Parkhill J et al. (1998) Deciphering the biology of *Mycobacterium tuberculosis* from the complete genome sequence. *Nature* **393**(6685):537–544.

37. Newton SM, Brent AJ, Anderson S et al. (2008) Paediatric tuberculosis. *Lancet Infect Dis* **8**(8): 498–510.

38. Boehme CC, Nabeta P, Hillemann D et al. (2010) Rapid molecular detection of tuberculosis and rifampin resistance. *N Engl J Med* **363**(11):1005–1015.

39. WHO (2014) Tuberculosis Diagnostics Xpert MTB/RIF Test. 2014. who.int/tb/publications/Xpert_factsheet.pdf?ua=1

40. Lawn SD, Mwaba P, Bates M et al. (2013) Advances in tuberculosis diagnostics: the Xpert MTB/RIF assay and future prospects for a point-of-care test. *Lancet Infect Dis* **13**(4):349–361.

41. Rangaka MX, Wilkinson KA, Glynn JR et al. (2012) Predictive value of interferon-γ release assays for incident active tuberculosis: a systematic review and meta-analysis. *Lancet Infect Dis* **12**(1):45–55.

ANSWERS

MCQ	Feedback
1. MDR tuberculosis is defined as which of the following? a. Resistant to prothionamide b. Resistant to rifampicin c. Resistant to moxifloxacin and rifampicin d. Resistant to rifampicin and isoniazid e. Resistant to rifampicin, isoniazid, and moxifloxacin	See definitions in Table 12.1.
2. First-line treatment for tuberculosis is: a. Rifampicin and isoniazid b. Rifampicin, isoniazid, and ethambutol c. Rifampicin, isoniazid, ethambutol, and pyrazinamide d. Rifampicin, moxifloxacin, ethambutol, and pyrazinamide e. Cycloserine, isoniazid, pyrazinamide, and moxifloxacin	Treatment with isoniazid provides early bactericidal activity to rapidly reduce the number of viable *Mycobacteria*. Rifampicin has a sterilizing activity, which will continue to kill *Mycobacteria* beyond the end of treatment. These two first-line drugs are the most effective in reducing the bacillary load and thus the infectivity of the patient. Without isoniazid and rifampicin treatment, for MDR-TB is substantially less effective and the infectious period is prolonged.
3. True or false. Which of the following is important in prevention of further MDR-TB cases? a. Positive-pressure rooms b. Negative-pressure rooms c. Use of linezolid d. Masks e. Isolation in a single room	a. False b. True c. False d. True e. True Environmental-control measures aim to reduce the concentration of airborne infectious particles in the environment by diluting contaminated air, removing infectious particles, and controlling airflow in patient areas. Isolation of infectious cases, preferably in a negative-pressure room is one of several measures that can be taken to prevent onward transmission. Another key intervention to reduce transmission is the use of personal protective equipment (PPE) for all healthcare staff and other contacts of infectious patients. Protective face masks with a capacity to filter 1-μ particles (for example, N95 or FFP3 respirator masks) should be worn. These masks need to be individually fitted, to ensure they provide the required protection.
4. Which of the following is not important in spread of MDR-TB? a. Vitamin D b. Droplet nuclei c. Coughing d. Overcrowding e. Poverty	TB is usually transmitted via airborne droplet nuclei, produced when the infected person coughs, sneezes, or sings, or during other procedures that generate aerosols from an individual with pulmonary TB. Those at highest infection risk are often close contacts, such as those living in the same household with the case or those in frequent prolonged contact, such as caregivers. TB disease is multifactorial and influenced by socioeconomic risk factors, such as poor quality or overcrowded housing conditions, poor air quality and barriers to accessing to health services. Populations living and working in congregate settings such as prisons, refugee camps, and homeless shelters have a particularly high risk of transmission. Individuals with social risk factors may be more likely to develop drug resistance as a result of difficulties in accessing services or adhering to treatment.
5. True or False. Which of the following are important in development of drug resistance? a. Compliance with treatment b. Inappropriate treatment c. HIV status d. Limited access to health services e. Smoking	a. True b. True c. False d. True e. False Drug resistance of *Mycobacteria* can occur from incomplete or interrupted treatment with first-line drugs. The development of drug-resistant TB strains is often associated with a history of failed or inappropriate treatment, from poor adherence, limited access to health services, or interruption of drug supply, or any combination of these.

MEASLES: ACHIEVING NATIONAL CONTROL OF A VACCINE PREVENTABLE INFECTION

CASE 13

Mary Ramsay and Subhadra Rajanaidu

Immunisation, Hepatitis and Blood Safety Department, Public Health England, London, UK.

In November 2012, the UK started seeing an increase of measles cases in 10- to 18-year-olds with a peak in 11-year-old children. Within the first quarter of 2013 alone, Public Health England confirmed a total of 741 cases in England and Wales. This was the highest number of monthly cases recorded since the enhanced surveillance system commenced in 1994. The outbreak in Wales was the largest since the introduction of the measles, mumps, and rubella (MMR) vaccine in 1988.

The reason for this rapid spread was high susceptibility in the 10- to 18-year-old cohort. Although MMR is part of the national childhood immunization schedule, coverage of the vaccine at two years of age fell from a high of around 92% to just below 80% in the early 2000s as a result of discredited safety concerns linking MMR vaccines to autism. Although coverage had fallen nationally, some areas had lower baseline coverage or experienced a more marked decline, or both. These areas included London and Swansea. Coverage in younger children had subsequently improved, but attempts to provide catch-up immunization to older children had been less effective.

As measles is highly infectious, with a basic reproduction number of between 15 and 20, a high level of immunity (>95%) is required to prevent transmission. Estimates suggested that up to around 10% of secondary school age children in the worst affected cohorts were susceptible to measles. This level of susceptibility, which may have been sufficient to prevent transmission in younger cohorts, was insufficient to prevent sustained transmission within secondary school settings, where closer contact patterns and higher numbers of contacts occur.

The increase in cases triggered a national investigation and response.

INVESTIGATION AND PREVENTION OF FURTHER CASES

The prevention of further cases required control measures to be taken around individual cases, prompt management of outbreaks, and improvement of national immunization uptake.

Control measures for individual cases

Measles is a notifiable disease in the UK. Local health protection teams in England are informed of cases by healthcare professionals and follow up each case to both reduce potential spread to at risk individuals and to provide accurate surveillance data to inform national control. The health protection team collects epidemiological and clinical data on each case, and all suspected cases are offered laboratory investigation, although action may precede the results of such testing. In England the main method of laboratory investigation is noninvasive testing using oral fluid for detection of measles-specific IgM and measles-specific RNA.

A risk assessment is then made by an experienced health protection professional based upon clinical and epidemiological features. Confirmed and epidemiologically linked cases or cases considered to be likely on the basis of this risk assessment require active contact tracing and are recommended to be kept away from school, nursery, or childcare workers for four days from the onset of rash. Cases unlikely to be measles are sent an oral fluid test with no further public health action required unless the diagnosis is confirmed.

Contacts are then assessed and interventions offered as appropriate. All contacts are assessed for susceptibility using a combination of age and vaccination history, and in some cases urgent IgG testing. Susceptible or likely susceptible contacts who are vulnerable to severe consequences of infection (infants, pregnant women, and immunocompromised individuals) are considered for post-exposure prophylaxis with an immunoglobulin product. Susceptible contacts who are potential sources of transmission in high-risk settings, such as susceptible healthcare workers, are considered for exclusion from work. Susceptible healthy contacts are offered post-exposure MMR vaccination ideally within three days of exposure.

In some cases, although the index case does not appear clinically or epidemiologically likely to be measles, a case of measles would have considerable public health impact, such as a case in a school or healthcare setting. In this type of case, urgent local laboratory testing of the index case may be required. The test would normally seek to detect viral RNA from oral fluid or throat swabs taken in the acute phase of the illness. As the virus is cleared quickly in acute infection, detection of viral RNA may not be possible even a few days after onset (particularly in throat swabs). A single negative RNA result is not, therefore, considered adequate for discarding cases as part of surveillance. Additional testing for measles-specific IgM in oral fluid should always follow; oral fluid should therefore be collected at the same time.

Control measures for outbreaks

Following evidence of local transmission an outbreak control team (OCT) meeting is called and an outbreak declared at the appropriate level. Because measles is highly infectious prior to onset of symptoms, within smaller settings such as schools and nurseries, most susceptible individuals would have already been exposed by the time an outbreak is recognized. The risk of further transmission can be assessed based on attack rate and the historical vaccine coverage in the setting. Vaccination sessions may be considered for all under-vaccinated individuals in the affected setting. A communication strategy should also be developed to inform primary care and secondary care colleagues, as well as schools and parents of the situation. Proactive press releases may also be released so that the outbreak is used as an opportunity to inform the public in the wider community. The main aim of this wider communication is to encourage better vaccine uptake in the community, usually through their primary care provider, and thus to prevent similar outbreaks in other settings. The latter strategy is likely to be more effective in preventing cases than in offering vaccination in a setting where an outbreak has already commenced. Improving herd immunity will also protect individuals at high risk, such as those with immunosuppression, who cannot be vaccinated. Early immunization of infants (aged 6–12 months) is only considered when the risk of exposure is considered extremely high.

Improvement of national immunization uptake

In April 2013, following the increase in cases in England, with several localized outbreaks in teenagers, a measles catch-up campaign was launched nationally in England and the devolved administrations. The aim was to ensure that at least 95% of 10- to 16-year-olds had received at least one dose of MMR, similar to the uptake in younger children. Around a million at-risk children were identified using general practitioner (GP) records, and letters were sent to their parents recommending that they get their children vaccinated. Vaccination took place at GP clinics with some areas setting up temporary vaccine clinics in schools, community centers, and similar locations.

Following the campaign launch, there was a monthly decrease in measles cases in England to 24 confirmed cases between October and December 2013 (equivalent to an annualized incidence in England of 1.8 per million). Case numbers have remained low up through the end of 2015. In addition, health services were urged to exploit key opportunities to sustain low susceptibility in the longer term such as offering MMR alongside the school leaving booster vaccination, with the human papillomavirus (HPV) school program, and when students transfer from primary to secondary school.

EPIDEMIOLOGY

Measles is one of the leading causes of mortality among young children around the world with 114,900 deaths reported in 2014, mostly in children under 5 years of age, despite the availability of a safe and cost-effective vaccine.

Measles is highly communicable with a basic reproduction number of between 15 and 20; the secondary attack rate among susceptible household contacts exceeds 90%. Because of this, a very high level of immunity in the population is required to stop the transmission of this virus. In most developed countries, a range of 90–95% distributed uniformly across geographic areas is suggested as the threshold immunity to be maintained to interrupt the transmission of measles. Levels of susceptibility need to be particularly high in settings with a high number of contacts, such as secondary schools.

Measles used to be considered a common childhood infection. In the London fever hospitals between 1911 and 1914, there were 9277 admissions in children under 5 years of age for measles, with a case fatality of 13.9%. This improved dramatically over the years, probably as a result of better nutrition, reduction in overcrowding, and improved treatment options. By 1960, there were around 100 deaths a year.

In 1968, a measles vaccine was introduced into the UK national childhood immunization schedule for the first time for children aged between one and two years (Figure 13.1). Coverage was poor at below 80% and there was limited impact on the transmission of the virus. This vaccine was then replaced by the combination MMR in 1988 for children 13–15 months with an improved implementation program, which achieved coverage levels over 90%. This immunization program resulted in a significant reduction in case notification, and incidence continued to fall to very low levels.

In 1994, in an attempt to prevent a predicted epidemic in school-aged children, an extensive catch-up campaign was undertaken, targeting over 8 million children from 5–16 years of age in England and Wales with a measles–rubella (MR) vaccine. By the end of the year,

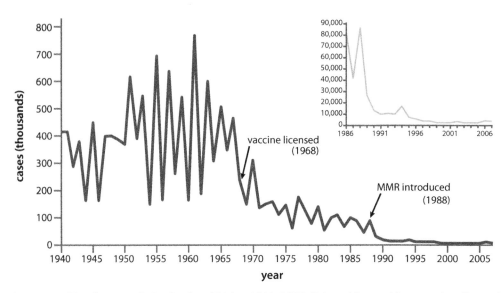

Measles cases in England and Wales, 1940–2007. (Adapted from public sector data, licensed under the Open Government Licence v3.0.)

92% of school-aged children had received the vaccine. After this, measles transmission was interrupted. To maintain elimination, by reducing the number of primary vaccine failures, a two-dose MMR schedule was introduced in 1996, with the second dose given at around 4 years of age.

Following a moderate reduction in vaccine coverage to below 80% in England, endemic transmission resumed in 2007, leading to large outbreaks in populations with low coverage, especially in areas within London. In 2009, following sustained efforts to improve coverage through a catch-up program in London primary schools begun in 2005, and a national catch-up campaign for children under 18 years of age in 2008, control improved again and infection rates remained at low levels during 2010.

Unfortunately, measles cases started increasing again in 2011, because of imported cases from Europe and limited transmission around these importations. A further rise in cases was observed in 2012 and continued into the first three months of 2013, with 2224 suspected and notified measles infections by April 2013. The age distribution of the cases in 2013 showed a peak in the 10- to 14- year-old age group, a shift when compared to the previous four years. Analysis of English cases by year of birth showed that, apart from children born in 2012 (who would be too young to be eligible for vaccination), the highest attack rate was in those born in 2000, corresponding with those who missed out on vaccination in the early 2000s, when concern around the discredited link between autism and the vaccine was widespread and MMR vaccination coverage fell nationally to less than 80%. Following the control measures described, including the catch-up vaccination campaign, case numbers in the UK fell to the equivalent of an annualized incidence of 1.8 per million and have remained low.

The reservoir of measles is infected humans and the route of transmission for measles is primarily person to person through respiratory droplets or when infectious secretions come into contact with a susceptible individual's mucosa, such as following transfer by hands. It is most infectious three days before the onset of rash, when the individual is experiencing nonspecific prodromal symptoms, such as fever, runny nose, cough, and red, watery eyes.

Measles continues to be infectious up to four days after the appearance of the rash. Because the early symptoms are nonspecific, infected individuals often continue to participate in social activities, facilitating transmission. There have also been transmissions that have occurred in primary healthcare settings when infected individuals seek health advice.

The measles virus is highly infectious in its droplet form where it can remain suspended in air for up to two hours if an infected person has coughed or sneezed. Susceptible individuals can become infected if they breathe the contaminated air or touch infected surfaces and then touch their eyes, noses, or mouths. The virus is however susceptible to strong ultraviolet light and desiccation.

PATHOLOGY

Measles is caused by a Morbillivirus, which is a member of the Paramyxoviridae family. It is an enveloped virus that can infect epithelial, reticuloendothelial, and white blood cells (epitheliatropic and immunotropic).

Some disease manifestations result directly from infection. In particular, the cough and coryza are related to infection of the respiratory mucosa. Other manifestations are related to the immune response against the infection. The characteristic measles rash onset coincides with the production of measles-specific cytotoxic CD8 T cells, soluble CD8, and interferon-gamma, and is postulated to be a cellular immune phenomenon. Immunologic abnormalities may persist after measles, resulting in increased susceptibility to secondary infections. *In vivo*, a delayed hypersensitivity response has been noted to be reduced with low natural killer (NK) cell activity and an increased IgE response. *In vitro* studies have shown a reduction in further lymphoproliferative responses, with many of these changes persisting for weeks despite apparent recovery from the infection.

DISEASE

The incubation period for measles ranges between 7 and 18 days, with an average of approximately 10 days from exposure to onset of fever. In immunosuppressed individuals, this it may increase up to 21 days.

There is often a prodromal phase with nonspecific symptoms of fever and malaise. The fever rises progressively, often exceeding 40°C. There is then the onset of a generalized characteristic maculopapular rash that usually starts on the head, spreading down to the trunk, and then to hands and feet. There may also be small raised white lesions in the mouth (Koplik spots). Although the latter are said to be diagnostic, they occur fleetingly before the skin rash and so the absence of Koplik spots is not helpful in excluding measles. Other associated complaints are cough, coryzal symptoms (for example, sneezing, nasal congestion, nasal discharge), and conjunctivitis. There is diarrhea in approximately 8% of cases.

Cases can be confirmed by laboratory investigation. In the UK, in line with World Health Organization (WHO) recommendations for nonendemic countries, all suspected measles cases are investigated through the detection of measles-specific IgM. In England, oral fluid samples are used for this investigation, and they can be taken by parents or patients and submitted to the national reference laboratory using the postal system. Samples should be taken as soon as possible after the onset of the rash or parotid swelling. Further confirmatory testing, characterization, and genotyping of measles RNA from confirmed cases are then undertaken by the WHO Global Specialised Reference Laboratory, followed by inclusion in the WHO global Measles Nucleotide Sequence (MeaNS) database.

In most developed countries, measles is usually a mild self-limiting disease in childhood, but can result in residual impairment and death in approximately 1–2 per 1000 cases. In children under five years of age, complications such as otitis media, bronchopneumonia, laryngotracheobronchitis, and diarrhea are relatively common with pneumonia occurring in 6% of cases. In older ages (over 20 years of age), there is an increased risk of pneumonia and acute encephalitis. Measles in pregnancy has an increased risk of premature labor, spontaneous abortion, and low-birth-weight infants.

Two other forms of encephalitis are observed after measles infection. Measles inclusion body encephalitis (MIBE) almost always occurs in patients with weakened immune systems, usually some weeks or months after mild or atypical measles, and is associated with high death rate. Subacute sclerosing panencephalitis (SSPE) is a rare degenerative neurological disease with an incidence of 5–10 cases per million reported measles cases and is hypothesized to arise from ongoing infection within the brain by the measles virus. In the first year of life, the risk of acquiring SSPE following measles is far greater than later on in life, with an average onset seven years after the initial infection (range 1 month–27 years). There is an insidious onset of neurological deterioration, progressing to abnormalities of behavior, intellect and motor control, seizures, and death.

Treatment and post-exposure prophylaxis: There is no specific treatment for measles. Rest, adequate fluid intake, and paracetamol or ibuprofen for symptomatic relief are recommended. Cases are requested to stay vigilant for potential complications and to seek help if concerned.

When given soon after exposure, both MMR vaccine and immunoglobulin may prevent or attenuate an attack of measles in a susceptible individual (post-exposure prophylaxis). In England, exposure is considered to be significant if the individual has had face-to-face contact or 15 min or more within the same room. The highest priority for contact tracing is to identify immunocompromised individuals, who should be rapidly assessed and, if necessary, admitted for intravenous immunoglobulin. Susceptible pregnant women and infants can be considered for human normal immunoglobulin given in the community, although a therapeutic dose cannot be feasibly achieved with intramuscular products. Susceptibility is assessed on the basis of age and vaccination history. Pregnant women who are not known to be fully vaccinated should be tested for antibodies against measles and only offered post-exposure prophylaxis if there is no evidence of immunity.

As the MMR vaccine is a live vaccine, it is contraindicated for pregnant and immunocompromised contacts. Post-exposure prophylaxis with MMR vaccine, however, can be offered to unimmunized contacts within three days of significant exposure to an individual infected with measles. As the attenuated virus is not transmitted, contacts of immunosuppressed individuals can be safely vaccinated. Although the effectiveness of this intervention is likely to be limited, it does ensure that they are protected against future exposure to measles, mumps, and rubella. Contact tracing of large numbers of immuncompetent individuals is of limited value and therefore of lower priority.

QUESTIONS

1. With respect to the features of measles, which one of the following statements is true?
   a. The likelihood of contracting measles in a susceptible population is low.
   b. Measles is a notifiable disease.
   c. Usual treatment of measles requires antibiotics.
   d. Primates and reptiles are the main reservoirs for the measles virus.

2. Regarding the clinical symptoms of measles in children, which one of the following statements is false?
   a. Measles causes a nonblanching rash that starts in the extremities.
   b. The initial symptoms of measles are nonspecific with malaise and fever.
   c. Koplik spots are always seen in cases of measles.
   d. Measles can cause very high fevers exceeding 40°C which can cause febrile seizures.

3. Which of the following groups should receive the MMR vaccination when exposed to an infected individual?
   a. Pregnant women in their third trimester
   b. An immunocompetent adult who lives in the same household as an immunecompromised individual
   c. A fully vaccinated colleague who works in the same open-plan area as the infected individual
   d. Newborn infants exposed to an infected individual within the same household

e. Healthy older contacts (born in the UK before 1970) who had measles in early childhood

4. Which of the following do not require exclusion from work or school?
   a. Case diagnosed with measles who works in a nursery
   b. Asymptomatic vaccinated sibling of an infected case
   c. A healthcare worker working within a pediatric unit who has been exposed to a confirmed case
   d. An infected case working as an IT specialist in an open-plan office

5. A local health protection team has just been informed of a second confirmed case of measles within a nursery setting within a five-day period. Which of these actions should the team take?
   a. Convene an outbreak control team to discuss any possible public health actions that need to be taken
   b. Determine whether any vulnerable individuals have been exposed to the case
   c. Arrange for vaccination for all susceptible contacts who are not pregnant or immunocompromised within three days
   d. Request that the nursery send a letter to parents informing of the cases within the nursery and current public health action
   e. All of the above

GUIDELINES

1. Department of Health (2013) Measles. In The Green Book. UK Department of Health, pp. 210–234.
2. NICE (2013) Scenario: Management of Measles. (cks.nice.org.uk/measles#!scenariorecommendation:2) [Accessed 29 June 2015].
3. Washington State Department of Health (2014) Measles (www.doh.wa.gov/Portals/1/Documents/5100/420-063-Guideline-Measles.pdf)[Accessed 29 June 2015].

REFERENCES

1. Bloch AB et al. (1985) Measles outbreak in a pediatric practice: airborne transmission in an office setting. *Pediatr* **75**:676–683.

2. Brown DW, Ramsay ME, Richards AF & Miller E (1994) Salivary diagnosis of measles: a study of notified cases in the United Kingdom, 1991–1993. *BMJ* **308**(6935):1015–1017.

3. CDC (2014) Transmission of Measles (www.cdc.gov/measles/about/transmission.html) [Accessed 29 June 2015].

4. CDC (2015) (www.cdc.gov/measles/about/signs-symptoms.html) [Accessed 29 June 2015].

5. Ehresmann KR et al. (1995) An outbreak of measles at an international sporting event with airborne transmission in a domed stadium. *J Infect Dis* **171**:679–683.

6. Griffin D (2010) Measles virus-induced suppression of immune responses. *Immunol Rev* **236**:176–189.

7. Griffin D, Cooper S, Hirsch R et al. (1985). Changes in plasma IgE levels during complicated and uncomplicated measles virus infections. *J Allergy Clin Immunol* **76**:206–213.

8. Griffin D, Ward B, Jauregui E & Johnson R (1990) Natural killer cell activity during measles. *Clin Exp Immunol* **81**:218–224.

9. Hutchins SS et al. (2001) National serologic survey of measles immunity among persons 6 years of age or older, 1988-1994. *MedGenMed* (**E5**):24.

10. Cherry, J.D. Measles. in: R.D. Feigin, J.D. Cherry (Eds.) Textbook of pediatric infectious diseases. 4th ed. WB Saunders Company, Philadelphia (PA); 2002: 2054–2074

11. Maino T et al. (2008) Measles virus infects both polarized epithelial and immune cells by using distinctive receptor-binding sites on its hemagglutinin. *J Virol* **82**(9)4630–4637.

12. Ramsay ME et al. (2003) The elimination of indigenous measles transmission in England and Wales. *J Infect Dis* **187**(Suppl 1):198–207.

13. Miller D (1964) Frequency of complications of measles, 1963. *Br Med J* **2**:75–78.

14. Morley M, et al. (1963) Measles in Nigerian children. A study of the disease in West Africa, and its manifestations in England and other countries during different epochs. *J Hyg Camb* **61**:115.

15. Orenstein W, Perry R & Halsey N (2004) The clinical significance of measles: A review. *J Infect Dis* **189**(Supp 1):S4–S16.

16. Ornoy A et al. (2006) Pregnancy outcome following infections by coxsakie, echo, measles, mumps, hepatitis, polio and encephalitis viruses. *Reprod Toxicol* **21**(4):446–457.

17. Public Health England (2006) Annual Cover Report: 2005/06. Summary of trends in vaccination coverage in the UK. (webarchive.nationalarchives.gov.uk/20140629102627/http://hpa.org.uk/webc/HPAwebFile/HPAweb_C/1194947367316).

18. Public Health England (2013) MMR Catchup Programme. (webarchive.nationalarchives.gov.uk/20140505192926/http://www.hpa.org.uk/Topics/InfectiousDiseases/InfectionsAZ/Measles/MMRCatchupProgramme/).

19. Public Health England (2014) Notifications (confirmed cases) 1995–2013 by annual quarter, England and Wales. (webarchive.nationalarchives.gov.uk/20140505192926/http://www.hpa.org.uk/web/HPAweb&HPAwebStandard/HPAweb_C/1195733808276).

20. Public Health England) 2014. Vaccine coverage and COVER. (webarchive.nationalarchives.gov.uk/20140629102627/http://hpa.org.uk/Topics/InfectiousDiseases/InfectionsAZ/VaccineCoverageAndCOVER/).

21. Ramsay M et al. (1994) The epidemiology of measles in England and Wales: rationale for the 1994 national vaccination campaign. *Commun Dis Rep CDR Rev.* **4**:R141–146.

22. Ramsay ME, Brugha R, Brown DW et al. (1998) Salivary diagnosis of rubella: a study of notified cases in the United Kingdom, 1991-1994. *Epidemiol Infect* **120**(3):315–319.

23. Ramsay, M (2013) Measles: the legacy of low vaccine coverage. *Arch Dis Child* **98**:752–754.

24. Remington PL et al. (1985) Airborne transmission of measles in a physician's office. *JAMA* **253**:1574–1577.

25. Salisbury DM et al. (1995) Measles campaign. *BMJ* **310**:1334.

26. Simpson R H (1952) Infectiousness of communicable diseases in the household (measles, chicken pox, and mumps). *Lancet* **2**:549–554.

27. Tamashiro V, Perez H & Griffin D (1987) Prospective study of the magnitude and duration of changes in tuberculin reactivity during complicated and uncomplicated measles.. *Pediatr Infect Dis J* **6**:451–454.

28. Van Binnendijk R et al. (1990) The predominance of CD8+ class 1 MHC-restricted cytotoxic T lymphocytes (CTL) in recovery from measles. *J Immnunol* **144**:2394–2399.

29. Washington State Department of Health (2014) Measles (www.doh.wa.gov/Portals/1/Documents/5100/420-063-Guideline-Measles.pdf).

ANSWERS

MCQ	Feedback
1. With respect to the features of measles, which one of the following statements is true? a. The likelihood of contracting measles in a susceptible population is low. b. Measles is a notifiable disease. c. Usual treatment of measles requires antibiotics. d. Primates and reptiles are the main reservoirs for the measles virus.	The measles virus is highly infectious and approximately 90% of susceptible individuals can become infected if exposed to an infectious case. Measles is a notifiable disease in the UK and all health care professionals are expected to inform the local health protection team when they diagnose a patient with the infection. Measles is usually a self-limiting disease and treatment is mainly supportive for symptomatic relief. Antibiotics however may be required for secondary bacterial infections. Humans are the only known reservoirs for the measles virus.
2. Regarding the clinical symptoms of measles in children, which one of the following statements is false? a. Measles causes a nonblanching rash that starts in the extremities. b. The initial symptoms of measles are nonspecific with malaise and fever. c. Koplik spots are always seen in cases of measles. d. Measles can cause very high fevers exceeding 40°C, which can cause febrile seizures.	Measles has a characteristic erythematous maculopapular rash, which usually starts in the head around the hairline, progressing down toward the trunk and limbs over three to four days. The initial symptoms of measles are nonspecific, however the individual is already infectious at this point. Koplik spots are small red spots with blueish-white centers. They appear in the mucous membranes of the mouth just before the characteristic rash appears and can be seen for an additional couple of days after the onset of the rash. Measles infection can cause extremely high temperatures, which can lead to febrile seizures.
3. Which of the following groups should receive the MMR vaccination when exposed to an infected individual? a. Pregnant women in their third trimester b. An immunocompetent adult who lives in the same household as an immunecompromised individual c. A fully vaccinated colleague who works in the same open-plan area as the infected individual d. Newborn infants exposed to an infected individual within the same household e. Healthy older contacts (born in the UK before 1970) who had measles in early childhood	The MMR vaccine is a live vaccine and therefore is contraindicated during all stages of pregnancy. Depending on the immune status of the pregnant woman, she may be offered human normal immunoglobulin. Individuals with a clear history of MMR vaccination are not offered a further vaccination. An exposure is considered significant when a susceptible individual is within the same room as an infected individual for 15 min or more. There may be considerations around the size of the room and ventilation, which will be part of the risk assessment conducted by the health protection team. The measles vaccine is not recommended for babies under 6 months. A newborn infant may have some protection from maternal antibodies, depending on the immune status of the mother. The infant may require human normal immunoglobulin (HNIG), should the mother be shown to be measles antibody negative. Measles circulated in the UK until the vaccine was introduced in 1968, so most older individuals are immune from natural exposure. Vaccination of these individuals is only considered if they are vulnerable or at risk of transmission (for example health care workers).
4. Which of the following do not require exclusion from work or school? a. Case diagnosed with measles who works in a nursery b. Asymptomatic vaccinated sibling of an infected case c. A healthcare worker working within a pediatric unit who has been exposed to a confirmed case d. An infected case working as an IT specialist in an open-plan office	All infected children and adults are excluded from school or work respectively for five days from the onset of the rash. Healthcare workers potentially exposed to a case are followed up by the health protection team for evidence of protection. For health care workers in high-risk settings, a lower level of exposure may be considered significant. Healthcare workers shown to have evidence of protection can continue to work but are requested to be vigilant for the following 18 days for symptoms of fever. Those who do not have satisfactory evidence of protection should be excluded from work until they are measles IgG positive.
5. A local health protection team has just been informed of a second confirmed case of measles within a nursery setting within a five-day period. Which of these actions should the team take? a. Convene an outbreak control team to discuss any possible public health actions that need to be taken b. Determine whether any vulnerable individuals have been exposed to the case c. Arrange for vaccination for all susceptible contacts who are not pregnant or immunocompromised within three days d. Request that the nursery send a letter to parents informing of the cases within the nursery and current public health action e. All of the above	As there are two cases that may be linked, an outbreak control team should be convened to discuss any further public health actions. All vulnerable individuals such as unvaccinated children, pregnant staff members, and immunocompromised children or adults should be identified to determine whether any further actions such as MMR vaccination or HNIG (immunoglobulin) should be prescribed. Appropriate communication to parents as well as primary and secondary care colleagues should be considered for raising awareness as well as managing parental concerns.

FOUR CASES OF MERS-COV

Jake Dunning[1] and Joseph Fitchett[2]

[1]National Infection Service, Public Health England, UK.
[2] Harvard T.H. Chan School of Public Health, USA.

A previously healthy 49-year-old man, a resident of Qatar, travelled to Saudi Arabia in August 2012. While there, he and his traveling companions developed rhinorrhoea and fever, which resolved spontaneously. On September 3, he sought medical care in Qatar for a cough and myalgia, for which he was prescribed empirical antibiotics. On September 8, he was admitted to the hospital in Qatar with a fever and hypoxia. A chest radiograph showed bilateral lower zone consolidation, and empirical treatment of community-acquired pneumonia was commenced (ceftriaxone, azithromycin, and oseltamivir). Within 48 hours, the patient had deteriorated and required invasive mechanical ventilation. He was transferred subsequently by air ambulance to London to receive specialist intensive care (Figure 14.1).

On arrival in London, the patient was severely hypoxic and required 100% oxygen with optimized pressure ventilation and inotropic support. Laboratory findings at the time of transfer revealed abnormally elevated C-reactive protein (CRP), creatinine, and procalcitonin levels. Antimicrobial cover was adjusted to meropenem, clarithromy-cin, and teicoplanin, with subsequent addition of colistin and liposomal amphotericin. On September 14, his renal function deteriorated and renal replacement therapy was commenced. By this stage, viral pneumonia was considered the most likely clinical diagnosis in this patient, despite the failure to detect several common respiratory viruses by laboratory testing. On September 20, the patient remained in a critical condition with acute respiratory distress syndrome (ARDS) and refractory hypoxaemia; extracorporeal membrane oxygenation (ECMO) support was initiated. On the same day, a report issued by ProMED-mail highlighted the detection of a novel coronavirus (subsequently named Middle East respiratory syndrome coronavirus, or MERS-CoV) in a patient from Saudi Arabia who had died in June 2012. On September 21, coronavirus RNA was detected in respiratory tract samples from the patient in London; sequencing performed the following day revealed that this was the same novel coronavirus identified in samples from the patient in Saudi Arabia. The patient continued on ECMO support for 231 days but developed pulmonary fibrosis as a result of MERS-CoV infection. He died on June 28, 2013.

IDENTIFICATION OF SUBSEQUENT CASES

A second unrelated UK case was identified on February 8, 2013 (Figure 14.2, case 1). The male patient was admitted to intensive care unit in the West Midlands and gave a history of recent travel to Saudi Arabia and Pakistan, 10 days prior to symptom onset. He had spent time in Mecca and Medina on a pilgrimage and returned to the UK on January 28, 2013. Prior to his return, he developed a fever and upper respiratory tract symptoms

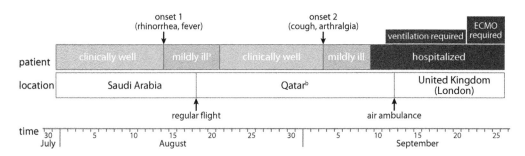

ECMO: Extracorporeal Membrane Oxygenation
ᵃ According to relatives of the patient.
ᵇ Contact with farm animals during stay (camels, sheep).

Figure 14.1. **Timeline of exposure and initial period of hospitalization.**

on January 24. By January 30, his symptoms had progressed and, following a visit to his general practitioner, he was admitted to hospital on January 31. The patient deteriorated and required invasive mechanical ventilation. The patient showed no sign of improvement and on February 5 was transferred to a tertiary centre in Manchester for ECMO treatment of ARDS. Influenza A(H1N1)pdm09 was detected in respiratory tract samples on February 1. Despite intensive care, administration of neuraminidase inhibitors to treat influenza, and empirical antibiotics, he failed to improve. Therefore, MERS-CoV infection was considered. On February 7, MERS-CoV was detected in a throat swab by reverse transcription polymerase chain reaction (RT-PCR), and confirmed by the reference laboratory on February 8. ECMO continued for 51 days, when the patient died of MERS-CoV pneumonitis with ARDS.

A total of 137 family, community, and healthcare-associated contacts were monitored closely for 10 days following their most recent potential exposure to the index case. Two secondary cases were confirmed by laboratory testing, but a further 135 contacts did not develop MERS-CoV infection. One of the secondary cases was a household contact with known immunosuppression (Figure 14.2, case 2), who went on to develop MERS-CoV pneumonitis resulting in death. The other secondary case (Figure 14.2, case 3) was

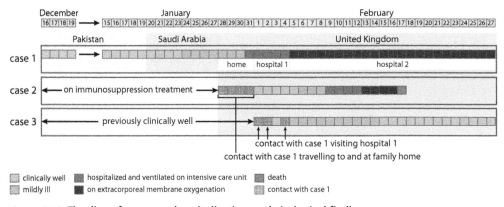

Figure 14.2. **Timeline of exposure, hospitalization, and virological findings.**

a relative who made three visits to the intensive care unit and experienced a mild, self-limiting respiratory illness. At the time, these cases provided the first examples of person-to-person transmission of MERS-CoV globally.

CLINICAL MANAGEMENT

As of 2016, there are no specific antiviral agents with proven efficacy for the treatment of MERS-CoV infection. Clinical management involves supportive measures such as rehydration, respiratory support, and treatment of any bacterial or viral co-infection. Experimental agents that have been used in cases of MERS-CoV including ribavirin and interferons have not demonstrated a consistent effect and data from clinical trials are lacking.

PREVENTION OF FURTHER CASES

Little was known about the virology or transmission dynamics of the novel coronavirus when the two UK index cases occurred. Seasonal coronaviruses circulate in the UK, but there was initial concern that the novel coronavirus, which is from the same family of beta coronaviruses as severe acute respiratory syndrome (SARS-CoV), might have the potential to spread easily between humans and cause significant morbidity and mortality. Therefore, a precautionary approach was taken initially to prevent and identify cases of related human-to-human transmission. The initial response was informed in part by measures taken globally during the SARS-CoV outbreak of 2003–2004.

When the hospitalized patients were found to be infected with MERS-CoV, they were isolated in negative-pressure single rooms (including when admitted to intensive care units). Personal protective equipment (PPE) for essential hospital staff included full-length fluid-repellent disposable gowns, disposable gloves, eye protection, and fit-tested high-filtration (FFP3) face masks. Additionally, strict hand hygiene was enforced (soap and water or alcohol gel, if hands were visibly clean). Visiting was discouraged, but was unavoidable for compassionate reasons. Visitors followed the same PPE and hand-hygiene protocol. Records were kept of all those who entered the side rooms.

Potential aerosol-generating procedures (AGP) were assessed for risk and only performed when essential. The number of staff present was restricted and access to the rooms was controlled during the procedures. Following any potential AGP, the negative-pressure rooms were to be left for 20 min and then cleaned according to protocol; these procedures had to be balanced with the need to provide ongoing care for the critically ill patients. Medical equipment remained in the room at all times and single-use equipment was preferred where possible. When reusable equipment had to be removed from a patient room for specific cleaning, it was contained prior to removal and then cleaned according to manufacturers' instructions. Mechanical ventilators were fitted with high-efficiency filters and closed-system suctioning was used.

Specific data on the effectiveness of disinfectants were not available for MERS-CoV at the time of the incidents. However, since all coronaviruses are lipid-enveloped RNA viruses, it was reasonable to apply disinfection policies for other enveloped RNA viruses to MERS-CoV. Thus, several existing hospital disinfectants and detergents could be used for cleaning environmental surfaces and equipment. Daily cleaning of the patient environment was performed, with staff wearing PPE and using dedicated or single-use cleaning materials. All waste (including excreta) was managed as potentially infectious clinical waste, double-bagged, and disposed of according to existing clinical waste protocols.

Fundamental to an effective public health response was the ability to identify and exclude MERS-CoV infection in suspected cases and contacts of the confirmed cases, using an appropriately sensitive and specific molecular test. Biological samples from confirmed cases were double-bagged with biohazard labelling and processed in laboratories at Containment Level 3. When samples needed to be transferred to other centres, enhanced Category B shipping measures were followed (Category A packaging and a Category B courier).

A national incident was declared and incident management was coordinated via an incident management team at the Health Protection Agency (HPA; succeeded by Public Health England). Advice and guidance was issued rapidly to healthcare professionals, the public, and travellers. The immediate, primary aim of the public health response was to identify and isolate confirmed cases, identify and test close contacts, and prevent onward transmission. Interim case definitions were developed to assist investigations. Close contacts included healthcare workers who attended confirmed patients without appropriate full PPE, household contacts, and any other individuals who had prolonged face-to-face contact with a symptomatic confirmed case. Contacts that developed respiratory illness within 10 days following most recent exposure were asked to self-isolate in their homes or were isolated in the hospital if admission was required. For the first UK case, 64 close contacts were identified (56 healthcare workers and 8 family members and friends). Thirteen healthcare-worker contacts developed respiratory symptoms and 10 submitted samples for respiratory virus detection by PCR; MERS-CoV was not detected in any of the contacts' samples.

A second and equally important priority was to raise awareness about the infection and how it might be acquired (especially by travellers to the Middle East), with the aim of preventing further individual infections and secondary transmission. All UK cases were reported to the World Health Organization (WHO) via the International Health Regulations (IHR) reporting mechanism. Following reporting of the UK cases and cases from other countries, the WHO IHR Emergency Committee met twice in July 2013 and ascertained that although MERS-CoV cases were of concern, the events did not constitute a Public Health Emergency of International Concern.

EPIDEMIOLOGY

Source

Coronaviruses infect and cause disease in a wide variety of animal species, including bats, birds, cattle, civet cats, dogs, mice, and pigs. Dromedary camels are believed to be the primary reservoir of MERS-CoV in terms of zoonotic transmission to humans, as suggested by phylogenetic analysis of MERS-CoV strains isolated from camels and infected humans. Coronaviruses that are similar, but not identical, to MERS-CoV have been detected in the faeces of bats in Asia, the Middle East, Europe, and Africa. To date, the complete MERS-CoV virus has not been detected in bat samples. However, one outbreak investigation study from Saudi Arabia reported that a 190-nucleotide sequence for a specific viral gene from a single bat sample was identical to the sequence for the same viral gene in a sample obtained from a related human index case. It is possible that bats are a zoonotic source of MERS-CoV or common ancestor coronaviruses, but bats are unlikely to transmit infection to humans directly.

Anti-MERS-CoV antibodies have been detected in sera obtained from the majority of adult camels sampled in studies across the Arabian Peninsula, Egypt, East African countries, and Nigeria. A lower seroprevalence has been observed in camels in Spain, the Canary Islands, Tunisia, and Kenya. Antibodies have not been detected in other domesticated animals (for example, goats, sheep, and cattle) from affected regions. Juvenile camels are less likely to have detectable antibodies but more likely to have detectable virus in nasal secretions or faeces as compared to adult camels. Some, but not all, of the community-acquired cases reported have been associated with camel husbandry or consumption of camel products, such as raw camel's milk, or both. Identical MERS-CoV isolates were obtained from an infected camel handler and camels with rhinorrhoea that the handler had been attending.

Transmission characteristics

Reports of suspected human-to-human transmission emerged following several hospital clusters in Saudi Arabia. Transmission was reported to have occurred in dialysis units, intensive care units, and hospital wards. There was wide variability in the numbers of individuals infected by an index case, raising the possibility of "super-spreading" events driving some of the transmission networks.

Limited human-to-human transmission has since been suggested by case clusters in several countries, including suspected nosocomial transmission clusters. Over half of the confirmed secondary cases reported to date have occurred in healthcare settings, often secondary to sub-optimal infection prevention and control measures. Sustained human-to-human transmission has not occurred in association with the majority of cases or clusters, but super-spreading events were observed following introduction of MERS-CoV to South Korea in 2015; this included secondary transmission to at least 70 people from a single patient treated in an emergency department. Similar super-spreading events were also reported during the 2003 SARS-CoV outbreak. MERS-CoV was introduced to South Korea originally by a traveller who had visited Saudi Arabia, leading to an outbreak that eventually involved 186 confirmed cases. Epidemiological investigations suggested that 83% of transmission events arose from multiple individual exposures to five patients with MERS-CoV pneumonia. Although secondary transmission to household contacts has been reported to have occurred in Saudi Arabia (approximately 5% secondary transmission rate), it can be difficult to exclude exposure to a common, nonhuman source of infection in community settings.

MERS-CoV can be detected in samples of the respiratory tract from laboratory confirmed cases, often with high viral loads. It is possible for virus to be detected in lower respiratory tract samples, but not in upper respiratory tract samples obtained at the same time, particularly when patients have established viral pneumonia and are critically ill. Precisely how human-to-human transmission can occur in hospital settings is unclear. It is possible that such transmission occurs via transfer of virus in respiratory droplets or virus on contaminated fomites to an individual's mucosal surfaces, and through inhalation of aerosolized virus. These potential routes have been suggested, in part, by epidemiological investigation of specific activities undertaken by healthcare workers caring for patients with MERS-CoV and SARS-CoV infections. The relative importance of each of these modes remains poorly understood, limiting our estimations of the likelihood of onward transmission in the wider population.

MERS-CoV RNA can be detected in respiratory tract secretions for several weeks following onset of symptoms, although it appears that levels of virus RNA typically decrease over time. In a study from Saudi Arabia, 13 patients and 13 infected contacts were sampled until a negative PCR result was obtained for lower- or upper-respiratory tract samples; at day 12, 30% of contacts and 76% of cases still had detectable viral RNA. MERS-CoV RNA can also be detected in stool and urine samples, albeit at lower frequencies and with lower amounts of virus compared to respiratory tract samples. MERS-CoV RNA can also be detected infrequently in blood samples from severe cases. Prolonged detection of MERS-CoV RNA for six weeks was detected in an asymptomatic healthcare worker who had cared for MERS patient in Saudi Arabia. However, it is unclear whether the continuing detection of viral RNA represented prolonged shedding of infectious virus, nor whether an asymptomatic individual poses a significant infection risk to others.

In a hospital outbreak in Saudi Arabia, the median incubation period for MERS-CoV was estimated to be 5.2 days, with over 95% of cases progressing to symptom onset within 12.4 days following infection. In the South Korean 2015 outbreak, one estimate of the mean incubation period was 6.3 days (95th percentile is 12.1 days). These incubation periods are notably shorter than those described during the SARS-CoV outbreak. However, several MERS cases were reported to have a more prolonged incubation period, up to 9–12 days, emphasising the difficulties faced by outbreak investigation teams in gathering critical information.

Although MERS-CoV RNA can be detected in air samples in camel barns and patient environments, it is unclear whether aerosolized virus is infectious and can lead to secondary infections via the airborne route. However, most public health agencies recommend measures to prevent possible airborne transmission when aerosol-generating procedures are performed, as a precautionary measure. Environmental sampling data from South Korea suggest that MERS-CoV can be detected (detection of viral RNA and potentially infectious virus) on hard surfaces, soft furnishings, medical equipment, air vents, and other objects in the rooms of affected patients. Additionally, virus has been detected in adjoining anterooms. Such environmental contamination appears to occur despite reported daily cleaning of the patient environment.

Survival in the environment

For influenza A(H1N1)pdm09 virus, a cool and dry environment is favourable for transmission, with warm and humid environments less favourable. This variability in environmental stability accounts for the differences in transmission dynamics with this particular pathogen. By comparison, MERS-CoV remains viable in the environment for a longer period than influenza A(H1N1) virus. In one study, influenza A(H1N1) virus was no longer viable after 4 hours under test conditions, whereas MERS-CoV was detected at 8, 24, and 48 hours, depending on environmental conditions. Furthermore, experimental aerosolization of MERS-CoV did not appear to decrease stability. Collectively, these findings increase the potential for contact and fomite transmission, and they are comparable with the environmental stability characteristics displayed by SARS-CoV. The ability for MERS-CoV to remain viable at high temperatures is an important character to have identified, particularly given its presence in the Middle East. Additionally, results of patient environment sampling studies are now beginning to emerge. In a study conducted during the South Korean outbreak of 2015, environmental sampling was conducted while four patients were hospitalized in isolation rooms, and also following discharge (up to 120 hours postdischarge). MERS-CoV could be isolated from swabs of multiple environmental surfaces, even after patient samples had become negative. There was evidence of significant virus contamination in the areas around patients' beds, as well as in ventilation grills and adjoining anterooms.

National and global epidemiology

Following the second UK case and associated cluster in 2013, no further cases of MERS-CoV infection have been detected in the UK. This is despite an established surveillance system for MERS in the UK and laboratory testing of patients in whom MERS is suspected. In particular, MERS-CoV has not been detected in those returning from Hajj and Umrah pilgrimages and who have compatible symptoms; by contrast, other respiratory viruses have been detected frequently in such individuals, including influenza virus and human rhinovirus.

Between September 2012 and May 2016, there were 1733 laboratory-confirmed cases of infection with MERS-CoV from 27 countries in total, including at least 628 related deaths (case fatality rate 36%). Most cases have occurred in the Middle East, predominantly in the Kingdom of Saudi Arabia. Other countries have experienced imported cases and secondary transmission of infection. Analysis of the first 133 cases from 9 countries suggested that 26% cases involved possible nosocomial transmission, including infection of 15 healthcare workers. In 2015, South Korea experienced the largest MERS-CoV outbreak outside the Middle East. Following introduction of infection by an individual who had visited several countries in the Middle East, South Korea experienced 186 cases and 38 deaths. Of those, 64 cases (34%) were believed to be nosocomial infections, including infection of 39 healthcare workers.

Although sporadic cases linked to acquisition from infected camels continue to be reported in the Middle East, most MERS cases to date have been associated with nosocomial outbreaks. MERS-CoV has not demonstrated pandemic potential yet, although this does not exclude the possibility of a pandemic in the future, particularly if the virus evolves to favour more efficient human-to-human transmission. Early estimates of the basic reproduction number (R_0) in 2013 suggested that the R_0 was between 0.60 and 0.69, which is below the prepandemic R_0 of SARS-CoV.

BIOLOGY

Coronaviruses, first discovered in the 1960s, are large enveloped, single-stranded, positive-sense RNA viruses classified into three genera (alpha-, beta- and gamma-coronavirus). Including MERS-CoV, there are six known human respiratory coronaviruses. Coronavirus 229E and OC43 are classically associated with the common cold or influenza-like illness. Strain NL63 is associated with upper respiratory tract infections, and HKU1 is more commonly associated with pneumonia. SARS-CoV, which caused 8096 cases and 774 deaths in 37 countries between 2002 and 2003, has not been seen globally since 2004. Coronavirus diversity arises from an error-prone RNA-dependent RNA polymerase, an unusually large genome for an RNA virus, and a high frequency of RNA recombination events.

The genomic sequence of MERS-CoV was published in 2012, shortly after the first reported infections in humans. The host-cell receptor, dipeptidyl peptidase 4 (DPP4; also known as CD26), was identified soon after. MERS-CoV belongs to lineage C beta-coronaviruses, and the closest known relatives are the bat coronaviruses HKU4 and HKU5. There is considerable phylogenetic diversity within MERS-CoV sequences, with the identification of two distinct clades (A and B). The rate of evolution of MERS-CoV was estimated at 1.6×10^{-3} substitutions per site per year, and the time of the emergence of the common ancestor of five MERS-CoV genomes was estimated to be mid-2011.

DISEASE

MERS-CoV is capable of causing severe respiratory failure, and most reported primary infection cases tend to be adults with severe viral pneumonia, acute respiratory distress syndrome, or both. Atypical presentations may occur if the infected individual is immuno-compromised; for example, an absence of respiratory tract symptoms despite radiological evidence of pneumonia. Individuals may deteriorate rapidly, requiring intubation within one week of onset of symptoms, which is five days earlier than observed in patients with SARS-CoV. Acute kidney injury is a recognized complication of infection. Gastrointestinal symptoms, such as diarrhoea and abdominal pain, can also occur, along with variable nonpulmonary organ dysfunction.

Most, but not all, individuals with MERS-CoV infection have evidence of fever on admission (Table 14.1), often associated with chills and myalgia. Hospitalized patients with lower respiratory tract involvement tend to have cough and shortness of breath at presentation, or they go on to develop these symptoms soon after initial presentation. By contrast, few hospitalized patients report upper respiratory tract symptoms, such as rhinorrhea and sore throat.

The reported clinical features, combined with reports of lymphopenia and other abnormalities of blood cell parameters, are reminiscent of SARS-CoV infection. In contrast to SARS-CoV infection, there has been a high rate of comorbidities in patients with MERS-CoV infection in Saudi Arabia (particularly diabetes mellitus and chronic renal failure). With

Table 14.1 **Symptoms and the proportion of patients that experience them**

Symptom	Proportion (%) of patients ($n = 47$)
Fever	98
Chills or rigors	87
Cough	83
Dyspnea	72
Myalgia	32
Diarrhea	26
Nausea	21
Vomiting	21
Sore throat	21
Abdominal pain	17
Hemoptysis	17
Chest pain	15
Headache	13
Rhinitis	4

relatively small numbers of infections, it is unclear whether there is a specific predisposition to infection in individuals with pre-existing medical conditions, or whether the current outbreak data reflects the background rates of comorbidities in the exposed population.

In addition to severe respiratory failure, it is important to note the case reports of less severe disease in secondary cases, including individuals within family and hospital clusters. Mild illness not requiring hospitalization has been described, including an infected family contact within the UK cluster of 2013. In a contact screening study of 3000 contacts in Saudi Arabia, seven healthcare workers were found to be infected; two of the contacts had asymptomatic infection and five had only mild upper respiratory tract symptoms. The spectrum of illness is further highlighted by case reports of a family cluster from Saudi Arabia, where the index case and a previously healthy household contact contracted severe respiratory failure and died, and two further cases experienced mild respiratory illness. Asymptomatic infection of individuals who had contact with infected camels has been detected through surveillance programs.

Although children can be infected, in a small case series of 11 children, only two of the children were symptomatic; this suggests that asymptomatic infection might be more common in paediatric populations. Limited case reports of infection in pregnancy suggest that MERS-CoV infection may be associated with adverse outcomes of pregnancy, including severe maternal disease and stillbirth. However, healthy and uninfected babies have been born to infected mothers who had severe MERS illness. Although patients with underlying chronic health conditions, such as diabetes mellitus, chronic kidney disease, cardiovascular disease, and immunosuppression, are well represented in hospital case series from the Middle East, it is not clear whether specific medical conditions increase the risk of infection with MERS-CoV. However, underlying cardiac disease and cancer were identified as independent risk factors for death in one large case series from Saudi Arabia. In an analysis of 159 MERS patients with known outcomes in the South Korean outbreak, the overall case fatality rate (CFR) was considerably lower than the CFR for all known MERS patients reported by Saudi Arabia at the time of the analysis (22% versus 44%, respectively). This discrepancy may be explained by the relatively low prevalence of underlying chronic health conditions in the Korean MERS patient population; subanalyses demonstrated a CFR of 14% in the 134 patients who did not have comorbidities, compared to a CFR of 64% in the 25 patients who did have comorbidities. A similar CFR of 60% was reported for a MERS population with comorbidities in Saudi Arabia.

Public Health England Case Definition for MERS-CoV Infection

Public Health England published updated case definition criteria for a possible case of MERS-CoV infection: Any person with severe acute respiratory infection requiring admission to hospital who fits all three of the following criteria:

- has symptoms of fever ($\geq 38°C$) or history of fever and cough
- shows evidence of pulmonary parenchymal disease (for example, clinical or radiological evidence of pneumonia or acute respiratory distress syndrome (ARDS)
- has symptoms that cannot be explained by any other infection or aetiology
  In addition, patients must meet at least one of the following criteria:
- have a history of travel to, or residence in, an area where infection with MERS-CoV could have been acquired in the 14 days before symptom onset
- have a close contact during the 14 days before onset of illness with a confirmed case of MERS-CoV infection while the case was symptomatic

- be a healthcare worker based in an intensive care unit (ICU) caring for patients with severe acute respiratory infection, regardless of history of travel or use of PPE
- be part of a cluster of two or more epidemiologically linked cases within a two-week period requiring ICU admission, regardless of history of travel

Note that the Public Health England case definition focuses on detecting infections causing viral pneumonia leading to hospitalization, rather than less common cases of mild illness that are unlikely to require hospitalization.

PATHOLOGY

Knowledge of the pathogenesis of MERS-CoV infection and the illness it causes is incomplete. Although the host receptor dipeptidyl peptidase 4 can be found on many different types of human cell, expression within the respiratory tract is confined to the lower respiratory tract. This might explain why human-to-human transmission has been relatively limited in community settings. It is possible that other, unidentified receptors exist.

Human autopsy data are limited. In one report, diffuse alveolar damage was evident, with abundant viral antigen detected in pneumocytes and lung epithelial cells but not in the kidneys and other organs (despite the patient having acute kidney injury). Nonhuman primate models of infection have revealed histological changes consistent with interstitial or bronchointerstitial pneumonia, with viral RNA detected throughout the respiratory tract and also conjunctival and tonsillar tissue, and sometimes other organs. There are *in vitro* animal model and human study data to suggest that the antiviral innate immune response to MERS-CoV may be poorly coordinated in the most severe cases, resulting in prolonged and exuberant inflammation, with ineffective activation of the adaptive immune response, and sometimes a failure of viral clearance despite the emergence of specific antibodies.

QUESTIONS

1. True or false: the following are used to prevent secondary transmission.
   a. FFP1 masks
   b. Restriction of staff numbers during endoscopy
   c. Eye protection
   d. Open-system suctioning
   e. Regular environmental decontamination

2. Which of the following is believed to be the primary zoonotic reservoir of MERS-CoV?
   a. Civets
   b. Cats
   c. Cattle
   d. Camels
   e. Bats

3. Which of the following are true concerning human–human transmission of MERS-CoV infections?
   a. Sexual contact is frequently associated with transmission.
   b. Most cases have been associated with nosocomial transmission.
   c. Virus can be detected in urine.
   d. Virus particles can be detected in air.
   e. Super-spreading events have not occurred.

4. Which of the following clinical features is not associated with MERS-CoV infection?
   a. Rigors
   b. Lymphopenia
   c. Dyspnea
   d. Acute kidney injury
   e. Microcephaly

5. Which of the following is not true for MERS-CoV?
   a. It is an error-prone DNA dependent polymerase.
   b. Its host cell receptor is CD26.
   c. It is a beta coronavirus.
   d. Expression of CD26 occurs in the lower respiratory tract.
   e. There is a high frequency of recombination events.

GUIDELINES

UK Guidelines

1. Public Health England (2016). Middle East respiratory syndrome coronavirus (MERS-CoV): clinical management and guidance. www.gov.uk/government/collections/middle-east-respiratory-syndrome-coronavirus-mers-cov-clinical-management-and-guidance

2. Health Protection Scotland (2015). Respiratory Infections: MERS-CoV. www.hps.scot.nhs.uk/resp/coronavirus.aspx

3. Public Health Wales (2017). Middle East Respiratory Syndrome Coronavirus (MERS-CoV). www.wales.nhs.uk/sitesplus/888/page/73600

International Guidelines

4. World Health Organization (2017). Middle East respiratory syndrome coronavirus (MERS-CoV). www.who.int/emergencies/mers-cov/en/

5. European Centre for Disease Prevention and Control (2017). Coronavirus Infections. ecdc.europa.eu/en/healthtopics/coronavirus-infections/Pages/index.aspx

6. Ministry of Health, Kingdom of Saudi Arabia (2017). MERS Health Staff Guidelines. www.moh.gov.sa/en/CCC/Pages/default.aspx

7. US Centers for Disease Control and Prevention (2016). Middle East Respiratory Syndrome (MERS). www.cdc.gov/coronavirus/mers/

REFERENCES

1. Zaki AM (2012) Novel coronavirus—Saudi Arabia: human isolate. *ProMED Mail* e143-144.

2. Bermingham A et al. (2012) Severe respiratory illness caused by a novel coronavirus, in a patient transferred to the United Kingdom from the Middle East, September 2012. *Euro Surveill* **17**:20290.

3. HPA UK Novel Coronavirus Investigation Team (2013) Evidence of person-to-person transmission within a family cluster of novel coronavirus infections, United Kingdom, February 2013. *Euro Surveill* **18**:20427.

4. Pebody RG et al. (2012) The United Kingdom public health response to an imported laboratory confirmed case of a novel coronavirus in September 2012. *Euro Surveill* **17**:20292.

5. Perlman S & Netland J (2009) Coronaviruses post-SARS: update on replication and pathogenesis. *Nat Rev Microbiol* **7**:439–450.

6. Haagmans BL et al. (2014) Middle East respiratory syndrome coronavirus in dromedary camels: an outbreak investigation. *Lancet Infect Dis* **14**:140–145.

7. Drosten C et al. (2015) An observational, laboratory-based study of outbreaks of middle East respiratory syndrome coronavirus in Jeddah and Riyadh, Kingdom of Saudi Arabia, 2014. *Clin Infect Dis* **60**:369–377.

8. Assiri A et al. (2013) Hospital outbreak of Middle East respiratory syndrome coronavirus. *N Engl J Med* **369**:407–416.

9. Guery B et al. (2013) Clinical features and viral diagnosis of two cases of infection with Middle East Respiratory Syndrome coronavirus: a report of nosocomial transmission. *Lancet* **381**:2265–2272.

10. Park HY et al. (2015) Epidemiological investigation of MERS-CoV spread in a single hospital in South Korea, May to June 2015. *Euro Surveill* **20**:1-6.

11. Cowling BJ et al. (2015) Preliminary epidemiological assessment of MERS-CoV outbreak in South Korea, May to June 2015. *Euro Surveill* **20**:7-13.

12. Chowell G et al. (2015) Transmission characteristics of MERS and SARS in the healthcare setting: a comparative study. *BMC Med* **13**:210.

13. Korea Centers for Disease Control and Prevention (2015) Middle East respiratory syndrome coronavirus outbreak in the Republic of Korea, 2015. *Osong Public Health Res Perspect* **6**:269–278.

14. Drosten C et al. (2014) Transmission of MERS-coronavirus in household contacts. *N Engl J Med* **371**:828–835.

15. Memish ZA, Assiri AM & Al-Tawfiq JA (2014) Middle East respiratory syndrome coronavirus (MERS-CoV) viral shedding in the respiratory tract: an observational analysis with infection control implications. *Int J Infect Dis* **29**:307–308.

16. van Doremalen N, Bushmaker T & Munster VJ (2013) Stability of Middle East respiratory syndrome coronavirus (MERS-CoV) under different environmental conditions. *Euro Surveill* **18**:1263–1264.

17. Breban R, Riou J & Fontanet A (2013) Interhuman transmissibility of Middle East respiratory syndrome coronavirus: estimation of pandemic risk. *Lancet* **382**:694–699.

18. Raj VS et al. (2013) Dipeptidyl peptidase 4 is a functional receptor for the emerging human coronavirus-EMC. *Nature* **495**:251–254.

19. Cotten M et al. (2013) Transmission and evolution of the Middle East respiratory syndrome coronavirus in Saudi Arabia: a descriptive genomic study. *Lancet* **382**:1993–2002.

20. Assiri A et al. (2013) Epidemiological, demographic, and clinical characteristics of 47 cases of Middle East respiratory syndrome coronavirus disease from Saudi Arabia: a descriptive study. *Lancet Infect Dis* **13**:752–761.

21. Memish ZA, Zumla AI & Assiri A (2013) Middle East respiratory syndrome coronavirus infections in health care workers. *N Engl J Med* **369**:884–886.

22. Public Health England (2015) MERS-CoV: public health investigation and management of possible cases. (www.gov.uk/government/publications/mers-cov-public-health-investigation-and-management-of-possible-cases) [cited 2016 21 May].

ANSWERS

MCQ	Feedback
1. True or false: the following are used to prevent secondary transmission	
a. FFP1 masks	a. False
b. Restriction of staff numbers during endoscopy	b. True
c. Eye protection	c. True
d. Open-system suctioning	d. False
e. Regular environmental decontamination	e. True
	Understanding of the precise routes of transmission of MERS-CoV is incomplete. However, standard, contact, and airborne precautions are recommended to prevent contact, airborne, and droplet transmission. FFP3 masks are recommended for all staff caring for patients, regardless of whether a potential aerosol-generating procedure is being performed. To reduce unnecessary potential exposures, only staff essential to care should attend the patient, including those performing medical procedures such as endoscopy. Because exposure of conjunctiva to MERS-CoV could result in infection, eye protection is recommended when attending a patient. Aerosol-generating procedures (AGPs) are believed to be associated with higher risk of transmission of MERS-CoV and should be avoided when possible. When AGPs cannot be avoided, risk to operators should be minimized, as, for example, by using closed rather than open suctioning. The optimal environmental cleaning schedule is unknown, but environmental sampling studies suggest that widespread contamination of the patient environment can occur; regular decontamination using appropriate agents is recommended.
2. Which of the following is believed to be the primary zoonotic reservoir of MERS-CoV?	The host range of coronaviruses is broad. However, for MERS-CoV, epidemiological and sampling studies (viral RNA and antibodies) suggest that camels are an important zoonotic reservoir for subsequent human infections, either through close contact with camels and their environments, or possibly by consumption or handling of raw camel products (such as unpasteurized camel milk or camel urine). It is possible that MERS-CoV or its forerunner virus passed into camels from bats, but to date only partial sequences from the MERS-CoV genome have been detected in bat fecal samples. Civets were proposed as a potential reservoir of SARS-CoV during the 2002–2003 outbreak in China, following detection of SARS-like coronaviruses in samples obtained from civets.
a. Civets	
b. Cats	
c. Cattle	
d. Camels	
e. Bats	
3. Which of the following are true concerning human–human transmission of MERS-CoV infections?	
a. Sexual contact is frequently associated with transmission.	a. False
b. Most cases have been associated with nosocomial transmission.	b. True
c. Virus can be detected in urine.	c. True
d. Virus particles can be detected in air.	d. True
e. Super-spreading events have not occurred.	e. False
	Although close contact with an infected case is a risk factor for infection, sexual transmission has not been demonstrated. Sporadic cases and small community clusters continue to occur, but most infections have been associated with hospital outbreaks. There is no evidence of sustained human-to-human transmission in the community. MERS-CoV can be detected in a variety of body fluids and excreta, although levels of virus tend to be greatest in respiratory tract secretions. MERS-CoV RNA and virus particles have been identified in air samples and swabs obtained from healthcare environments associated with confirmed cases. There are indications from these studies that infectious virus may be present, although their role in nosocomial transmission has not been determined. Super-spreading events have been described for MERS-CoV outbreaks, most notably in the 2015 outbreak in South Korea.

MCQ	Feedback
4. Which of the following clinical features is not associated with MERS-CoV infection? a. Rigors b. Lymphopenia c. Dyspnea d. Acute kidney injury e. Microcephaly	Most hospitalized patients with MERS-CoV infection experience fever, often accompanied by rigors or chills. Like many acute severe viral infections, lymphopenia is a common symptom at presentation. Lymphocytosis has also been described, but is unusual. Dyspnea is also common symptom in hospitalized patients, and is more likely in those with lower respiratory tract involvement (pneumonia, ARDS, or both). Acute kidney injury is a common feature in critically ill MERS patients. Microcephaly has not been described, but stillbirth and premature birth have been associated with infection during pregnancy.
5. Which of the following is not true for MERS-CoV? a. It is an error-prone DNA dependent polymerase. b. Its host cell receptor is CD26. c. It is a beta coronavirus. d. Expression of CD26 occurs in the lower respiratory tract. e. There is a high frequency of recombination events.	MERS-CoV is a beta coronavirus with an error-prone RNA-dependent polymerase. This, together with an unusually large viral genome and a high frequency of genetic recombination events, helps explain coronavirus diversity. CD26 (also known as DPP4) is a recognized, functional human receptor for MERS-CoV. Experiments have shown that MERS-CoV infects lower respiratory cell lines much more easily than upper respiratory tract cell lines. The tissue tropism of MERS-CoV appears to be particularly broad, compared to other coronaviruses that infect humans.

A CASE OF HOSPITAL-ACQUIRED MRSA

Wael Elamin[1] and John Holton[2,3]

[1]Mid Essex Hospital NHS Trust & Queen Mary University of London.
[2]Department of Natural Sciences, School of Science & Technology, University of Middlesex.
[3]National Mycobacterial Reference Service - South, Public Health England.

A 42-year-old female presented to the accident and emergency department with a left-sided weakness and was admitted to the hospital. A computerized tomography (CT) scan of her brain showed a space-occupying lesion. Five days into her admission, the patient had a brain biopsy of her space-occupying lesion. Her condition deteriorated, and she required a decompressive craniotomy and admission to the intensive therapy unit (ITU), where she was intubated and required mechanical ventilation and central venous access. In order to monitor her intracranial pressure, an intracranial pressure probe (ICP) was inserted in theatre. Her methicillin-resistant *Staphylococcus aureus* (MRSA) triple-site screening swabs on admission to the ITU were negative. The patient required further surgical interventions, and a ventriculoperitoneal (VP) shunt was inserted. She remained in the ITU for three weeks and was later discharged to the general ward. On admission to the ward, a repeat MRSA triple-site screening swab was reported as being positive. Then, 48 hours later, the patient spiked a temperature of 39°C and further dropped her Glasgow coma scale (GCS). The doctor was asked to review her. Her examination was unremarkable for the cardiovascular and respiratory system, but he noted small blistering on the right forearm at the peripheral vascular access site. Blood cultures were taken from the patient and she was started on the vancomycin for a possible cellulitis caused by MRSA. The next morning, the microbiologist called, informing the clinical team that the patient's blood culture was growing Gram-positive cocci in clusters in both the aerobic and anaerobic bottles, probably indicative of MRSA. The organism was fully identified 24 hours later as MRSA.

INVESTIGATION OF THE CASE

MRSA bloodstream infections are considered preventable and avoidable healthcare-associated infections. Legislation in several developed countries (the UK included) has set a zero tolerance target for MRSA bloodstream infections. In the UK post-infection review (PIR) of all cases is mandatory, with the aim of identifying factors and learning from the events. Toolkits that aid in performing the review are readily available online.

The PIR is undertaken by data collected in relation to a case of MRSA from a questionnaire. The data collected include a description of the incident and associated clinical factors and outcome; the location where the incident took place; the patient demographics, risk factors, comorbidities, involvement of any medical devices, and a description of skin integrity; the results of screening and suppression regimen; antibiotic use; results of any relevant audits and hand hygiene compliance; organizational issues such as staff-to-patient ratio;

information given to the patient or family and any outcomes from the investigation. The incident is summarized and this latter sent to Public Health England (PHE) via the data capture system.

The PIR replaces the Root Cause Analysis previously used in the United Kingdom (UK). The current Guidelines for the PIR have been updated (April 2014) to include an arbitration process if there is a difficulty in assigning the case to a specific organization. This change reflects the fact that some patients move from one healthcare practitioner to another in a shorttime scale. The arbitration process is handled by the regional medical and nursing directors. Further amendments have been to increase the time for completion of the PIR to 14 days and add the option to assign the case to a third party, when it is not possible to know where the infection was acquired or if the case could not be assigned to a specific organization. The assignment to a third party reflects the institutional complexity of some cases.

A PIR reflects the zero tolerance approach to MRSA infections and ensures a consistent approach to dealing with MRSA bloodstream infections.

CLINICAL MANAGEMENT

Screening for MRSA is usually performed by polymerase chain reaction (PCR) to detect the *mecA* gene. Alternatively, it is possible to detect the presence of MRSA by the use of oxacillin or cefoxitin (the latter being a more effective inducer of the *mecA* gene) as a disk or by incorporation into agar when incubated at 33°C. Additionally there are various commercially available chromogenic agars that will detect the presence of MRSA or a latex agglutination assay for Penicillin Binding Protein 2a (PBP2a).

On identification of the positive MRSA screening swabs, the patient was transferred to a side room, and decolonization was attempted as per the hospital's standard protocol with chlorhexidine washes and nasal mupirocin. Following the phone call from the microbiology laboratory, the antibiotics were changed to vancomycin (aiming for a trough level of 10–15 mg/L). The peripheral cannula was removed, and an echocardiogram was requested on the patient. As the patient had a drop in her Glasgow coma (GCS), a CT brain was requested, and this showed a possible collection, which was drained neurosurgically. Intrathecal vancomycin was also administered. The patient remained on antibiotics for a total of four weeks following drainage of the intracerebral collection.

PREVENTION OF FURTHER CASES

Infection prevention and control policies for the management of MRSA are universal in hospitals in the UK. Therefore, following the hospital policy, the patient was moved to a negative-pressure isolation room with a single bed to prevent onward transmission in the ward. A notice was placed in the room door to ensure all the staff were aware of the need to take isolation precautions. The door of the isolation room was kept closed at all times. Attendants having direct contact with the patient wore disposable gloves and a disposable apron. Hand hygiene was the single most important factor in reducing transmission, and handwashing before and after each patient contact was mandatory. The effectiveness of handwashing, using soap and water or alcohol-based hand rubs, has been shown in a landmark study.

The remaining patients who shared the bay with the index case were all screened for MRSA from three sites: the nose, groin, and throat. No further cases were identified. The index patient was commenced on chlorhexidine washes and mupirocin nasal ointment in an attempt to supress the colonization and reduce the bioload, thereby lowering the risk of the patient becoming infected and transmitting MRSA to others in the hospital.

Healthcare-acquired MRSA is potentially a preventable disease. Good infection prevention and control practices have been shown to reduce the risk of infections. The decline of MRSA bloodstream infections once legislation was introduced to record the rates of infection in the UK was an important component of monitoring the effectiveness of the preventative measures.

EPIDEMIOLOGY OF THE ORGANISM

MRSA was first described in the early 1960s, with outbreaks reported soon after its discovery. The organism is currently widespread in both healthcare (HA-MRSA) settings and the community (CA-MRSA) with five (of six) different clones predominating globally. The epidemiology is markedly different between HA-MRSA and CA-MRSA.

HA-MRSA was previously known as nosocomial MRSA and is defined as an MRSA infection occurring more than 48 hours following hospital admission, or within 12 months of exposure to healthcare facilities. Infections with MRSA in hospitals are usually associated with severe invasive disease, and the isolates tend to have multidrug resistance and usually carry the staphylococcal cassette chromosome *SCCmec* type 2 cassette. The prevalence of HA-MRSA ranges from <1% in Scandinavia to 40% in countries such as Japan and Israel. MRSA has been associated with most types of healthcare-acquired infections. This results in part from the organisms' ability to form biofilms in invasive devices.

Several risk factors are associated with HA-MRSA infections. These include previous MRSA colonization, prolonged hospital admissions, long-term facility stay, intensive care admission, haemodialysis, and antibiotic use. Patients with MRSA infections have higher mortality and morbidity rates and longer hospital stay in comparison to their counterparts with methicillin-sensitive *Staphylococcus aureus* (MSSA) infections.

CA-MRSA is defined as an MRSA infection in a patient who lacks specific risk factors for healthcare exposure. It is often associated with skin and soft tissue infections (SSTI). Groups with high-intensity physical contact are more at risk of infections. Strains of CA-MRSA are usually sensitive to non-beta-lactam classes of antibiotics. Most strains usually carry *SCCmec* type IV or V and frequently carry the Panton–Valentine leukocidin gene, which enhances its virulence.

There has been a gradual decrease in the rates of invasive HA-MRSA infections between 2005 and 2013 owing to introduction of screening and isolation.

Patients colonized with MRSA are at an increased risk of infections following invasive procedures.

The primary site of carriage for *S. aureus* is the anterior nares, although the organism can be isolated from other body sites, with some individuals only colonized in the throat. Most infants are colonized shortly after birth, usually with the same strains as their mothers. Carriage decreases with age from 40–60% at 2 months to 21–28% at 6 months. There is also a marked change of carriage between the ages of 10 and 20 years, with those under 20 years having higher persistent carriage.

Overall, MRSA admission prevalence from the NOW study in the United Kingdom was found to be 1.5% (national one-week prevalence audit of MRSA Screening—2013).

There are three different patterns of carriage.

- Persistent carriage: Approximately 20% of healthy people are asymptomatic persistent nasal carriers of *S. aureus.* They usually carry the same strain for prolonged periods of time. Persistent carriers also have a higher load and higher risk of infection.
- Intermittent carriage: Approximately 30% of healthy people are intermittent carriers.
- Noncarriers: Approximately 50% of the population are classified as noncarriers.

Several risk factors have been associated with MRSA infections. These include:

- Antibiotic use: Selective pressure secondary to antibiotic use is believed to have led to MRSA, as there is a strong correlation between antibiotic use and colonization with MRSA, in particular quinolones and cephalosporins.
- HIV infection: HIV was found to be an important risk factor for colonization and infections with MRSA (OR 3.3).
- Haemodialysis: Patients undergoing dialysis via long-term central venous catheters (CVC) have an increased risk of infections from MRSA. Dialysis patients have a 100-fold increased risk over the general population.
- Long-term care facilities: Several studies have shown an increased prevalence of MRSA colonization among long-term care residents.

It is believed that the most common route of transmission of MRSA is by the contaminated hands of healthcare workers. Most studies that evaluated compliance with hand hygiene have shown improvements in the infection rates. A recent Cochrane systematic review failed to identify studies that assess the effectiveness of wearing gloves in reducing transmission. Transmission may also occur through contact with contaminated fomites and contaminated surfaces. Transmission may also occur via contaminated medical equipment such as stethoscopes.

BIOLOGY OF THE ORGANISM

Staphylococci are Gram-positive cocci that can be divided into strains that produce coagulase and those that don't. Coagulase-producing strains include *S. aureus* and the *S. hyicus-intermedius* group, although *S. aureus* is the most clinically important. The remainder of the staphylococci are coagulase negative. Staphylococci grow in clusters (appearing microscopically as bunches of grapes). There are over 40 different named coagulase negative staphylococci. Taxonomically, *Staphylococcus* spp. are closely related to *Gamella* and *Bacillus.* Staphylococci are facultative anaerobes, catalase positive, oxidase negative, resistant to bacitracin, sensitive to furazolidone, and can grow in the presence of a 6.5% concentration of sodium chloride. *S. aureus* produces a wide range of toxins: cytolysins (alpha, beta, gamma, delta toxins), enterotoxins (A–G), toxic shock syndrome toxin (TSS-1), leukocidin (Panton–Valentine leukocidin, PVL), and exfoliatin toxin.

Current global MRSA has developed from the acquisition of a staphylococcal chromosomal cassette *mec* (*SCCmec*) by *S. aureus* from another staphylococcal species, shortly after the introduction of methicillin, making MRSA resistant to all beta-lactam antibiotics. This occurred in the late 1950s to early 1960s, and since then MRSA has spread worldwide and acquired other resistance markers. The *mec* cassette is located on the chromosome of *S. aureus* and carries a *mecA* gene, which codes for a novel penicillin-binding protein (PBP2a) to which beta-lactam antibiotics bind poorly, yet the enzyme serves its physiological function in cell-wall synthesis for the organism.

Next to the *mecA* gene are two regulatory elements, *mecR1* and *mecI*. The former is a signal transduction protein, and the latter a regulator of transcription. They are similar to two regulatory proteins of a beta-lactamase system (blaZ) to such an extent that the regulatory elements of blaZ are able to induce *mecA*. *MecA* is not readily induced and an isolate may be *mecA* positive yet sensitive to methicillin because *mecI* tightly controls the expression of PBP2a, although mutations can occur in *mecI* leading to constitutive expression of PBP2a.

The *mec* cassette can be divided into 11 groups (I–XI) based upon the sequence of the *mecA* complex carrying the *mecA* gene, and the *ccr* region, which codes for recombination enzymes. These enzymes can excise and re-integrate the *mec* cassette into other locations. Type I and III are usually found in HA-MRSA, whereas IV and V are found in CA-MRSA. PVL is often associated with Type IV and strain EMRSa-15 is a Type IV SCCmec.

The *mecA* complex can itself be divided into six (A, B, C1, C2, D, E) subtypes based on the arrangement of its component genes and the ccr complex into eight types (1–8).

Studies on the environmental survival of MRSA showed that vinyl and plastic had the highest survival rate and wood had the lowest. Organic matter increased the survival and high relative humidity (~50%) reduced survival compared to low relative humidity (RH) of 16%. When tested in food, *S. aureus* survived well in high salt concentrations, was relatively heat tolerant, and more affected by weak acids than strong acids. MRSA and MSSA survived equally well in marine environments at lower temperatures (13°C versus 20°C) and survived better in marine environments compared to fresh water. In real-life environments (a dentist's chair) MRSA concentrations reduced by 90% within 15 min but were still detected after 4 months, implying a risk of spreading throughout the community.

DISEASE

S. aureus is a pyogenic infection (producing pus) that can produce a number of toxins. It can cause both superficial skin infections (folliculitis, boils, and impetigo), deep organ infections (osteomyelitis, septic arthritis, abscesses, endocarditis, and pneumonia), and septicaemia. The toxin-induced illnesses are toxic shock syndrome, scalded skin syndrome, and gastroenteritis.

The coagulase-negative staphylococci are opportunist organisms, usually affecting an individual who is in some way compromised, although *S. saprophyticus* is associated with cystitis in women and *S. lugdunensis* can cause invasive skin disease in otherwise healthy persons. Coagulase-negative staphylococci, typically *S. epidermidis*, is associated with prosthetic joint infections, prosthetic heart valve infection, and intravenous-line-associated infections.

PATHOLOGY

Acute inflammation

S. aureus induces an acute inflammatory reaction, which occurs when foreign material enters the host. Bacterial components, such as the peptidoglycan of the cell wall of *S. aureus*, and host products, for example tumour necrosis factor (TNF) and interleukin-1 (IL-1), are released by immune and epithelial cells in response to the organism. Foreign material is recognized by toll-like pattern receptors (TLR) and TNF/IL-1 activate their appropriate cell receptors leading to the up-regulation of NFkB, a transcription factor that in turn switches on a series of genes that ultimately lead to release of chemokines. The net effect of these cell-signalling changes is to induce increased blood flow to the affected area. The blood vessels then become leaky and host cells, principally neutrophils, are recruited into the extravascular space and move toward the organism. The neutrophils phagocytose the organism but, in addition, the release of phenol-soluble modulins from *S. aureus* induces cytolysis of neutrophils with the release of reactive oxygen radicals into the interstitium, causing further host-cell damage.

Toxin-induced damage

TSST-1 is a superantigen that stimulates a broad range of T-cells to release TNF and IL-1, leading to a generalized acute inflammatory reaction and shock. This leads to the toxic shock syndrome. *S. aureus* enterotoxins (SEA-SE/U2) are superantigens and have a similar mechanistic effect to TSST-1, but induce food poisoning with rapid onset of nausea, vomiting, and sometimes diarrhoea. Alpha toxin is a pore-forming toxin, which induces cell necrosis by allowing intracellular contents to leak out of the affected cells, causing local host cell death. PVL is associated with necrotizing pneumonia and severe invasive skin or bone infections. It induces cytolysis of cells, particularly neutrophils, thus abrogating the host innate immune response. Exfoliatin A and B produced by *S. aureus* affects cadherin intercellular adhesion proteins and their loss leads to separation of the upper skin layers, with desquamation of the epidermal layer. This illness is called scalded skin syndrome.

QUESTIONS

1. True or false: The following are associated with *Staphylococcus aureus*:
    a. Oxidase-positive reaction
    b. Production of coagulase
    c. Catalase-positive reaction
    d. PVL
    e. *SCCmec*

2. True or false: Resistance to beta-lactams in *Staphylococcus aureus* is associated with:
    a. Enzymatic destruction of penicillin
    b. Production of PBP2a
    c. Infections only occurring in hospitals
    d. Use of oxacillin to detect resistance
    e. Mandatory reporting

3. True or false: The following toxins are found in *Staphylococcus aureus*:
   a. Vacuolating cytotoxin
   b. Leukocidin
   c. Alpha toxin
   d. Cytolethal distending toxin
   e. Superantigens

4. True or false: The following diseases are linked with *Staphylococcus aureus*:
   a. Toxic shock syndrome
   c. Haemolytic uremic syndrome
   d. Lady Windermere syndrome
   e. Scalded skin syndrome

5. True or false: The following are associated with MRSA:
   a. Screening
   b. Nasal carriage
   c. Petting zoo
   d. Dialysis
   e. Beta-lactamases

GUIDELINES

1. PHE (2006) Screening for Methicillin-resistant *Staphylococcus aureus* (MRSA) colonisation: A strategy for NHS trusts: a summary of best practice.
2. NHS England (2013) Guidance on the reporting and monitoring arrangements and post infection review process for MRSA bloodstream infections from April 2014 (version 2).

REFERENCES

1. O'Hara P et al. (2008) A Geographic variant of the Staphylococcus aureus Panton-Valentine leukocidin toxin and the origin of community associated methicillin-resistant *S. aureus*. *USA300 J Infect Dis* **197**:184–194.
2. Kallen AJ et al. (2010) Health care associated invasive MRSA infections, 2005–2008. *JAMA* **304**:641–647.
3. Dantes R et al. (2013) National burden of invasive methicillin resistant Staph aureus infections USA. *JAMA* **173**:1970–1978.
4. Hidron AI et al. (2005) Risk factors for colonization with methicillin-resistant Staphylococcus aureus (MRSA) in patients admitted to an urban hospital: emergence of community-associated MRSA nasal carriage. *Clin Infect Dis* **41**:159–166.
5. Stone ND et al. (2012) Surveillance definitions of infections in long-term care facilities: revisiting the McGeer criteria. *Infect Control Hosp Epidemiol* **33**:965–977.
6. Longtin Y et al. (2014) Contamination of stethoscopes and physicians' hands after a physical examination. *Mayo Clin Proc* **89**:291–299.
7. Pittet D et al. (2000) Effectiveness of a hospital-wide programme to improve compliance with hand hygiene. Lancet **356**:1307–1312.
8. Simor AE et al. (2007) Randomized controlled trial of chlorhexidine gluconate for washing, intranasal mupirocin, and rifampin and doxycycline versus no treatment for the eradication of methicillin-resistant *Staphylococcus aureus* colonization. *Clin Infect Dis* **44**:178–185.
9. Wassenberg MWM et al. (2010) Rapid screening of methicillin-resistant Staphylococcus aureus using PCR and chromogenic agar: a prospective study to evaluate costs and effects. *Clin Microbiol Infect* **16**:1754–1761.
10. Coates T (2009) Nasal decolonization of Staphylococcus aureus with mupirocin: strengths, weaknesses and future prospects. *J Antimicrob Chemother* **64**:9–15.

ANSWERS

MCQ	Feedback

1. True or false: The following are associated with
   Staphylococcus aureus:
   a. Oxidase-positive reaction
   b. Production of coagulase
   c. Catalase-positive reaction
   d. PVL
   e. *SCCmec*

    a. False
    b. True
    c. True
    d. True
    e. False
S. aureus is catalase-positive and oxidase-negative. It produces coagulase, and some *S. aureus* produce a toxin called Panton–Valentine leukocidin. Some *S. aureus* carry the *mac* cassette, making the organism resistant to beta-lactam antibiotics.

2. True or false: Resistance to beta-lactams in
   Staphylococcus aureus is associated with:
   a. Enzymatic destruction of penicillin
   b. Production of PBP2a
   c. Infections only occurring in hospitals
   d. Use of oxacillin to detect resistance
   e. Mandatory reporting

    a. False
    b. True
    c. False
    d. True
    e. True
Resistance to beta-lactams in *S. aureus* results from the production of a novel penicillin-binding protein PBP2a to which beta-lactam antibiotics do not bind very efficiently. It is not a result of destruction of the antibiotic by beta-lactamases. Infections occur both in hospitals and in the community. The presence of this resistance is usually detected in the laboratory by resistance to oxacillin, by a chromogenic agar, or by the polymerase chain reaction to detect *mecA*.

3. True or false: The following toxins are found in
   Staphylococcus aureus:
   a. Vacuolatin, cytotoxin
   b. Leukocidin
   c. Alph toxin
   d. Cytolethal distending toxin
   e. Superantigens

    a. False
    b. True
    c. True
    d. False
    e. True
Vacuolating cytotoxin is produced by *Helicobacter pylori*. Cytolethal distending toxin is produced by *Campylobacter jejuni*. *S. aureus* produces Panton–Valentine leukocidin, alpha toxin (one of the enterotoxins produced by *S. aureus*), and both the toxic shock syndrome toxin TSST-1 and enterotoxins are superantigens.

4. True or false: The following diseases are linked with
   Staphylococcus aureus:
   a. Toxic shock syndrome
   b. Necrotizing pneumonia
   c. Haemolytic uremic syndrome
   d. Lady Windermere syndrome
   e. Scalded skin syndrome

    a. True
    b. True
    c. False
    d. False
    e. True
S. aureus can cause toxic shock syndrome by secretion of TSST-1 and was initially recognized in association with tampons, but it can occur in either gender. It can cause a necrotizing pneumonia from secretion of PVL and scalded skin syndrome from exfoliatin. Haemolytic uremic syndrome is cause by verotoxin, which is found in *E. coli*, and Lady Windermere syndrome is voluntary suppression of the cough reflex, leading to retained secretions and a *Mycobacterium avium* chest infection.

MCQ	Feedback

5. True or false: The following are associated with MRSA:
    a.   Screening
    b.   Nasal carriage
    c.   Petting zoo
    d.   Dialysis
    e.   Beta-lactamases
	a. True
	b. True
	c. False
	d. True
	e. False
	S. aureus is found in the nose, axilla, and groin (carrier sites), and screening is an important method of controlling infection and onward transmission of MRSA by giving an appropriate suppression regimen to the colonized patient and isolating the patient in a single room. Risk factors for MRSA are residence in a long-term care facility, admission to hospital, prior use of antibiotics, haemodialysis, and HIV infection. Resistance to beta-lactams is not a result of the presence of beta-lactamases. Petting zoos are a risk factor for infection with verotoxin-producing organisms, leading in some cases to the haemolytic uremic syndrome.

NEISSERIA GONORRHOEAE WITH HIGH-LEVEL RESISTANCE TO AZITHROMYCIN

CASE 16

Monica Desai,[1] Helen Fifer,[2] and Gwenda Hughes[3]

[1]Consultant Epidemiologist, Department of HIV and STIs, Centre for Infectious Disease Surveillance and Control, National Infection Service, Public Health England, UK.
[2]Consultant Microbiologist, National Infection Service, Public Health England, UK.
[3]Consultant Scientist and Head, STI section, Department of HIV and STIs, Centre for Infectious Disease Surveillance and Control, National Infection Service, Public Health England, UK.

Between November 2014 and March 2015, five patients presented to a sexual health clinic in England and were diagnosed with gonococcal infection that was highly resistant to azithromycin. Contact tracing identified three additional cases. High-level azithromycin resistance is a rare phenotype previously observed only sporadically in the UK and elsewhere. Whole genome sequencing was performed on seven of the isolates, and it demonstrated that they were identical. The isolates remained susceptible to ceftriaxone, ciprofloxacin, and spectinomycin, and all patients were treated successfully.

This case is based upon an outbreak reported in England in the Sexually Transmitted Infections Journal in 2015.

INVESTIGATION OF THE OUTBREAK

The investigation focused on transmission characteristics, sexual network mapping, and microbiological characteristics, including resistance testing. In Case 16, we outline the basic tenets of an outbreak investigation for sexually transmitted infections (STI), with a focus on gonorrhoea. Managing an outbreak of gonorrhoea needs to consider not just case finding, but also the susceptibility profile to ensure that there is no onward transmission of a potentially resistant organism. It is vital to work collaboratively for the successful management of any outbreak, as illustrated in this example of gonorrhoea.

Detecting an STI outbreak

Typically, an STI outbreak is recognized by the observed number of cases over a defined period that exceeds the number expected in a given community (that is, exceedance). The expected number of cases can be estimated by automated statistical simulation tools using previous years' data with allowance for variation. Initially the objectives are to confirm the diagnosis, confirm that there is an increase in cases, and determine that this is a real and significant increase that represents an outbreak. Once confirmed, an outbreak control team (OCT) is formed. The first step in any outbreak is to provide a case and outbreak definition. In the case of gonorrhoea, the case definition may be based on clinical or microbiological definitions or both (as in this case). However, other outbreak definitions may include linked cases, such as those in close sexual networks, or the burden on healthcare systems, such as the inability of current clinic capacity to manage the increase in cases.

It is then important to identify cases and contacts, to determine the nature of the outbreak, and collect appropriate specimens.

Special characteristics of STI outbreaks

Sexually transmitted infections have certain special features that need to be considered when managing an outbreak. First, patients may wish to remain anonymous and do not want their health records shared with other health professionals. Early treatment of the index case and prompt identification and treatment of contacts is important to prevent onward transmission. Co-infection with other sexually transmitted infections is common. Anyone diagnosed with gonorrhoea should have a full STI screen, including screening for human immunodeficiency virus (HIV) as indicated in the BASHH guidelines, and be treated promptly as appropriate. Other measures may be required to control the outbreak and will require a multidisciplinary approach. These include contact tracing and other interventions to improve infection detection and treatment, such as locally expanded STI screening. Interventions that foster behavioural change can also help reduce transmission, especially in high-risk sexual networks. Finally, transmission not only depends upon transmissibility of the organism, but on the transmission dynamics within the sexual network. As a result, outbreaks involving dense high-risk sexual networks characterized by multiple and concurrent sexual partnerships may develop rapidly, whereas those in networks where partner turnover is lower may develop slowly (Figure 16.1). The speed of development of the outbreak will also depend upon the duration of infectiousness of the organism.

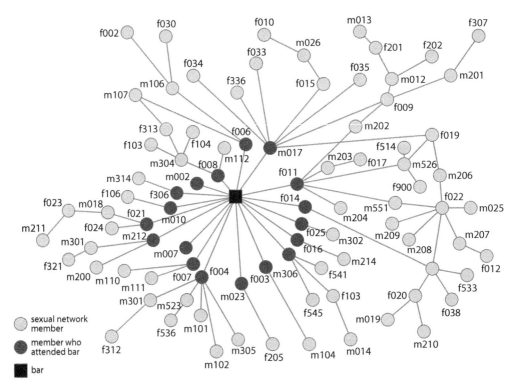

Figure 16.1. **Network members viewed by their connection through a bar associated with gonorrhoea acquisition.** A prefix to the unique identifier of m designates a male and f indicates a female sexual partner. (From De P, Singh AE, Wong T et al. [2004] *Sex Transm Infect* 80:280–285. With permission from BMJ Publishing Group Ltd.)

Roles and responsibilities in outbreak investigation

Effective management of an STI outbreak needs an effective multidisciplinary team. Key stakeholders include the affected local authorities (LAs), Public Health England (PHE), the British Association of Sexual Health and HIV (BASHH), and certain third-sector (voluntary) organizations. The OCT is usually, but not always, chaired by the local Consultant in Communicable Disease Control (CCDC) or Consultant in Health Protection (CHP), and would usually include representation from the local sexual health clinic team (genitourinary medicine (GUM) physicians, nurses, and health advisors), the PHE Health Protection Team (HPT), and PHE Field Epidemiology Services (FES). For larger outbreaks infection specialists from the National Infection Service, the Director of Public Health (DPH), National Health Service (NHS), or PHE microbiologists, third-sector organizations such as the Terrence Higgins Trust, and a designated press officer from PHE or the affected local authorities or both. The roles and responsibilities are set out in Table 16.1.

Table 16.1. **Key roles and responsibilities in managing STI outbreaks**

Role	Responsibilities
Outbreak Control Team Chair–usually CCDC or CHP	Directing and coordinating overall management of outbreak
	Ensuring communication between members of the OCT and other parties
	Declaring outbreak over, with OCT
CCDC/Consultant in Health Protection	Identifying outbreak through routine surveillance
	Providing local epidemiological support
	Liaising with commissioning authority to ensure resources are available to investigate outbreak
	Maintaining heightened surveillance of infection during (and if required after) outbreak to evaluate interventions
	Developing materials for training from outbreak lessons
	Auditing management of outbreak (in liaison with local GUM)
	Providing a link between public health and GUM
PHE Health Protection and National Infection Service	Undertaking a national review of surveillance data and providing guidance on whether the observed increase is an outbreak or caused by other factors
	Providing technical and expert advice on incident management, and investigation tools
	Identifying outbreaks through routine surveillance
	Providing expert epidemiological support
	Providing advice and support on local outbreak research
	Providing specialist advice on evaluation of control measures
PHE or NHS Microbiology	Identifying outbreaks through routine surveillance
	Providing expert advice to OCT on interpretation of microbiological data, specimen collection, specialist diagnostic methods, and outbreak control methods
	Facilitating prompt analysis and reporting of clinical samples and further testing as appropriate at reference laboratories
Communications Manager	Advising on media management (for example, being proactive versus reactive)
	Preparing media messages in collaboration with OCT chair
	Liaising with communications managers of other involved stakeholders

Service Commis-sioner	Ensuring funding of surge capacity or change to service delivery required to manage outbreak
Sexual Health Network	Coordinating sexual health promotion and provide links where more than one region in the network is involved

Regular communication of standards, available epidemiological data, local plans, and up-to-date professional networks are essential to facilitate the effective management of an outbreak.

TESTING FOR *N. GONORRHOEAE*

Nucleic acid amplification tests (NAATs) for the detection of *N. gonorrhoeae* have become the method of choice for testing asymptomatic individuals for urethral or endocervical infection. NAATs have an advantage over conventional culture methods in that they offer accurate diagnosis (high sensitivity and specificity), and they produce a result within a working day, which results in shorter turnaround times. They can be used with noninvasive specimens (urine or self-taken vulvo-vaginal swabs) without the need for an invasive examination or for chaperoning, which facilitates testing both in GUM clinics and in primary care. Detection of *N. gonorrhoeae* in extragenital sites, the rectum, and the pharynx, using NAATs is now known to be significantly more sensitive than culture, and NAATs are the test of choice at these sites in men who have sex with men and other high-risk individuals. There is the potential to cross-react with commensal *Neisseria* species that may be present at significant levels at these sites, particularly in the pharynx. It is therefore essential that reactive specimens from the rectum and pharynx are confirmed by supplementary testing using a different nucleic acid target from the original test.

In males, a first-pass urine (FPU) is the preferred sample for NAAT testing. A urethral or meatal swab should be used for microscopy and culture. Rectal and pharyngeal swabs can be taken for NAATs and culture, and their use should be directed by clinical history and sexual risk. Men who have sex with men who report oral–anal or digital–anal contact should also be considered for rectal and oral swabs based on sexual history. In females, vaginal or endocervical swabs can be used for NAAT testing for gonorrhoea and are equally sensitive. Vaginal swabs can be self-taken or taken by a clinician. Endocervical and urethral swabs are required for microscopy and culture. Urine has poor sensitivity for detecting *N. gonorrhoeae* in women. Rectal and pharyngeal swabs should be directed by history or if the patient is presenting as a sexual contact of someone with gonorrhoea.

In all cases of gonorrhoea that are diagnosed by NAAT, culture should be taken if possible before antibiotics are given. This practice allows antibiotic susceptibility and resistance to be identified. It is important to screen for other STIs including HIV.

The use of NAATs allows the simultaneous detection of *N. gonorrhoeae* and *Chlamydia trachomatis* from the same sample. There has been an increased use of dual NAATs among microbiology laboratories in recent years. Screening in populations with a prevalence below 1% will have a low positive predictive value (PPV), and most reactive test results are likely to be false positives. The latest gonorrhoea testing guidelines in England have been updated to include an algorithm to estimate local PPVs, which can be used to inform decisions on whether testing for gonorrhoea using NAATs is appropriate and, when done, whether supplementary testing is recommended for confirmation.

CLINICAL MANAGEMENT

Treatment should be initiated under any of the following situations: Identification of intracellular Gram-negative diplococci on microscopy of a smear from the genital tract, a positive culture of *N. gonorrhoeae* from any site, a positive NAAT for *N. gonorrhoeae* from any site, or a recent sexual partner of confirmed case of gonococcal infection.

Recommended treatment is currently ceftriaxone 500 mg intramuscularly as a single dose with azithromycin 1 g oral as a single dose. Azithromycin is recommended as cotreatment, regardless of the results of chlamydia testing, to delay onset of widespread cephalosporin resistance. There is some evidence to support improved eradication of pharyngeal gonorrhoea when using azithromycin with cephalosporin therapy. Treatment for individuals infected with HIV is the same as for those who are HIV negative.

Alternative regimens include cefixime 400 mg oral single dose, spectinomycin 2 g intramuscular, and other single dose cephalosporins. Quinolones are not recommended for empirical treatment of gonorrhoea because of high prevalence of resistance. For up to date guidance, refer to the BASHH Gonorrhoea Treatment Guidelines.

Treatment of complicated infections, such as gonococcal pelvic inflammatory disease, epidiymo-orchitis, conjunctivitis, and disseminated infection, can be found in the BASHH 2011 guidelines and US-specific guidance in the Centers for Disease Control (CDC) 2010 guidelines.

Co-infection with chlamydia is common. Clinic studies suggest that co-infection with chlamydia is as high as 50% those diagnosed with gonococcal infection, whereas all of those with gonococcal infection in a recent British population-based survey were co-infected with chlamydia. All adults with gonorrhoea should be tested for HIV and be offered a full STI screen including for chlamydia.

Follow-up may be of use to ensure symptoms have resolved, compliance with treatment, and to ask about adverse reactions. The possibility of re-infection should be explored through a full sexual history and advice on safer sexual behaviour given. Patients should be encouraged to notify partners so that they may be tested and treated if needed.

Test-of-cure (TOC) is recommended for all cases to detect emerging resistance and antibiotic failure. Ideally all patients should be offered a TOC, but if a targeted approach needs to be taken, those with persisting symptoms or signs, those with pharyngeal infection, and those who received treatment other than first-line treatment should be prioritized. Note that infection following treatment could be the result of re-infection.

There is a lack of good evidence to guide timing of TOC. In general, in those with persisting symptoms or signs, a test with culture should be performed at least 72 hours after completing therapy. In those who are asymptomatic, a NAAT test two weeks after completion of therapy is currently recommended. However, studies are being conducted to determine whether this period is sufficiently long to avoid detection of dead organisms when using NAATs.

PREVENTION OF FURTHER DISEASE

Primary prevention aims to modify sexual risk-taking behaviours to prevent infection. Public health strategies to prevent infection include general and targeted health promotion campaigns and targeted outreach work for specific high-risk populations, which highlight the importance of correct and consistent condom use for those with new or casual sexual partners. Some campaigns may also highlight the risks associated with concurrent sexual partnerships.

Secondary prevention aims to reduce the onward transmission from those already infected. An important element of this, which is especially important in STI outbreak control, is effective partner notification. All individuals infected with *N. gonorrhoeae* should be encouraged to undertake partner notification themselves or through health advisors. Typically, male partners who have symptomatic urethral infection are asked to notify all partners with whom they have had sexual contact in the past two weeks or the last partner if longer ago. Patients with infection at other sites or who have asymptomatic infection should notify all partners within the past three months. Sexual partners should be offered a full HIV/STI screen including testing and epidemiological treatment for gonorrhoea. Patients with gonorrhoea are also advised to abstain from sexual intercourse until they and their partner or partners have completed treatment to avoid re-infection.

During an outbreak, strategies to interrupt the onward transmission of infection include:

- active case surveillance that will involve collection of detailed case information through enhanced surveillance, case interviews, social and sexual network investigation, and monitoring of effectiveness of partner notification. Focused research studies and in-depth interviews can elucidate the social context of the local epidemic and is of particular value with marginalized populations.
- microbiological investigation, which may involve collection of specimens and the use of typing methods to aid the mapping of transmission.
- control measures that will include health promotion interventions, partner notification, publicity campaigns to raise awareness in the community, and among health-care professionals, and outreach clinics.

Follow-up of the control measures introduced is vital. This can include evaluation of process measures, such as the proportion of the target population accessing the intervention, coverage of intervention delivery, and coverage of health promotion interventions. However, the impact of control measures on the primary outcome is most important. The primary outcome measure is the number of reported cases, and effective control should result in a decline in numbers. If incidence has been high and the outbreak large, stability in case numbers suggests the infection may have become endemic in the population.

NATURAL HISTORY AND EPIDEMIOLOGY

The primary sites of gonorrhoea infection are the mucous membranes of the urethra, endocervix, rectum, pharynx, and conjunctiva. Sexual intercourse with an infected partner is the main risk factor for gonorrhoea. A single episode of unprotected vaginal intercourse has a female-to-male transmission risk of 20%. This may increase to up to 80% after several exposures. The risk of transmission from an infected man to a female partner is approximately 50–70% per episode of unprotected vaginal intercourse.

In contrast to rectal and pharyngeal infection in men and gonococcal infection in women, urethral infection in men has a short incubation period, and together with prominent symptoms leading to prompt treatment, means that the duration of the infectious period is short. These factors, together with the relatively poor transmission efficiency through heterosexual intercourse, means that high rates of sex partner change are required to sustain transmission in a population. Consequently, gonorrhoea tends to be concentrated in population groups with more complex sexual networks. A key infection control measure is to ensure treatment of the sex partners of infected people, who often will not spontaneously seek healthcare.

Globally, gonococcal infections account for 106 million of the estimated 498 million new cases of curable STIs each year. In the United States, there are an estimated 700,000 new gonococcal infections each year, making it the second most common reported bacterial STI.

In England in 2016, there were 36,244 new diagnoses of gonorrhoea reported to Public Health England, representing 10% of all STI diagnoses made at sexual health centres in England. Men who have sex with men (MSM) are disproportionately affected, accounting for 22,408 of the infections of gonorrhoea in 2015, an increase of 21% on the previous year. Gonorrhoea was the most commonly diagnosed STI among MSM in 2015; 10% (2188) had multiple site infections, 15% (3400) were infected only in the pharynx, and 25% (5570) were diagnosed with rectal infection.

Almost 50% of all UK gonorrhoea cases are diagnosed in London, and infection is strongly associated with deprived urban areas. The key groups at risk are MSM, black Caribbean populations, and young heterosexuals with complex sexual networks. Young heterosexual adults aged 15–24 account for 43% (7440/13,514) of gonorrhoea diagnosis in the UK and 42% in Europe. In Europe, age-specific rates are highest among 20–24 year olds (36 per 100,000 people). The male-to-female ratio of the rate of diagnosis was 2.7:1 in Europe and 2.6:1 in the UK.

Antimicrobial resistance in gonorrhoea

In England and Wales, surveillance of resistance in gonorrhoea is provided through the Gonococcal Resistance to Antimicrobials Surveillance Programme (GRASP). GRASP, which was established in 2000, is now coordinated by Public Health England. The GRASP collection combines laboratory and clinical data on gonococcal isolates diagnosed in 26 collaborating GUM clinics covered by 24 laboratories. Isolates are collected from consecutive patients with gonorrhoea over a three-month period each year. In Europe, antimicrobial resistance is monitored in 21 countries through a sentinel surveillance program called the European Gonococcal Antimicrobial Surveillance Programme (Euro-GASP), and is coordinated by the European Centre for Disease Prevention and Control (ECDC).

In 2010, GRASP data showed that there was a consistent drift in decreased susceptibility to cefixime, detected predominantly in isolates from MSM, and occasional reports of treatment failure began to be reported in the UK, Europe, and Japan. In 2011, Euro-GASP reported that 7.6% of isolates were resistant to cefixime and for the first time detected isolates with decreased ceftriaxone susceptibility. Therefore, in 2011 BASHH recommended that first-line therapy should be changed before the 5% resistance level was reached for cefixime, in order to delay the accumulation of treatment failure and extend the useful life of the cephalosporins. The UK treatment guidelines were changed to intramuscular ceftriaxone at the increased dose of 500 mg in combination with oral azithromycin 1g as first-line therapy.

The threat of untreatable multidrug-resistant gonorrhoea is of global concern, and the World Health Organization (WHO) Global Action Plan and ECDC Response Plan have been published to provide strategic advice. The GRASP Action Plan aims to raise awareness and inform national guidelines. Case definitions have been recommended for confirmed and probable treatment failure. Therapeutic failure to ceftriaxone is still rare but is documented, and in 2016, the world's first documented case of treatment failure to dual ceftriaxone and azithromycin therapy was reported in England.

In addition, the outbreak of high-level azithromycin resistant gonorrhoea, described earlier, has continued to spread across England. This outbreak presents a significant threat to the current front-line dual therapy for gonorrhoea, as it renders the azithromycin component ineffective. Since November 2015, an increasing number of cases are known to be in men who have sex with men, and concern remains about the potential for rapid spread among this group.

BIOLOGY

N. gonorrhoeae is a nonmotile, non-spore-forming, Gram-negative coccus that characteristically grows in pairs (diplococci). It is a fastidious organism with complex growth requirements; specimens must be cultured on an enriched medium with an added iron source, which is then incubated in a carbon dioxide-enriched environment with high humidity. The medium also contains antimicrobial agents, which are selective to suppress the overgrowth of normal microbiota, while allowing growth of *N. gonorrhoeae*. Colonies are identified by Gram staining and oxidase testing. Any oxidase-positive, Gram-negative diplococci are further identified using biochemical and immunological tests. The isolation of a viable organism also allows antimicrobial susceptibility testing to be performed. The sensitivity of culture depends on collection of the appropriate specimen, provision of a high-quality isolation medium, and good transportation between the collection of the sample and the laboratory.

Traditionally, strains have been characterised either by serotyping, which is based upon variations in the major outer membrane protein (porin), or by auxotyping, which identifies the strain's growth requirements. Molecular methods allow greater discrimination between strains. Isolates can be discriminated from different sources and can be used to link sexual contacts and short transmission chains and to identify local clusters. *N. gonorrhoeae* multi-antigen sequence typing (NG-MAST) differentiates strains according to variations in two hypervariable loci, *porB* and *tbpB*, and determines a sequence type from the two-allele combination. It is highly reproducible, allows high throughput, and produces transferable data. Whole genome sequencing offers even greater discrimination between strains and is likely to replace NG-MAST in the near future.

Resistance may be mediated by chromosomes or plasmids. *N. gonorrhoeae* has developed increasing minimum inhibitory concentrations (that is, decreasing susceptibility) followed by frank resistance to the antimicrobial classes most commonly used for therapy, including penicillins, tetracyclines, macrolides, and fluoroquinolones, thus progressively reducing available therapeutic options.

DISEASE

The most common clinical presentation in men is acute urethritis a few days after unprotected vaginal or anal intercourse. Of genital infection in men, 95% will have symptoms, commonly dysuria and a purulent penile discharge. Untreated gonococcal infections infrequently progress locally (periurethral abscess, epididymitis, or prostatitis).

Endocervical infection is the most common clinical presentation of gonorrhoea in women. Half of women with cervical infection are asymptomatic. Most infected women who develop symptoms will do so within 10 days. Symptoms include mucopurulent vaginal discharge, dysuria, and intermenstrual bleeding. The cervical mucosa may be erythematous and friable with a purulent exudate. Complications include Bartholin abscesses, salpingitis, pelvic inflammatory disease (occurs in 10–20% of women with cervical gonorrhoea), ectopic pregnancy, infertility, and perihepatic inflammation (Fitz-Hugh-Curtis syndrome).

N. gonorrhoeae can be isolated from the rectum in up to 40% of women and in a similar proportion of MSM with uncomplicated gonorrhoea. The rectum is often the only infected site in MSM, whereas in women, perineal contamination with vaginal discharge is more common. Rectal infection is usually asymptomatic but can cause acute proctitis. Almost all pharyngeal infections are asymptomatic but may cause pharyngitis.

Disseminated gonococcal infection classically presents with polyarthritis–dermatitis syndrome or septic arthritis. Only half of patients with disseminated infection have positive blood or synovial cultures, but 80% have positive cultures from mucosal sites. More rarely, patients may present with endocarditis, osteomyelitis, or meningitis.

Babies born through infected birth canals may develop ophthalmia neonatorum, a severe purulent eye discharge with periorbital oedema that occurs within a few days of birth.

PATHOLOGY

N. gonorrhoeae is an obligate human pathogen. It is never found as a commensal, although a proportion of those infected, particularly women, may remain asymptomatic. It is transmitted almost exclusively by sexual contact, but may also cause ophthalmic infection in neonates by contact with an infected birth canal during labour. Gonorrhoeal infection is generally limited to superficial mucosal surfaces lined with columnar epithelium. It primarily affects the mucous membranes of the urethra and cervix, and less frequently those of the rectum, oropharynx, and conjunctivae.

Attachment of gonococci to the surface of columnar epithelial cells is mediated by pili, which are filamentous outer membrane appendages composed of multiple repeating protein subunits. Piliated variants are better at attaching to mucosal surfaces and experiments have found them to be more pathogenic. A single strain of *N. gonorrhoeae* may have pili with great antigenic variability. Other outer membrane proteins such as Opa (opacity-related proteins) increase adherence, and certain Opa variants also promote cell invasion.

Attachment to mucosal cells is followed within 48 hours by cell invasion. This process involves the protein porin that occurs in two main antigenic forms, PorA and PorB. Porin inserts itself in the host cell membrane and forms transmembrane channels that allow endocytosis. Strains expressing PorA are associated with resistance of *N. gonorrhoeae* to serum bactericidal activity and are more likely to disseminate. Disseminated gonococcal isolates are also able to evade killing by human complement.

QUESTIONS [Q8]

1. Which of the following would be an appropriate definition for an outbreak of gonorrhoea?
   a. The number of cases of gonorrhoea in Pool County over a one-month period exceeds the number that would normally be expected.
   b. Clinicians are concerned that there have been more cases of gonorrhoea presenting to the genitourinary medicine clinic than they would normally expect over a one-month period.
   c. A new resistant strain of gonorrhoea is reported by the local laboratory in two specimens.
   d. Two sexually linked cases of primary gonorrhoea have occurred within a sexual network.

2. Which of the following would be included as members of an outbreak control team for an outbreak of gonococcal disease?
   a. Consultant in Communicable Disease Control (CCDC)
   b. Microbiologist
   c. Sexual health clinician
   d. Public Health Consultant
   e. Communications Officer
   f. All of the above

3. Which of the following is the preferred first-line test when testing asymptomatic men for urethral gonorrhoea?
   a. Microscopy using a urethral swab
   b. Culture using a urethral swab
   c. Nucleic acid amplification test (NAAT) of first-pass urine
   d. None of the above

4. Screening for gonorrhoea in low-prevalence populations using NAATs is not recommended. What is the main reason for this?
   a. The test is too expensive.
   b. Culture methods are more sensitive than NAATs.
   c. NAATs cannot be used for male urine specimens.
   d. NAATs have low specificity in low-prevalence populations.
   e. NAATs have low positive predictive value in low-prevalence populations.

5. *N gonorrhoeae* has developed resistance to which classes of antibiotics?
   a. Penicillins
   b. Cephalosporins
   c. Tetracyclines
   d. Macrolides
   e. All of the above

GUIDELINES

1. www.bashh.org/documents/3920.pdf.

2. www.hpa.org.uk/webc/HPAwebFile/HPAweb_C/1214553002033.

3. www.cdc.gov/std/program/outbreak.pdf.

4. whqlibdoc.who.int/publications/2007/9789241563475_eng.pdf.

5. www.uptodate.com/contents/epidemiology-pathogenesis-and-clinical-manifesta-tions-of-neisseria-gonorrhoeae-infection.

6. www.who.int/reproductivehealth/publications/rtis/9789241505895/en/.

7. Simms et al. (2017) Managing outbreaks of Sexually Transmitted Infections Operational Guidance. Public Health England (in press).

8. Centers for Disease Control and Prevention, Program operations guidelines for STD prevention: outbreak response plan.

9. World Health Organization (2007) Global strategy for the prevention and control of sexually transmitted infections: 2006–2014: breaking the chain of transmission.

10. National Health Service, England and Wales (1974). National Health Service (Venereal Diseases) Regulations 1974.

11. Department of Health (2013) Sexual Health: Clinical Governance.

12. Health Protection Agency, British Association for Sexual Health and HIV (2010). Guidance for gonorrhoea testing in England and Wales 2010.

13. Bignell CJ (2004) BASHH guideline for gonorrhoea. *Sex Transm Infect* **80**(5):330–331.

REFERENCES

1. Chisholm SA et al. (2015) An outbreak of high-level azithromycin resistant *Neisseria gonorrhoeae* in England. *Sex Transm Infect.* **54**(9):3812-3816.

2. Annan T, Hughes G, Evans B et al. (2010) Guidance for managing STI outbreaks and incidents, Health Protection Agency.

3. Whiley DM, Tapsall JW & TP (2006) Nucleic acid amplification testing for *Neisseria gonorrhoeae*: an ongoing challenge. *J Mol Diagn* **8**(1):3-15.

4. Health Protection Agency (2010) Detection of *Neisseria gonorrhoeae* using molecular methods. National Standard Method QSOP 62.

5. Bignell C & Fitzgerald M (2011) UK national guideline for the management of gonorrhoea in adults, 2011. *Int J STD AIDS* **22**(10):541-547.

6. British Association for Sexual Health and HIV (2014) Standards for the management of sexually transmitted infections (STIs).

7. Hughes G, Field N, Folkard K et al. (2014) Guidance for the detection of gonorrhoea in England, Public Health England.

8. Sathia L et al. (2007) Pharyngeal gonorrhoea– is dual therapy the way forward? *Int J STD AIDS,* **18**(9):647-648.

9. Centers for Disease Control and Prevention (2015) 2015 STD Treatment Guidelines.

10. Fifer H & Ison C (2014) Nucleic acid amplification tests for the diagnosis of *Neiserria gonorrhoeae* in low-prevalence settings: a review of the evidence. *Sex Transm Infect* **90**(8):577-579.

11. Sonnenberg P et al. (2013) Prevalence, risk factors, and uptake of interventions for sexually transmitted infections in Britain: findings from the National Surveys of Sexual Attitudes and Lifestyles (Natsal). *Lancet* **382**(9907):1795-1806.

12. Hooper RR et al. (1978) Cohort study of venereal disease. I: the risk of gonorrhea transmission from infected women to men. *Am J Epidemiol* **108**(2):136–144.

13. Lin JS et al. (1998) Transmission of *Chlamydia trachomatis* and *Neisseria gonorrhoeae* among men with urethritis and their female sex partners. *J Infect Dis* **178**(6):1707–1712.

14. Wiesner PJ et al. (1973) Clinical spectrum of pharyngeal gonococcal infection. *N Engl J Med* **288**(4):181–185.

15. European Centre for Disease Prevention and Control, Annual Epidemiological Report (2013) Reporting on 2011 surveillance data and 2012 epidemic intelligence data. *ECDC*, Stockholm.

16. Public Health England (2016) Sexually transmitted infections and chlamydia screening in England, Health Protection Report.

17. World Health Organization (2012) Global action plan to control the spread and impact of antimicrobial resistance in *Neisseria gonorrhoeae*.

18. European Centre for Disease Prevention and Control (2012), Response plan to control and manage the threat of multidrug resistant gonorrhoea in Europe. *ECDC*, Stockholm.

19. Health Protection Agency (2013) Gonococcal Resistance to Antimicrobials Surveillance Programme (GRASP) Action Plan for England and Wales: Informing the Public Health Response.

20. Fifer H et al. (2016) Failure of dual antimicrobial therapy in treatment of gonorrhea. *N Engl J Med* **374**(25):2504–2506.

21. Public Health England (2016) Outbreak of high level azithromycin resistant gonorrhoea in England. Health Protection Report.

22. Grad YH et al. (2014) Genomic epidemiology of *Neisseria gonorrhoeae* with reduced susceptibility to cefixime in the USA: a retrospective observational study. *Lancet Infect Dis* **14**(3): 220–226.

23. Lewis, D.A., The Gonococcus fights back: is this time a knock out? Sex Transm Infect, 2010. **86**(6): p. 415-21.

24. Platt R, Rice PA & McCormack WM (1983) Risk of acquiring gonorrhea and prevalence of abnormal adnexal findings among women recently exposed to gonorrhea. *JAMA* **250**(23): 3205–3209.

25. Weel JF, Hopman CT & van Putten JP (1991) In situ expression and localization of *Neisseria gonorrhoeae* opacity proteins in infected epithelial cells: apparent role of Opa proteins in cellular invasion. *J Exp Med* **173**(6):1395–1405.

26. Knapp JS et al. (1984) Serological classification of *Neisseria gonorrhoeae* with use of monoclonal antibodies to gonococcal outer membrane protein I. *J Infect Dis* **150**(1):44–48.

ANSWERS

MCQ	Feedback
1. Which of the following would be an appropriate outbreak definition for an outbreak of gonorrhoea? a. The number of cases of gonorrhoea in Pool County over a one-month period exceeds the number that would normally be expected. b. Clinicians are concerned that there have been more cases of gonorrhoea presenting to the genitourinary medicine clinic than they would normally expect over a one-month period. c. A new resistant strain of gonorrhoea is reported by the local laboratory in two specimens. d. Two sexually linked cases of primary gonorrhoea have occurred within a sexual network.	A disease outbreak is the occurrence of cases of a disease in excess of what would normally be expected in a defined community, geographical area, or season. An outbreak may occur in a restricted geographical area or can extend over a larger area, and it can last a few days, weeks, or several years. The WHO also regards a single case of communicable disease that has been absent for a long time from a population, or caused by an agent that was not previously recognized in that community or area, or emergence of a previously unknown disease, as an outbreak. Hence, option c could be used as an alternative outbreak definition, but it would require more detail in the description. (www.who.int/topics/disease_outbreaks/en/)
2. Which of the following would be included as members of an outbreak control team for an outbreak of gonococcal disease? a. Consultant in Communicable Disease Control (CCDC) b. Microbiologist c. Sexual health clinician d. Public Health Consultant e. Communications Officer f. All of the above	The key stakeholders in an outbreak control team for a sexually transmissible infection are sexual health clinicians, CCDC in their health protection role, public health, microbiology, communications officer, and can also include nonstatutory organizations such as the National AIDS Trust and regional or national health protection or epidemiology, depending on the extent of the outbreak, sexual health commissioner, and sexual health network coordinator. It is vital to have a multidisciplinary team to effectively manage the outbreak. Convening an outbreak team is a major step in the preliminary phase of the outbreak.
3. Which of the following is the preferred first-line test when testing asymptomatic men for urethral gonorrhoea? a. Microscopy using a urethral swab b. Culture using a urethral swab c. Nucleic acid amplification test (NAAT) of first-pass urine d. None of the above	A first-pass urine (FPU) is the preferred sample for NAAT testing. Asymptomatic men should be tested for gonorrhoea using FPU NAAT as first line. A urethral swab processed for microscopy and culture is recommended for use in symptomatic patients. The NAAT platform allows for dual detection of *N. gonorrhoeae* and *Chlamydia trachomatis* from the same sample. However, a suitable testing algorithm must be in place to avoid high rates of false positive test results. The testing algorithm should give a minimum positive predictive value (PPV) of 90%. In almost all cases, this will require the use of a supplementary NAAT with a different nucleic acid target to confirm the result, even in settings where gonorrhoea prevalence exceeds 1%.
4. Screening for gonorrhoea in low-prevalence populations using NAATs is not recommended. What is the main reason for this? [Q10] a. The test is too expensive. b. Culture methods are more sensitive than NAATs. c. NAATs cannot be used for male urine specimens. d. NAATs have low specificity in low-prevalence populations. e. NAATs have low positive predictive value in low-prevalence populations.	NAATs are more sensitive than culture methods and can be used on male but not female urine specimens. NAAT platforms allow the simultaneous dual detection of *N. gonorrhoeae* and *Chlamydia trachomatis* from the same sample at no extra cost. With the widespread availability of dual NAAT technology, increased opportunistic testing for gonorrhoea is occurring in low-prevalence settings where chlamydia screening is offered. Screening in populations with a prevalence below 1% will have a low positive predictive value (PPV) and most reactive test results are likely to be false positives, suggesting unselected screening would be of limited public health benefit. Note that it is the positive predictive value and not the sensitivity or specificity of a test that varies with the prevalence of the disease.
5. *N gonorrhoeae* has developed resistance to which classes of antibiotics? a. Penicillins b. Cephalosporins c. Tetracyclines d. Macrolides e. All of the above	*N gonorrhoeae* has developed resistance to all classes of antibiotics. It is of grave concern that during the latest decade *N. gonorrhoeae* has developed resistance to the injectable extended-spectrum cephalosporin ceftriaxone, the last option for empiric first-line treatment, whilst retaining resistance to previously recommended antimicrobials. In 2010, in response to the fear that gonorrhoea may become untreatable, dual antimicrobial therapy (ceftriaxone 250–500 mg plus azithromycin 1–2 g) was introduced as recommended first-line treatment in the UK, USA, Europe, and Australia.

INCREASED NUMBER OF INFECTIONS WITH *PLASMODIUM* SPP DURING A PERIOD OF SOCIOPOLITICAL INSTABILITY

CASE 17

Geraldine A O'Hara,[1] Peter L Chiodini[2]

[1]London School of Hygiene and Tropical Medicine, UK.
[2]Hospital for Tropical Diseases and London School of Hygiene and Tropical Medicine, UK.

During a period of sociopolitical instability in Rwanda and Burundi, increasing numbers of patients were admitted to treatment centres with symptoms of malaria infection. They complained of headache, fatigue, fevers, rigors, and aching muscles and joints. Fever pattern was not a reliable indicator of malaria and some patients with severe malaria were afebrile.

INVESTIGATION OF THE CASES

The following investigations were undertaken on the patients in this region, subject to available resources: a full blood count, liver function tests, urea and electrolytes, and in selected cases a glucose level test. If evidence of haemolysis was present, haptoglobin levels, lactic dehydrogenase, and a reticulocyte count were requested. If neurological symptoms were present, a lumbar puncture (LP) was performed.

Specific tests for malaria were rapid diagnostic dipstick tests (RDT) plus thick and thin blood films for species identification (Table 17.1).

In resource-rich countries, the diagnostic tests for malaria include blood films and RDTs, plus polymerase chain reaction (PCR) assays in reference centres.

CLINICAL MANAGEMENT

Any febrile illness occurring after exposure to a region where malaria is endemic should prompt consideration of malaria as a diagnosis. Presentation of malaria varies with infecting species, age, and immunity.

Table 17.1. **Phenotypic appearance of different malarial parasites**

Phenotypic appearance	P. falciparum	P. ovale	P. vivax	P. malariae
Presence of early forms	+	–	–	–
Red blood cell (RBC) with multiple parasites	+++	+	++	+
Age of infected RBC	All ages	young	young	old
Schüffner's dots	–		+	–
James' dots		+		

Symptoms of malaria are nonspecific and include fever, chills, myalgia, malaise, fatigue, headache, cough, anorexia, nausea, vomiting, abdominal pain, diarrhoea, and arthralgia.

Malaria presents along a continuum from mild to severe disease, and clinical management is influenced by the species causing infection and the existence of naturally acquired immunity.

Infection with *Plasmodium falciparum* carries the highest risk of a fatal outcome but severe illness is also recognised to occur with *P. knowlesi* and *P. vivax*. *P. ovale* and *P. malariae* usually cause milder disease.

Attention should be paid to local and national guidelines before prescribing treatment for malaria. When treating uncomplicated malaria, the main aim is to cure the infection as rapidly as possible by eliminating the parasite and thus preventing progression to severe disease. Management of severe malaria additionally requires expert supportive treatment of complications.

Resistance to antimalarials has been documented for *P. falciparum* within a wide geographical area. Resistance has been observed for amodiaquine, chloroquine, mefloquine, quinine, and sulfadoxine-pyrimethamine. Delayed parasite clearance of *P. falciparum* by artemisinin derivatives has been confirmed in parts of Southeast Asia. *P. vivax* exhibits resistance to sulfadoxine-pyrimethamine in many areas; resistance to chloroquine is found in Indonesia, Papua New Guinea, Timor-Leste, and other parts of Oceania. There are also reports of resistance from Brazil and Peru.

Treatment of falciparum malaria requires combination therapy. Two or more blood schizonticidal compounds with independent modes of action are administered simultaneously with the aim of avoiding or slowing down the development of drug resistance.

Current World Health Organization (WHO) guidelines deploy artemisinins as the cornerstone of malaria therapy as a component of artemisinin combination therapy (ACT). Artemisinins eliminate the blood stages of all *Plasmodium* species and have the fastest parasite clearance times of any antimalarial. Importantly, artemisinins are active against gametocytes, so their widespread usage can reduce malaria transmission.

The ACTs currently recommended for treatment of uncomplicated falciparum malaria are:

- artemether plus lumefantrine
- artesunate plus amodiaquine
- artesunate plus mefloquine
- artesunate plus sulfadoxine-pyrimethamine
- dihydroartemisinin plus piperaquine

In severe malaria, parenteral drug treatment should be commenced without delay and prompt care to manage life-threatening complications of the disease is also vital. Supportive measures (for example, oxygen, ventilatory support, renal support, and cardiac monitoring) should be instituted as required. Other ancillary therapies should be administered as needed (for example, anticonvulsants, intravenous glucose and fluids, antipyretics, antibiotics, and blood transfusion). A lumbar puncture should be considered in unconscious patients to rule out concomitant bacterial meningitis.

PREVENTION OF FURTHER CASES

Many different strategies are used to prevent malaria, and the choice often depends on whether travellers or residents of endemic areas are being targeted.

Minimizing the risk of travellers' malaria is based on the ABCD of malaria prevention.

- **A** is awareness of risk. Travelers to areas where malaria is endemic should be counselled about the extent and distribution of malaria, a potentially life-threatening infection, in the area to be visited and given an individual risk-assessment based on their medical history, including medication.
- **B** is bite prevention. All travellers should deploy mosquito bite avoidance measures, including using an insect repellent such as DEET (*N,N*-diethyl-meta-toluamide), covering up with loose-fitting clothing, long sleeves, long trousers, and socks, avoiding being outside at peak mosquito biting time (dusk through to dawn), using insecticide-treated bed nets, and using room protection with a knockdown insecticide.
- **C** is chemoprophylaxis. Travellers to malaria endemic areas should be considered for chemoprophylaxis. Note that chemoprophylaxis should be given to people who may have been born in endemic areas and are returning there for a visit. This group is classed as visiting friends and relatives (VFR). A VFR is a migrant, ethnically distinct from the majority population of the country of residence, who returns to his or her country of birth. The partial immunity of these individuals wanes rapidly, and VFRs display similar risk patterns for malaria acquisition as travellers not previously exposed to malaria do. Indeed, VFRs in the US are eight times more likely to be diagnosed with malaria than other tourist travellers. Similarly, in the UK, higher rates of malaria are exhibited in VFR travellers to West Africa compared to non-VFR tourists visiting the area.

Recommendations for chemoprophylaxis against malaria must be individualized for each traveller. They depend on local patterns of transmission, resistance to antimalarials, potential drug toxicities, and interactions with existing medication, weighed against the overall likelihood of acquisition of malaria.

It is beyond the scope of this discussion to cover antimalarial chemoprophylaxis in detail, and recommendations differ among different national expert bodies (see Guideline 1 for the PHE Guidelines for malaria prevention in travellers from the UK).

- **D** is diagnose promptly and treat without delay. Malaria must be considered in every ill patient who has returned from the tropics in the previous year, and fever on return from the tropics should be considered to be malaria until proven otherwise. Suspected cases should be investigated by obtaining a blood film diagnosis as a matter of urgency.

In areas endemic for malaria, prevention strategies vary slightly. Intermittent preventive therapy or intermittent preventive treatment (IPT) involves the administration of antimalarials to at-risk subjects such as infants (IPTi), children (IPTc), schoolchildren (IPTsc), and pregnant women (IPTp). IPT is designed to reduce disease burden at a population level.

In IPTp, the major aims are to reduce maternal malaria episodes, maternal and foetal anaemia, placental parasitaemia, low birth weight, and neonatal mortality. The WHO recommends that in areas of Africa with moderate-to-high malaria transmission, a full treatment course of sulfadoxine-pyrimethamine (SP) is offered to all pregnant women at each scheduled antenatal care visit except during the first trimester. The WHO-recommended schedule is four antenatal care visits. Although resistance to SP is increasing in Africa, the preventive efficacy persists in IPTp. Alternative drugs are being investigated for IPTp.

IPTi similarly aims to reduce clinical malaria, anaemia, and severe malaria by administering full therapeutic courses of antimalarial medicines to infants. Current WHO recommendations are that in areas of sub-Saharan Africa with moderate-to-high malaria transmission, a treatment course of SP is delivered to all infants at 8 to 10 weeks, 12 to 14 weeks, and 9 months of age (corresponding to the schedule of the Expanded Programme on Immunization).

Given that the high burden of malaria-associated morbidity and mortality in Africa occurs in children aged 3–59 months, particularly in the rainy season, IPT has been extended to children. Monthly amodiaquine and sulfadoxine-pyrimethamine should be administered to children aged 3–59 months from the start of the yearly transmission season to a maximum of four doses during the malaria transmission season.

Prevention of mosquito bites remains a mainstay in avoidance of malaria infection in malaria-endemic areas. Mosquito nets provide a physical barrier to *Anopheles* spp., reducing the chance of bites. Insecticide-treated bednets (ITN) treated with pyrethroid insecticides, are preferred as mosquitoes encountering them are killed by the insecticide component.

An insecticide-treated net is a mosquito net that repels mosquitoes, by disabling, killing, or both, any mosquitoes that come into contact with insecticide on the netting material. There are two categories of ITNs: conventionally treated nets and long-lasting insecticidal nets . A conventionally treated net is a mosquito net that has been treated by dipping in a WHO-recommended insecticide. To ensure its continued insecticidal effect, the net should be treated again after three washes, or at least once a year. A long-lasting insecticidal net (LLIN) is a factory-treated mosquito net made with netting material that has insecticide incorporated within or bound around the fibres. The net must retain its effective biological activity without re-treatment for at least 20 WHO standard washes under laboratory conditions and three years of recommended use under field conditions.

ITNs have been shown to reduce malaria-related morbidity and mortality, reducing the number of *P. falciparum* and *P. vivax* malaria cases where used by approximately 50%. In trials performed in areas of moderate to high transmission ITNs have been shown to reduce the all-cause mortality in children under five years by about 20%. Use of ITNs in sub-Saharan Africa has shown increases in mean birth weight, reduction in low birth weight, and reduction in miscarriage rates. In addition, the mosquitocidal effects exhibited by ITNs mean that community members not sleeping under ITNs are offered a degree of protection as ITNs reduce vector burden. WHO recommends sufficient distribution of LLINs to achieve full coverage of populations at risk of malaria.

Historically, spraying the inside surfaces of houses with a residual insecticide was the main method of malaria control in endemic areas—this is referred to as indoor residual spraying (IRS). This approach relies on *Anopheles* spp. remaining on the wall or ceiling for a sufficient amount of time to acquire a lethal dose of insecticide. In 2006, the WHO recommended that where appropriate, IRS should be a central part of national malaria control strategies after meta analysis showed that a combination of LLINs and IRS had an additive reduction in risk compared to either strategy alone.

EPIDEMIOLOGY

Malaria transmission is a function of the interactions among the anopheline mosquito vector, *Plasmodium* spp., human hosts, and the environment.

Transmission dynamics are affected by the number of vectors (density), the human-biting habits (indoors or outdoors), the efficiency of biting, and the longevity of the female anopheline vectors. Only around 25 of the more than 400 anopheline species are good vectors. *Anopheles* mosquitoes breed in water and each species has its own breeding preference. For example, some prefer shallow collections of fresh water, such as puddles, rice fields, and hoof prints, explaining why different species predominate in different areas. Transmission is more intense in places where the mosquito lifespan is longer (giving the parasite sufficient time to complete its development inside the mosquito). Vectors prefer to bite humans rather than other animals, rest and feed within dwellings, and exhibit high breeding capacity. For example, in sub-Saharan Africa, *Anopheles gambiae* mosquitoes, which are efficient hosts of *P. falciparum*, can deliver up to 120 infective bites per person per year, a major reason why approximately 90% of the world's malaria deaths are in Africa.

Climatic conditions, such as rainfall patterns, temperature, and humidity, affect the number and survival of mosquitoes. In many places, transmission is seasonal, with the peak during and just after the rainy season. Malaria epidemics can occur when climate and other conditions suddenly favour transmission in areas where people have little or no immunity to malaria. They can also occur when people with low immunity move into areas with intense malaria transmission, for instance, to find work or as refugees.

Malaria now occurs exclusively in tropical and subtropical regions. In 2015, the WHO estimated there were 214 million cases of malaria, a decline of 18% from the year 2000. Over the same period, estimated deaths fell by 48% to 438,000 in 2015, of which 306,000 were in children under 5 years old.

Risk of infection with malaria parasites is vector dependent, that is, related to proliferation of *Anopheles* mosquitoes, which in turn is linked to environmental factors such as rainfall, temperature, or altitude, coupled with factors such as farming practices that allow collection of stagnant water where mosquito larvae can mature. Even within relatively small geographical areas, the incidence of malaria can vary widely, depending on local factors, and outbreaks can be predicted by monitoring rainfall patterns. Ecological analysis is valuable, but the overall models of transmission and severity of malaria remain complex, and environmental features may not have the same effect in different settings.

The nature of clinical malaria depends greatly on the background level of acquired immunity in the host, which is affected by the pattern and intensity of malaria transmission in the area of residence (see Table 17.1).

Malaria endemicity was defined in 1950 by the WHO and based upon palpable spleen rates in children two to nine years of age.

- hypoendemic (≤10%), very intermittent transmission

- mesoendemic (11 to 50%), regular seasonal transmission
- hyperendemic (51 to 75%), intense, but with periods of no transmission during dry season
- holoendemic (≥75%), transmission occurs all year long

The activity of the anopheline vector of malaria provides the basis for calculating the entomological inoculation rate (EIR), the number of infectious female anopheline bites per person per year. In holo- and hyperendemic areas (for example, certain regions of tropical Africa or coastal New Guinea where there is intense *P. falciparum* transmission), people may receive more than one infectious mosquito bite per day.

Broadly speaking, an EIR of <10 per year is low transmission, 10 to 49 per year is intermediate, and ≥50 per year is a high-transmission area. However, it is important to remember that the methods used to calculate EIR do not take into account ecological, demographic, and socioeconomic differences within a population.

Constant, year-round infection is termed stable transmission, generally occurring in areas with EIRs of ≥50 per year; in such areas, most malaria infections in adults are asymptomatic because frequent exposure throughout life enables development of protective immunity against disease although not against infection (Figure 17.1). Development of immunity to malaria is often referred to as premunition—the ability of the immune system to respond to parasites without eliminating the infection entirely. Premunition involves both B and T cells responding to frequent antigen exposure from infective bites. It is acquired relatively rapidly, is only partially effective, and is short-lived. However, it

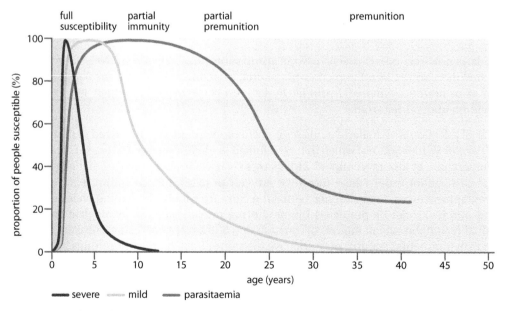

Figure 17.1 Relation between age and malaria severity in an area of moderate transmission intensity. With repeated exposure protection is acquired, first against severe malaria (red line), then against illness with malaria (blue line), and, much more slowly, against microscopy-detectable parasitaemia (purple line). (From White N, Pukrittayakamee S, Tinh Hien T et al. [2014] *Lancet* **383**:723–735. With permission from Elsevier.)

is highly effective at preventing severe disease and occurs in hyper- and holoendemic areas. In areas where transmission is low, highly seasonal, or focal, immunity is not acquired or only partially acquired, and symptomatic disease occurs at all ages; this is termed unstable transmission. Malaria behaves like an epidemic disease in areas with unstable transmission. Even in areas with stable transmission, epidemics may develop; climactic instability such as heavy rains following a period of drought, migration of refugees from a nonmalarious region to an area of high transmission, and disruption in malaria control and treatment will alter the epidemiological picture.

Malaria infection may also be transmitted across the placenta, via blood transfusion or solid organ transplantation, and via contaminated medical equipment, although these modes of transmission are far more rare.

BIOLOGY OF THE ORGANISM

The life cycle of malaria is shown in Figure 17.2.

Infection with malaria is initiated when sporozoites are injected with the saliva of a feeding mosquito. Within 60 minutes the circulatory system delivers sporozoites to the liver and hepatocyte invasion occurs. Once within a hepatocyte, the parasite undergoes asexual replication, known as pre-erythrocytic schizogony. This asexual reproductive cycle culminates in the production of merozoites, which are released into the bloodstream.

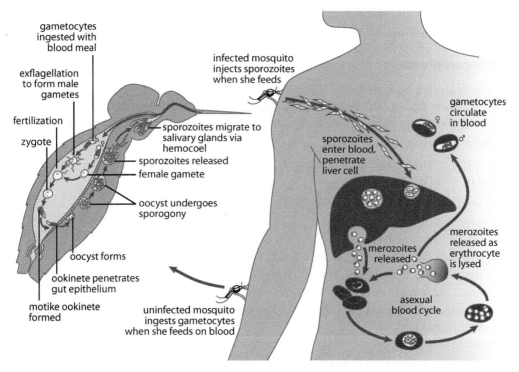

Figure 17.2. **Malaria life cycle.** (Adapted from Su X, Hayton K & Wellems TE [2007] *Nat Rev Genet* **8**:497–506. With permission from Macmillan Publishers Ltd.)

Merozoites invade erythrocytes and the parasite enlarges, producing a trophozoite. Ingestion of host cytoplasm and digestion of hemoglobin into amino acids occurs as the enlarging parasite becomes increasingly metabolically active. Finally, nuclear division occurs, resulting in a schizont-containing merozoites.

The infected erythrocyte ruptures releasing the merozoites that invade erythrocytes and thus initiate the erythrocytic cycle.

A few merozoites do not develop to schizonts but differentiate into gametocytes, which are ingested by female mosquitoes during feeding. In the mosquito's midgut, the gametocytes develop into male or female gametes and fertilization takes place, resulting in the formation of ookinetes. The ookinetes penetrate the mosquito gut wall and attach to the external gut membrane. Oocysts are formed and produce sporozoites that then migrate to the mosquito salivary glands.

In *P. vivax* and *P. ovale* infections, a small number of liver stage parasites do not progress to schizonts immediately, but become hypnozoites, dormant stages that are capable of reactivation weeks to months (occasionally more than a year) after the primary infection. For this reason, adequate treatment of *P. vivax* or *P. ovale* requires elimination of both blood and liver stages of the parasite.

DISEASE

P. falciparum malaria

The WHO criteria for the definition of severe malaria consist of a combination of clinical findings and abnormalities on laboratory testing. Note that malaria can progress rapidly, especially in children, and uncomplicated infection can become life-threatening in a short space of time.

Adults and children with severe malaria present differently. Children with severe disease often experience convulsions, coma, hypoglycaemia, metabolic acidosis, severe anaemia, and defects in cognition. Adults tend to experience severe jaundice, acute renal failure, and acute pulmonary oedema more frequently than children.

The WHO lists the following clinical signs of severe malaria (WHO Guidelines for the treatment of malaria, 2nd ed):

- impaired consciousness (including unarousable coma)
- prostration, that is, generalized weakness so that the patient is unable to sit, stand, or walk without assistance
- multiple convulsions: more than two episodes within 24 hours
- deep breathing and respiratory distress (acidotic breathing)
- acute pulmonary oedema and acute respiratory distress syndrome
- circulatory collapse or shock, systolic blood pressure <80 mm Hg in adults and <50 mm Hg in children
- acute kidney injury
- clinical jaundice plus evidence of other vital organ dysfunction
- abnormal bleeding

Corresponding to clinical abnormalities are recognized laboratory findings associated with severe malaria. These include anaemia, haemoglobinuria, renal impairment, hypoglycaemia, metabolic acidosis, hyperlactatemia, and pulmonary oedema (radiological).

In *P. falciparum* infection the level of parasitaemia is a risk factor for death, but this relationship is strongly influenced by the presence or absence of antimalarial immunity. In low-transmission areas, where naturally acquired immunity is low, mortality from *P. falciparum* malaria increases with parasite densities over 100,000 per μL (~ 2% parasitaemia). Higher parasite densities may be well tolerated in areas of high transmission, where the population is frequently exposed to *P. falciparum* and subsequently exhibits some immunity.

Neurological complications of malaria range from abnormal posturing or impaired consciousness, nystagmus, conjugate gaze palsies, and opisthotonus, to seizures, coma, and death. Other risk factors for cerebral malaria include pregnancy, poor nutritional status, HIV infection, history of splenectomy, and host genetic predisposition. In children, it is important to realise hypoglycaemia from to malaria may play a significant role in disturbance of consciousness.

In both *P. falciparum* and *P. vivax* malaria severe anaemia (haemoglobin <5 g/dL, packed cell volume <15% in children; <7 g/dL, packed cell volume < 20% in adults) can ensue. Anaemia can be the result of repeated low-level infections or can occur rapidly in infection with high parasite densities. Haemoglobinuria, the presence of free haemoglobin in the urine, results from excessive intravascular haemolysis.

Severe anaemia, cerebral malaria, and hypoglycaemia can all provoke impaired tissue perfusion which may result in acidosis (plasma bicarbonate < 15 mmol/L) and hyperlactatemia (lactate > 5 mmol/L). The increased metabolic demands found in severe malaria infection coupled with low glycogen stores can cause hypoglycaemia (glucose < 2.2 mmol/L or < 40 mg/dL); this may be exacerbated by the use of quinine as treatment. Impaired tissue perfusion can also result in renal impairment, although haemoglobinuria may also contribute to tubular damage (serum creatinine < 265 μmol/L).

In severe malaria, capillary permeability may be come deranged causing respiratory distress. Differentiating between acute respiratory distress syndrome (ARDS) and pulmonary oedema is extremely difficult, and both can occur in falciparum or vivax malaria. In falciparum malaria, mortality associated with late complications can be as high as 80%.

Bacteraemia occurs at higher frequency in severe malaria, particularly in children. Historically, an association existed between severe falciparum malaria and bloodstream infection with nontyphoidal *Salmonella* spp.

Malaria infection in pregnancy requires special consideration as it poses significant health risks to both mother and foetus, including maternal anaemia, miscarriage, low birth weight, congenital infection, and maternal death. The prevalence of malaria in pregnancy is related to intensity of transmission—in holoendemic areas women have a higher frequency of malaria episodes. A feature of falciparum malaria in pregnancy is the unique ability of parasitized red blood cells to sequester in the intervillous space of the placenta. These parasites express a specific class of variant surface antigens that mediate adhesion of parasite-infected erythrocytes to chondroitin sulfate A on the syncytiotrophoblast lining the intervillous space.

Parasitized erythrocytes adhere to placental villi and induce accumulation of inflamma-tory leukocytes, followed by syncytial degradation, increased syncytial knotting, and, in rare cases, localized destruction of villi. Importantly, pregnancy selects for parasite clones that have the ability to adhere to the placenta. Partial immunity to these clones develops with successive pregnancies.

P. vivax infection

Severe vivax malaria may present with symptoms similar to those of severe P. falciparum malaria and can be fatal although overall mortality rates for vivax malaria are far lower. Severe disease occurs in low-transmission areas, including India and South America; elsewhere, severe disease is often a reflection of chloroquine resistance.

P. knowlesi infection

P. knowlesi is a newly emerging pathogen of humans in Southeast Asia. Previously only recognised to infect macaques, use of molecular techniques such as PCR have been crucial to its recognition—predominantly because P. knowlesi is often indistinguishable from P. malariae on blood film examination. Whereas the schizogonic cycle of P. falciparum takes 48 hours, that of P. knowlesi takes 24 hours, so its faster cycle can result in rapidly increasing parasite densities, severe disease and death in some individuals. Manifestations of severe P. knowlesi infection are similar to those of severe falciparum malaria. In South-east Asia, patients with severe symptoms should have P. knowlesi considered as a diagnosis.

P. malariae and P. ovale infection

Neither species causes severe malaria and signs of severe disease should prompt consid-eration of mixed infection (with a more pathogenic species), misidentification of the para-sites present, or an alternative diagnosis.

PATHOLOGY

Lysis of parasitized erythrocytes at the completion of schizogony releases merozoites and waste products, including red cell membrane fragments, hemozoin pigment, and glycosylphosphatidylinositol. Waste products activate macrophages and endothelial cells to secrete cytokines and inflammatory mediators such as tumour necrosis factor, interferon-γ, IL-1, IL-6, IL-8, macrophage colony-stimulating factor, and lymphotoxin, as well as superoxide and nitric oxide. Systemic manifestations of malaria, for example, headache, fever and rigors, nausea, and vomiting, can be attributed to these cytokines. In severe malaria, it is postulated that cytokines and small molecules become unregulated and lead to a systemic inflammatory response syndrome-like state.

Parasitization of erythrocytes by P. falciparum causes development of structural abnormal-ities that increase the cytoadherence of red cells. Parasite-related proteins, such as PfEMP-1 or 2 or RESA, are expressed on the surface of erythrocytes as sticky knobs and affected erythrocytes bind to a wide variety of cell types. Adhesion to vascular endothelial cells in capillary beds causes sequestration of erythrocytes, partial vessel obstruction, endothelial cell breakdown, and local inflammation. A microaerophilic environment develops as blood supply is impaired, which favours parasite growth and development while also avoiding splenic clearance of parasitized cells and peripheral immune responses. Cerebral malaria is an extreme manifestation of sequestration. Furthermore, parasitized erythrocytes can adhere to non-infected erythrocytes—a phenomenon known as rosetting, which causes further sequestration and capillary sludging and removes functioning red cells from the circulation. As a result of the sequestration of the growing parasites in the deeper vasculature, only the ring-stage trophozoites of *P. falciparum* are generally seen circulating in the peripheral blood, whereas the more mature trophozoites and schizonts are bound in the deep microvasculature and thus seldom seen on peripheral blood examination.

QUESTIONS

1. Which one of the following drugs is a key component in the treatment of severe malaria?
   a. Mefloquine
   b. Artesunate
   c. Ciprofloxacin
   d. Lumefantrine
   e. Amodiaquine

2. True or false: The following patients are at particular risk of a serious life-threatening infection with malaria:
   a. Children under 12 years
   b. Travelers visiting friends and family
   c. First time visitors
   d. Pregnant women
   e. Men

3. Which of the following is important in reducing mosquito bites?
   a. Insecticide-impregnated nets
   b. House spraying
   c. Eradication of snails
   d. Insecticides
   e. Artesunate

4. Which of the following spleen rates relates to a holoendemic situation?
   a. A 10% spleen rate
   b. A 25% spleen rate
   c. A 50% spleen rate
   d. A 60% spleen rate
   e. an 80% spleen rate

5. Which of the following relates to the definition of the entomological inoculation rate per year?
   a. The number of male mosquitoes
   b. The number of female mosquitoes
   c. The number of mosquito larvae
   d. The number of breeding sites
   e. The number of infectious female bites

GUIDELINES

1. Chiodini PL, Patel D, Whitty CJM & Lalloo DG (2015) Guidelines for malaria prevention in travellers from the United Kingdom, 2015. Public Health England. www.gov.uk/government/publications/malaria-prevention-guidelines-for-travellers-from-the-uk

2. World Health Organisation (2010) Guidelines for the treatment of malaria, 2nd ed. www.searo.who.int/entity/malaria/documents/Glo_treatment_guide/en

3. World Health Organisation (2013) Management of severe malaria—A practical handbook, 3rd ed. www.who.int/malaria/publications/atoz/9789241548526/en

4. WHO Global Malaria Programme. World Malaria Report 2015. www.who.int/malaria/publications/world-malaria-report-2015/report/en

REFERENCES

1. White NJ et al. (2014) Malaria. *Lancet* **383**:723–735.

2. Vythilingam I et al. (2016) Current status of *Plasmodium knowlesi* vectors: a public health concern. *Parasitol* 1–9.

3. WWARN Gametocyte Study Group (2016) Gametocyte carriage in uncomplicated *Plasmodium falciparum* malaria following treatment with artemisinin combination therapy: a systemic review and meta analysis of individual patient data. *BMC Med* 14 (doi: 10.1186/s12916-016-0621-7).

4. Birkholtz LM et al. (2016) Discovering new transmission-blocking antimalarial compounds—challenges and opportunities. *Trends Parasitol* S1471–4922 (doi: 10.1016/j.pt.2016.04.017).

5. Alpern JD et al. (2016) Personal protection measures against mosquitoes, ticks and other arthropods. *Med Clin North Am* **100**(2):303–316.

6. Ndenga BA et al. (2016) Malaria vectors and their blood meal sources in an area of high bed net ownership in the western Kenya highlands. *Malar J.* **15**:76 (doi: 10.1186/s12936-016-1115-y).

7. Dantzler KW et al. (2015) Ensuring transmission through dynamic host environments : host pathogen interactions in *Plasmodium* sexual development. *Curr Opin Microbiol* 17–23 (doi: 10.1016/j.mib.2015.03.005).

8. Hansen DS & Schofield L (2010) Natural regulatory T cells in malaria: host or parasite allies. *PLoS Pathog* **6**:e1000771 (doi:10.1371/journal.ppat.1000771).

9. Shigidi MM et al. (2004) Parasite diversity in adult patients with cerebral malaria: a hospital based case control study. *Am J Trop Med Hyg* **71**:754–757.

10. Soni R, Sharma D & Bhatt TK (2016) *Plasmodium falciparum* secretome in erythrocyte and beyond. *Front Microbiol* **7**:194 (doi: 10.3389/fmicb.2016.00194).

ANSWERS

MCQ	Feedback

1. Which one of the following drugs is a key component in the treatment of severe malaria?
    a. Mefloquine
    b. Artesunate
    c. Ciprofloxacin
    d. Lumefantrine
    e. Amodiaquine
	Artemisinin combination therapy is the mainstay of malaria treatment as per WHO guidelines. Artemisinins eliminate the blood stages of all *Plasmodium* species and have the fastest parasite clearance times of any antimalarial.

2. True or false: The following patients are at particular risk of a serious life-threatening infection with malaria:
    a. Children under 12 years
    b. Travelers visiting friends and family
    c. First time visitors
    d. Pregnant women
    e. Men
	a. True
	b. False
	c. False
	d. True
	e. False
	Children and pregnant women, especially primigravida, are most at risk.

3. Which of the following is important in reducing mosquito bites?
    a. Insecticide-impregnated nets
    b. House spraying
    c. Eradication of snails
    d. Insecticides
    e. Artesunate
	Only insecticide-treated nets prevent mosquito bites. House spraying and insecticides reduce the local burden of mosquitos. Artesunate is effective treatment and also can reduce ongoing transmission, as it is gametocidal. Snail eradication is a control strategy for schistosomiasis.

4. Which of the following spleen rates relates to a holoendemic situation?
    a. A 10% spleen rate
    b. A 25% spleen rate
    c. A 50% spleen rate
    d. A 60% spleen rate
    e. an 80% spleen rate
	Malaria endemicity was defined in 1950 by the WHO and based upon palpable spleen rates in children two to nine years of age.
	• hypoendemic ($\leq$10%), very intermittent transmission
	• mesoendemic (11 to 50%), regular seasonal transmission
	• hyperendemic (51 to 75%), intense, but with periods of no transmission during dry season
	• holoendemic ($\geq$75%), transmission occurs all year long

5. Which of the following relates to the definition of the entomological inoculation rate per year?
    a. The number of male mosquitoes
    b. The number of female mosquitoes
    c. The number of mosquito larvae
    d. The number of breeding sites
    e. The number of infectious female bites
	The activity of the anopheline vector of malaria provides the basis for calculating the entomological inoculation rate (EIR), the number of infectious female anopheline bites per person per year.

A CLINICAL INCIDENT LINKED TO PRION-ASSOCIATED DISEASE

John Holton

Department of Natural Sciences, School of Science & Technology, University of Middlesex.
National Mycobacterial Reference Service - South, Public Health England.

A 75-year-old patient underwent ophthalmic surgery involving the retina to correct visual impairment. Six weeks later, he developed a tremor and a degree of cognitive impairment. Over the following weeks the symptoms progressed. A probable diagnosis of Creutzfeldt–Jakob disease (CJD) was eventually made based on the symptom complex and laboratory investigation. Evidence from his case notes was consistent with symptoms of CJD prior to the operation. It was recognized that there was a risk that reusable surgical instruments could have become contaminated and that there was a possibility of onward transmission of prion disease to other surgical patients.

INVESTIGATION OF THE INCIDENT

The first step in this incident is to establish the level of certainty regarding the diagnosis in the index case and ascertain what sort of prion disease is likely to be involved. Diagnosis of probable CJD is based primarily on clinical presentation. Laboratory investigation is used to exclude other causes as well as to detect the surrogate marker, the 14-3-3 protein in the cerebrospinal fluid (CSF). There are also assays for the direct detection of PrP^{Sc} in the CSF, including protein misfolding cyclic amplification (PMCA) assay and real-time quaking-induced conversion (RT-QUIC). More recently, aggregation of gold nanoparticles by beta-pleated sheet proteins had been used to detect prion proteins in the CSF. Confirmation is provided by histopathological and immunological investigation of brain tissue. Brain biopsies are rarely undertaken during life as they will usually not affect prognosis, and, as the changes are diffuse, a biopsy may miss the relevant area. Confirmation may therefore be post mortem. The classical histopathological features are neuronal loss, spongiform changes, and astrocytic gliosis. In addition, the use of antibodies 3F4 and KG9 provide immunohistochemical confirmation. If a diagnosis of prion disease is possible, it is necessary to assess the risk from the tissues involved in the surgery. Quarantine any reusable instruments pending confirmation of the diagnosis. Prior to storage, the instruments should be decontaminated. Determine the fate of any disposable instruments and confirm if they were incinerated.

In this case, some reusable instruments will already have been used on other patients. Undertake a review to ascertain which patients have been exposed. As well as instruments used during the retinal surgery, the review should also include any endoscopes or reusable laryngoscopes used during anesthesia. For high-risk (retina) and medium-risk (tonsil) tissue, the first ten patients or two patients, respectively, should be identified. After the diagnosis of CJD is confirmed, inform the patients identified as being in contact with the contaminated instruments that they are at an increased risk of CJD and advise them not to donate blood or organs. They will be placed on the at-risk register and may need counselling.

Undertake a root-cause analysis to determine the details of the incident and whether or not the transmissible spongiform encephalopathy (TSE) guidelines were followed. Inform the local Health Protection Team (HPU) of the details of the incident. The local HPU will then manage the public health response. Inform the National CJD Research and Surveillance Unit.

CLINICAL MANAGEMENT

Patients with CJD may suffer from a number of generalized symptoms (see the disease section below) affecting coordination, vision, and memory. There is currently little that can be done for these patients other than symptomatic relief to mitigate the severity of the symptoms. Early attempts at therapeutic interventions, such as the development of specific antibodies targeting the abnormal prion protein, linked to a peptide that facilitates crossing the blood-brain barrier. Other pharmacological agents under investigation for prion disease are an antihistamine-like compound, astemizole; an immunosuppressive drug, tacrolimus; arylpiperazine compounds; and dendromers, which are synthetic polymers. They act by reducing the amount of protein, preventing the conversion to an abnormal isoform, or suppressing inflammation. Several other drugs have been assessed, for example, pentosan polysulfate, and thus far none of them can be regarded as a therapy for CJD.

PREVENTION OF FURTHER CASES

In the surgical context, prevention of onward transmission of TSE relies on identifying patients who are at risk of prion disease before the surgery, using appropriate infection control precautions during the procedure and suitable decontamination or disposal of instruments.

Screening should be routine before surgical or endoscopic procedures and should employ a standardized questionnaire that identifies patients at risk of prion disease. Risk factors that need to be determined are:

1. Family history of prion disease

2. Receipt of human-derived growth hormone or human-derived gonadotrophin

3. History of neurosurgical procedure, in particular a dura mater graft prior to 1992

4. Receipt of a large volume (>50 units) of blood or repeated (>20 occasions) blood transfusion

5. Notification that they have been exposed to a prion disease through other iatrogenic routes

If the responses are all negative, most surgeries can proceed routinely. For surgeries involving high risk tissues, the surgeon should also review the patient's notes and ensure that the patient is showing no clinical signs of prion disease.

Clinical waste containing high- or medium-risk tissue from the patient should be incinerated but, if any of the responses are affirmative, the patient should be considered as at risk of having prion disease and enhanced infection prevention and control measures should be employed during the surgery. The patient should also be referred to the appropriate care pathway for information or investigation as appropriate.

Table 18.1. **Tissue infectivity**

Level of infectivity	System	Tissue type
High	Nervous	Brain, spinal cord, implanted dura mater grafts pre-1992 (UK), cranial nerves, intracranial part of other cranial nerves, cranial nerve ganglia, optic nerve
	Visual	Posterior eye, posterior hyaloid face, retina, retinal pigment epithelium, choroid, subretinal fluid, optic nerve
	Endocrine	Pituitary gland
Moderate	Lymphoreticular	Spleen, lymph nodes, tonsils, thymus
	Gastrointestinal	Jejunum, ileum appendix, large intestine
	Nervous	Spinal ganglia
	Other	Adrenal, blood vessels, nasal mucosa
	Body fluids	Blood
Low	Reproductive	Testes, prostate, semen, ovary, uterus
	Musculoskeletal	Heart, pericardium, skeletal muscle
	Visual	Anterior eye, cornea
	Nervous	Peripheral nerves
	Other	Gingiva, dental pulp, trachea, skin, fat, thyroid
	Body fluids	Milk, saliva, sweat, tears, nasal mucus, urine, faeces

Infection prevention and control procedures for performing surgery on at-risk patients or probable or confirmed prion disease cases are based on the infectivity of the tissues involved in the procedure, as indicated in Table 18.1. Collection of routine preoperative samples such as blood, urine, sputum, and other low-risk materials require no additional precautions other than normal universal precautions. Specimens sent to the laboratory should be labelled as biohazard. Low-risk samples can be disposed of as normal clinical waste.

The operation should be at the end of the day's list and the minimum safe number of healthcare staff should be involved. Personal protective equipment (PPE) should be worn, consisting of fluid-repellent gown, gloves, mask, and goggles. If the patient is symptomatic, the personal protective equipment should be single use. Single-use disposable equipment should be used wherever possible. Where reusable instruments are required, those that are in contact with low-risk tissue must be kept separate from those in contact with medium- or high-risk tissue.

Postoperatively the handling of the instruments depends on the tissues involved. Note that there are more medium-risk tissues if the patient has variant Creutzfeldt–Jakob disease (vCJD) than other forms of CJD. The requirements for handling of surgical instruments are necessary because standard decontamination procedures do not eliminate prion proteins. Some enhanced decontamination agents have increased activity against prion proteins, but they are unpredictable in efficacy and also generally impractical for use on surgical instruments. Examples include 96% formic acid applied for 1 h, 20,000 ppm sodium hypochlorite exposure for 1 h, or autoclaving at 134–137°C for 18 min. Novel approaches such as the use of a thermostable protease to degrade the prion protein have been shown in studies to be potentially effective, though they are not widely available.

Single-use instruments in contact with high- or medium-risk tissue should be sealed in flammable containers and incinerated. Instruments used on high- or medium-risk tissue should be washed free of material by an operator wearing personal protective equipment and with the instrument held under water to prevent aerosol generation. After drying, they should be placed on an impervious instrument tray and sealed in a flammable container. If the case is definite, the container should be incinerated. If the case is currently probable or at risk, the instruments should remain quarantined in a secure location. If the case is subsequently confirmed to have prion disease, the unopened box should be incinerated. If the diagnosis is refuted, the instruments can be processed as normal and reused. Instruments used only on low-risk tissues can be processed as usual.

EPIDEMIOLOGY

Prion diseases are rapidly progressive fatal neurodegenerative conditions that are caused by an abnormal version of a normal membrane protein: PrP^c. Both animals and humans can be affected; in animals the equivalent diseases are called transmissible spongiform encephalopathies (TSE). A wide range of animals is affected: sheep (scrapie); cattle (bovine spongiform encephalopathy; BSE); mink (transmissible mink encephalopathy; TME); deer, elk, moose, and reindeer (chronic wasting disease; CWD); cats (feline spongiform encephalopathy; FSE), and various ungulates (exotic ungulate encephalopathy; EUE). Genetic human prion diseases are Gerstmann–Straussler–Scheinker (GSS) disease; fatal familial insomnia (FFI); and familial Creutzfeldt–Jakob Disease (fCJD) caused by specific mutations in the PrP gene. Sporadic prion diseases are sporadic Creutzfeldt–Jakob Disease (sCJD) and fatal sporadic insomnia (FSI), caused by spontaneous conversion of the normal PrP^c to the abnormal PrP^{Sc} (Figure 18.1). Infectious human prion diseases include kuru, associated with transmission through cannibalism in the Fore people of New Guinea. The practice of cannibalism has now been prohibited.

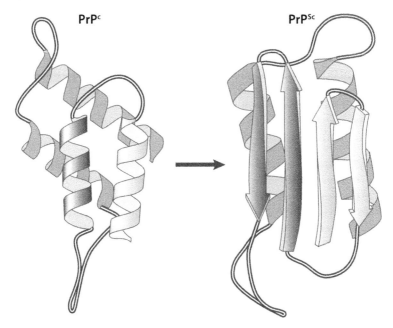

Figure 18.1 **The conversion of normal prion (PrP^c) to abnormal prion (PrP^{Sc}).** (From Hofkin BV [2017] Living in a Microbial World, 2nd ed. Garland Science.)

In the UK between 1990 and 2014, there have been 1926 deaths caused by CJD: 1514 sCJD; 161 fCJD; 177 vCJD; and 76 iatrogenic Creutzfeldt–Jakob disease (iCJD).

The variant type of CJD was first reported in 1995 in the UK and linked to the consumption of meat from cattle that were suffering from BSE. Since 1995 in the UK, there have been 177 deaths from vCJD, principally between 1996 and 2003, with a subsequent reduction in numbers. In the UK, histological analysis of tonsil tissue and appendix tissue in anonymized population studies before and after the recognition of vCJD have been performed to assess the risk and likely public health problems associated with vCJD. Examination of 32,441 appendix samples from 41 hospitals demonstrated that 16 were positive for abnormal PrP. This finding indicates an overall prevalence of 493 per million population. The prevalence from 1941 to 1960 (before the BSE outbreak) was not significantly different from those born between 1961 and 1985. The rate of carriage of PrP^{Sc} in the population is much higher than the number of cases of vCJD, suggesting a potential problem in future years.

There is no evidence that prion diseases are transmitted by social or nursing contact, or by any route other than direct inoculation with, or ingestion of, infected tissues. Patients do not require isolation. There is no evidence of vertical transmission based on one study of 125 children born to a parent with vCJD. Iatrogenic transmission is a continuing concern with over 250 cases reported worldwide. In the UK between 1990 and 2014, there have been 76 deaths from iCJD. Iatrogenic CJD has been associated with dura mater grafts, the use of growth hormone prepared from the pituitary of infected individuals, blood transfusions, and contact with PrP^{Sc}-contaminated surgical instruments.

BIOLOGY

The normal prion protein (PrP^c) is a glycosylphosphatidylinositol (GPI) anchored membrane glycoprotein found in lipid rafts in cell membranes and consists of 208 amino acids. It is highly expressed in nervous tissue with lower expression in lymphoid tissue. The protein is cycled through the endoplasmic reticulum, where it is glycosylated and then inserted into the cell membrane, where it resides until the protein is internalized and degraded in the endolysosomal pathway by endogenous proteases.

Normal PrP protein seems to have at least two functions. It is involved in developing and maintaining the myelin sheath by signalling the Schwann cell to cover the axon with myelin. PrP^c null mice, in which the protein is missing from the neuron, develop a chronic demyelinating peripheral neuropathy that can be reversed by expression of the protein in neurons but not in the Schwann cells. A second function is related to strengthening synaptic connections in the hippocampus that involve activation of protein kinase A; in null mice, synaptic connections are decreased and protein kinase C is expressed. PrP^c may also be involved in long-term memory. Many other functions have been ascribed to PrP^c including hemopoetic stem cell renewal, inducing or preventing apoptosis, and relieving oxidative stress.

DISEASE

Prion diseases have a progressive and often fatal course. The classical presentation is mental deterioration, which may affect concentration, sleep, mood, and cognitive functions, accompanied by myoclonus. However, presentations are diverse and include cerebellar syndromes, spasticity, and visual deficits. In sCJD, the time course of the illness is 8 months after clinical presentation with symptoms dominated by ataxia, failing cognition, and visual symptoms leading to blindness. In vCJD the onset of disease occurs at an earlier age (26 years on average) but has a more prolonged course before death (15 months on average). Psychiatric and sensory symptoms predominate. Prion diseases are invariably fatal once clinical symptoms appear.

PATHOLOGY

Sporadic CJD has diverse neuropathological features. Genetic testing of the *prnp* gene demonstrates a polymorphism at codon 129 (ATG/GTG) leading to either a methionine or valine in the protein. Classification of sCJD can be made based upon this polymorphism (MM, MV, or VV) combined with different isoforms of the PrP^{Sc} protein. Type 1 isoform is 21 kDa and nonglycosylated; Type 2A is 19 kDa and monoglycosylated; Type 2B is diglycosylated. The most common subtype is MM1 (57%). Other subtypes are MM2A (7%), MV1 (6%), MV2A (14%), VV1 (2%), and VV2A (14%). The subtyping scheme correlates with the neuropathological appearance and relates to the size and distribution of the vacuoles and the pattern of PrP distribution. The neuropathological appearance for the MM2A subtype varies depending if it is thalamic (MM2T) or cortical (MM2C).

The neuropathological features and polymorphism of vCJD are different from sCJD, in that vCJD has a more homogenous neuropathology, illustrated by prominent plaques, more severe spongiform changes, and a characteristic distribution of the PrP^{Sc} protein. Also only the MM subtype at position 129 is found. The isoform of the protein is different from that in sCJD (Type 2B). Additionally, the PrP^{Sc} protein is found outside the central nervous system (CNS) in lymphoreticular tissue—tonsil, lymphocytes, appendix, spleen, gut-associated lymphoid tissue (GALT)—and muscle.

The PrP^{c} protein is composed of alpha-helices and its key biophysical property is a change to a beta-pleated sheet (see Figure 18.1). In this abnormal form (PrP^{Sc}), the key biochemical property is the carboxy-terminal end of the protein, which becomes resistant to proteolytic cleavage and the protein self-aggregating, which in effect precipitates in the tissues. It is not understood how the conformational change occurs nor how precipitation occurs, although two methods have been suggested: either a template-directed misfolding induced by PrP^{Sc}, or an equilibrium model in which PrP^{Sc} naturally coexists with PrP^{c} with the equilibrium toward PrP^{c}. Over time or in the presence of some other factor the PrP^{Sc} begins to aggregate, precipitate, and induce the disease.

Finally, the mechanism or mechanisms whereby PrP^{Sc} induces neuropathological changes and the actual clinical aspects of the disease are not understood. Interestingly, PrP^{c}-knockout mice have several subtle physiological changes and introduction of PrP^{Sc} does not induce disease. On the other hand, the presence of PrP^{c} in the cytosol induces neurodegeneration but in the absence of PrP^{Sc}.

Recently a new prion disease has been identified called variable protease-sensitive prionopathy (VPSPr). VPSPr occurs in all *prnp* genotypes and has limited transmission dynamics. Histopathologically, the appearance is different from sCJD showing microplaques in the thalamus and cerebellum. Biochemically the protein is not as resistant to protease digestion as other forms of CJD and the protease-resistance varies with the *prnp* genotype. A prudent approach would be to treat patients as if they had CJD.

QUESTIONS

1. Which of the following is not important in investigating exposure to PrP^{Sc}?
   a. Determination of the type of CJD in the index case
   b. Whether the index case has had growth hormone in the past
   c. Whether the person is a heavy smoker
   d. Whether any reusable instruments have been used on the index case
   e. The family history of the index case

2. True or False: The following are important in preventing transmission of PrP^{Sc} during surgery:
   a. A screening questionnaire prior to surgery
   b. Special precautions when handling urine samples
   c. Placing the operation first on the list
   d. Use of disposable equipment where possible
   e. Incineration of reusable instruments used on retinal tissue in confirmed CJD

3. Which of the following is linked to consumption of the brain?
   a. iCJD
   b. Kuru
   c. vCJD
   d. GSS
   e. sCJD

4. Are the following statements true or false concerning PrP^C?
   a. It is a membrane bound protein.
   b. It is resistant to proteinase K hydrolysis.
   c. It is an abnormal protein conformer.
   d. It may be involved with memory.
   e. It may be involved with myelin metabolism.

5. Are the following statements true or false concerning CJD?
   a. Polymorphism occurs at position 129 in the *prnp* gene.
   b. Two isoforms of PrP^{Sc} can be found.
   c. The most common polymorphism is VV 1.
   d. The normal isomer is the beta-pleated sheet.
   e. In vCJD PrP^{Sc} can be found in tonsils.

GUIDELINES

1. WHO Infection Control Guidelines for Transmissible Spongiform Encephalopathies (1999). WHO/CDS/CSR/APH/2000.3
2. www.gov.uk/government/publications/guidance-from-the-acdp-tse-risk-management-sub-group-formerly-tse-working-group.
3. www.nice.org.uk/IPG196. Patient safety and reduction of risk of transmission of Creutzfeldt-Jakob disease (CJD) via interventional procedures (IPG196).

REFERENCES

1. Communicable Disease Network Australia Infection Control Guidelines for the Prevention of Transmission of Infectious Diseases in the Health Care Setting (2004) Ch. 31.
2. Rutala WA & Weber DJ (2001) Creutzfeldt-Jakob disease: recommendations for disinfection and sterilization. *Clin Infect Dis* **32**:1348–1356.
3. Berberidou C et al. (2013) Homogenous photocatalytic decontamination of prion infected stainless steel and titanium surfaces. *Prion* **7**(6):epub.

4. Secker TJ et al. (2012) Doped diamond-like carbon coatings for surgical instruments reduce protein and prion-amyloid biofouling and improve subsequent cleaning. *Biofouling* **28**:563–569.

5. Aguzzi A et al. (2013) The immunobiology of prion disease. *Nat Rev Immunol* **13**:888–902.

6. Aguzzi A et al. (2007) Insights into prion strains and neurotoxicity. *Nat Rev Mol Cell Biol* **8**:552–561.

7. Kovacs GG & Budka H (2008) Prion disease: from protein to cell pathology. *Am J Pathol* **172**:555–565.

8. Torrent J et al. (2015) The volumetric diversity of misfolded prion protein oligomers revealed by pressure dissociation. *J Biol Chem* (jbc.M115.661710).

9. Skinner PJ (2015) Treatment of prion disease with heterologous prion proteins. *PLOS One* (doi: 10.1371/journal.pone.0131993).

10. Asante EA et al. (2015) Transmission properties of human PrP102L prions challenge the relevance of mouse models of GSS. *PLOS Pathol* (doi: 10.1371/journal.ppat.1004953).

11. Cassard H et al. (2014) Evidence for zoonotic potential of ovine scrapie prions. *Nat Commun* (doi: 10.1038/ncomms6821).

ANSWERS

1. Which of the following is not important in investigating exposure to PrP^{Sc}?
   a. Determination of the type of CJD in the index case
   b. Whether the index case has had growth hormone in the past
   c. Whether the person is a heavy smoker
   d. Whether any reusable instruments have been used on the index case
   e. The family history of the index case

The type of case and family history is important epidemiologically. Individuals who have had growth hormone, because it was prepared from human sources and thus may carry prions, are in a risk category. Reuseable instruments used should be quarantined until the diagnosis is confirmed and can be used on the same patients. Otherwise the instruments should be incinerated.

2. True or False: The following are important in preventing transmission of PrP^{Sc} during surgery:
   a. A screening questionnaire prior to surgery
   b. Special precautions when handling urine samples
   c. Placing the operation first on the list
   d. Use of disposable equipment where possible
   e. Incineration of reusable instruments used on retinal tissue in confirmed CJD

   a. True
   b. False

   c. False
   d. True
   e. True

A screening questionnaire is used before surgery to quantify the risk the patient has of being TSE positive, which in turn determines how the case should be treated. Urine is not a high-risk specimen. If possible operations on confirmed or suspect cases should be last on the list. Because it is very difficult to decontaminate instruments used on potential or actual cases, it is best to use single-use instruments wherever possible to reduce the risk of onward transmission. If contaminated instruments are not to be used on the same patient, they should be incinerated.

3. Which of the following is linked to consumption of the brain?
   a. iCJD
   b. Kuru
   c. vCJD
   d. GSS
   e. sCJD

The only TSE to be linked to cannibalism is Kuru, which occurred in Papua New Guinea.

4. Are the following statements true or false concerning PrP^{C}?
   a. It is a membrane bound protein.
   b. It is resistant to proteinase K hydrolysis.
   c. It is an abnormal protein conformer.
   d. It may be involved with memory.
   e. It may be involved with myelin metabolism.

   a. True
   b. False
   c. False
   d. True
   e. True

PrP^{c} is the normal glycosylated membrane-bound protein that is sensitive to digestion. The abnormal nondigestible conformer is PrP^{Sc}. The function of the normal protein is thought to be related to long-term memory and physiologically seems to be involved in maintaining the myelin sheath and synaptic connections.

5. Are the following statements true or false concerning CJD?
   a. Polymorphism occurs at position 129 in the *prnp* gene.
   b. Two isoforms of PrP^{Sc} can be found.
   c. The most common polymorphism is VV 1.
   d. The normal isomer is the beta-pleated sheet.
   e. In vCJD PrP^{Sc} can be found in tonsils.

   a. True

   b. True
   c. False
   d. False
   e. True

Polymorphism occurs at position 129. There are two isoforms Type 1 and Type 2, although the latter can be subdivided into 2A and 2B. There are three possible polymorphisms: MM, MV, and VV; the polymorphism MM is the most common. The normal protein is an alpha-helix and the abnormal protein undergoes a conformational change to a beta-pleated sheet. The abnormal protein has been found in tonsils.

AN OUTBREAK OF *PSEUDOMONAS AERUGINOSA* IN A NEONATAL INTENSIVE CARE UNIT

Trupti A. Patel,[1] Michael Kelsey[2]

[1]Royal Free Hospital NHS Trust, London, UK.
[2]The Whittington Hospital, London, UK.

A low-birth-weight preterm infant was transferred to an English neonatal intensive care unit for respiratory support. *Pseudomonas aeruginosa* was cultured from an eye swab a week after admission. The neonatal unit had 34 cots: 12 intensive care cots, 6 high-dependency cots, and 16 special care cots.

Four days later, *P. aeruginosa* was found in a blood culture from a two-week-old neonate, who was in a cot situated at the opposite end of the unit. This baby had been admitted since birth with problems related to prematurity and necrotizing enterocolitis. He had a femoral central venous catheter, from which the blood culture had been taken, and an arterial catheter.

One week later a third baby with congenital hydrocephalus, admitted for ventricular shunt insertion, developed *P. aeruginosa* ventriculitis postoperatively. At this stage an outbreak was declared. Following these infections, 14 patients developed symptomatic infection over a six-week period before the outbreak was declared over: four had bacteraemia and multiorgan failure, three of whom died, six had ventilator-associated pneumonia, and four had ear infections or conjunctivitis.

INVESTIGATION OF THE INCIDENT

Alert organisms are microorganisms that have the potential to cause harm and disease in individuals and that can cause an outbreak of infection in a hospital environment. Since *P. aeruginosa* was an alert organism in this unit, the infection prevention and control (IPC) team began an investigation to determine if the first two cases were linked. An outbreak in a hospital will usually be declared by the Consultant Microbiologist or the Director of Infection Prevention and Control (DIPC) and the establishment of an Outbreak Control Team (OCT) should follow immediately. Key roles and responsibilities of each member are listed in Table 19.1. Outbreaks can occur either with a single strain (usually from a point source) or from an increase in the incidence of infection caused by one or more organisms, usually indicating a breakdown of infection prevention protocols.

Table 19.1. **Key roles and responsibilities in managing *P. aeruginosa* outbreaks**

Stakeholder	Responsibilities
OCT Chair (usually DIPC or Consultant Microbiologist)	Direct and coordinate overall management of outbreak Ensure communication between members of the OCT and other parties Responsible with OCT for declaring outbreak over
Consultant Microbiologist	Present microbiological information relating to the outbreak Identify resources to enable microbiological testing to be undertaken speedily and efficiently and to report on this to the OCT Provide advice and guidance on the microbiological aspects of the investigation and control of the outbreak Arrange microbiological testing of relevant patient and environmental samples and to arrange molecular typing at reference laboratory Provide results of all testing to the OCT Assist clinical colleagues with treatment protocols
Consultant in Public Health Medicine (or Healthcare Epidemiologist)	Provide local epidemiological support Liaise with commissioning authority to ensure resources are available to investigate outbreak Maintain heightened surveillance of infection in other local healthcare establishments during (and if required after) outbreak to evaluate interventions Develop materials for training from outbreak lessons Audit management of local outbreak
Public Health England Health Protection representative (or equivalent outside of England)	National review of surveillance data and guidance on whether observed increase is a local outbreak or involves other hospitals Technical and expert advice on incident management and investigation tools Advice and specialist microbiological support including molecular typing of isolates
Infection Prevention and Control representative	Provide results of investigations into control processes such as hand hygiene audits and cleaning records Provide surveillance data on colonization rates of *P. aeruginosa*, if applicable Ensure application of, and compliance with, the evidence-based guidelines for preventing healthcare-associated infections Ensure best practice advice relating to handwashing basins is followed to minimize the risk of *P. aeruginosa* contamination Observe and record clinical practice in relation to water sources and identify any hazardous procedures or deviation from good practice Coordinate control measures in preventing further cases
Clinical Lead for affected area	Support the investigation of the OCT by providing resources within the affected unit as needed
Water Safety Group* (WSG) representative / Estates and Facilities Manager (or equivalent from engineering and maintenance)	Provide the most recent copy of the hospital's water safety plan (WSP) Provide information on any recent microbiological hazards investigated by the WSG Provide records of any recent changes to the water-distribution system Provide routine maintenance and monitoring records, including water temperatures and microbiological testing results Develop incident protocols Advise on remedial action when water systems or outlets are found to be contaminated Produce a report in appropriate and timely manner to OCT Provide records of cleaning of affected area
Relevant healthcare professionals	Feedback information from OCT meetings to physicians and nurses responsible for care of patients Alert the OCT of any presumptive cases Facilitate implementation of infection prevention and control measures during the outbreak
Administrative support	Take minutes and action notes at each meeting; produce and distribute a timely written record of the meeting to the group members
Communications manager	Advise on media management (for example, proactive versus reactive) Draft media messages in collaboration with clinical lead Liaise with communications managers of other involved stakeholders

*Every healthcare institution is mandated to have a WSG that is a multidisciplinary group formed to undertake the commissioning and development of the WSP. The WSP provides a risk-management approach to the microbiological safety of water and establishes good practices in local water usage, distribution, and supply (see Guideline 1).

There are two key factors to establish initially: identification and typing of the organism, and identification of the source. In this case, the first two isolates were sent to the laboratory and were identified as *P. aeruginosa*. Typing, in this case by variable number of tandem repeats (VNTR), demonstrated the isolates were different strain types with different VNTR profiles and an outbreak was not declared.

Once the third case was identified, an outbreak was declared. Control processes for minimizing the risk of *P. aeruginosa* infection were investigated, particularly the examination of hand hygiene audits and routine environmental investigations related to the hospital water system. Estates and facilities staff (responsible for engineering and maintenance) should have kept accurate records and diagrams of the layout and operational manuals of the whole water system. That information would help identify areas of water stagnation such as dead legs (plumbing cul-de-sacs). Training of maintenance staff also needs to be deemed adequate, and maintenance and estates cleaning schedules checked. Environmental samples were collected from appropriate areas such as water outlets, showerheads, ice machines, mop heads, and equipment in contact with neonates, such as incubators.

In addition, a plan for screening was made to identify those who were colonized but not necessarily infected. Such screening would normally involve nose and throat swabs, endotracheal swabs, and rectal swabs or a stool specimen. A database containing the results of these investigations and clinical characteristics was created. Further demographic information of admission and transfer dates, bed movements, and contact with water and high-risk equipment, such as ventilator equipment, was gathered.

An additional 14 symptomatic infections were identified during the outbreak investigation. The source of the outbreak, after these investigations, including typing using both VNTR and pulsed field gel electrophoresis (PFGE), was found to be a contaminated water tap that had also been used for washing respiratory equipment. However, patient-to-patient transmission by healthcare workers also played a significant role.

Continuous communication between members of the OCT and with those affected in clinical areas is paramount to preventing further cases of infection. Information disseminated to the media and general public should be the responsibility of the Communications Manager. The decision of taking a proactive or reactive stance will depend on the nature and extent of the outbreak. Press statements should be prepared and reviewed by the OCT. Consideration should be given of the method of communication with affected patients, parents, and caregivers, depending on the audience and level of understanding.

CLINICAL MANAGEMENT

Individual cases should be managed with appropriate antimicrobials such as third- and fourth-generation cephalosporins, ureidopenicillin, carbapenems, fluoroquinolones, or aminoglycosides after discussion with the microbiologists. If the isolate is multidrug-resistant or part of a suspected outbreak strain, then cases will need to be isolated in side rooms or in cohorts. In a neonatal unit, it is often sufficient to isolate infected patients within a designated area of the ward because nursing is provided on a one-to-one basis and infants are contained within incubators. Staff should be reminded, however, of the importance of hand hygiene to prevent further transmission. Generally, with a limited number of single-rooms, these should be reserved for patients who have conditions that facilitate transmission of infectious material to others (for example, draining wounds, stool incontinence not contained within diapers, and uncontained secretions) or for those who are at increased risk of acquisition and adverse outcomes (for example, immunosuppression).

Debate continues as to whether combination therapy with more than one active agent is beneficial for the treatment of bacteraemia. Recent data, however, have failed to demonstrate a benefit of using more than one agent to which the isolate is susceptible.

Data on *P. aeruginosa* bacteraemias in England demonstrates the proportion of isolates that were resistant to piperacillin and tazobactam increased by 3% (from 6 to 9%) between 2008 and 2012, whereas resistance to imipenem (13%), meropenem (9%), ceftazidime (6%), ciprofloxacin (9%), and gentamicin (4%) reduced slightly or remained static. There are regional as well as international differences in susceptibility patterns, and local data should be used to inform treatment.

In recent years, there has been a rapid increase in the incidence of infections and colonization by multidrug-resistant strains (resistance to three or more classes), extensively drug-resistant strains (resistance to all but one or two classes), or even pandrug-resistant strains. A proportion of these are acquired carbapenemase (metallo-β-lactamase) producers, organisms that produce β-lactamases that hydrolyse carbapenems. These are of major concern as carbapenems are often viewed as the last resort therapeutically, and outbreaks of infection caused by these organisms are difficult to treat and are associated with a high mortality. Because these organisms are extremely drug resistant, few therapeutic options exist but include polymyxin B and E (colistin) and fosfomycin. Susceptibility to an aminoglycoside may be retained in some isolates.

PREVENTION OF FURTHER CASES

An augmented care unit is one in which water quality must be of a higher standard than that provided by the supplier. (Water supplied from the public mains, or a private company, to healthcare premises must comply with current legislation and is considered to be potable, although it is unusual for this water to be entirely free from aquatic organisms.) Most care that is documented as augmented will be in settings where medical procedures render the patients susceptible to invasive disease from environmental and opportunistic pathogens, such as *P. aeruginosa*. Such patient groups include those who are immunosuppressed, such as bone marrow transplant recipients, those cared for in units where organ support is necessary (critical care, renal, respiratory—including cystic fibrosis units), or those with extensive breaches in their dermal integrity (burns patients). (See Guideline 3.)

Given that pseudomonads form part of the ecosystem of water, it is not surprising that they enter human water distribution systems. From a legal perspective, although water suppliers are obliged to deliver water to healthcare premises free of prescribed organisms such as *Escherichia coli*, *Enterococcus faecalis*, and *Clostridium perfringens* (indicating animal or human faecal contamination), no such standards have been set for *P. aeruginosa*. As long as water suitable for drinking is delivered "wholesome at the time of supply," in that it is not unacceptable because of taste or odour, there is no limitation on the supply of *P. aeruginosa* (Guideline 5).

Central to water-outlet testing methods is the concept that although most bacteria are trapped within a biofilm on plumbing materials throughout the water system, the biofilm constantly generates bacteria that are released in free-floating individual cells (planktonic forms). In order to maximize the recovery of free-floating planktonic bacteria that cause infection, water samples from outlets should be taken during a period of no use (at least two hours or preferably longer) or low use. A preflush and postflush sample should be taken. If the source of *P. aeruginosa* is located at or near the outlet, the postflush sample will give lower bacterial counts than the preflush sample. If contamination is upstream, flushing will not affect bacterial counts.

Water outlets that should be sampled are those that supply water that is in direct contact with patients, is used to wash staff hands, or is used to clean equipment that will have contact with patients. Satisfactory water for augmented care units should have no detectable *P. aeruginosa*. Counts between 1 and 10 colony forming units (cfu) per 100 mL require retesting and referral to the WSG to perform a risk assessment on water use in the unit. Counts of >10 cfu per 100 mL are not satisfactory and these outlets should be removed from service if preflush counts are higher than postflush counts. If both preflush and postflush samples from a particular outlet are >100 cfu per 100 mL and other nearby outlets have no or low counts, it shows that the single outlet is heavily contaminated despite the high postflush count. If the sampling indicates that the water services are the problem, then most outlets would possibly be positive and other points in the system could then be sampled to assess the extent of the problem.

Other than sampling water outlets, environmental samples should also be taken from in and around areas that have contact with tap water, including swabs of handwashing basin surfaces, patient equipment such as incubators, humidifiers, milk bank pasteurizer, and bottle warmers, as well as samples from ice machines and water coolers. Environmental isolates should also be sent for molecular typing as it is likely that more than one strain type will be present and the results can then be compared with the isolates obtained from clinical samples.

If a particular water outlet has been taken out of service because of contamination, daily flushing while the outlet is out of use is necessary to prevent water stagnation and exacerbation of the contamination. Replacing contaminated taps with new taps may not provide a permanent solution. When replacement taps are fitted, consideration should be made for the installation of removable taps that are easy to use, may be easily dismantled, and allow filters to be attached to the spout. Entire handwashing basins may need to be removed if found to be heavily contaminated, especially if they are underused, because recolonization is more likely to occur.

Flow straighteners (a device inserted into the spout outlet of a tap to modify flow, take out turbulence, and create an even stream of water) should be permanently removed if practical. If not, they will need regular removal and cleaning. Flexible hoses should be replaced with plastic or metal ones.

It may be necessary to disinfect the hot- and cold-water distribution systems that supply the unit, although hyperchlorination is not effective against established biofilms.

Patient contact equipment such as incubators, humidifiers, nebulizers, and respiratory equipment found to be heavily contaminated need to be removed from use and appropriately disinfected and retested before use. The requirement for continued use of such equipment should also be assessed.

All staff with responsibility for cleaning should be adequately trained and made aware of the importance of high standards of cleanliness. Additional training and increased supervision should be given if a specific area does not maintain the expected standard of cleanliness. Staff should be reminded that handwashing basins should only be used for hand hygiene and should not be used for disposal of body fluids, environmental cleaning fluids, or washing or storage of patient equipment. All handwashing stations should be used regularly and all taps flushed regularly. Alcohol gel dispensers should be located away from handwashing basins at point-of-care and should be prefilled single-use bottles.

From 2011 to 2012, there were several well-publicized outbreaks of *P. aeruginosa* infections in neonatal care units in Northern Ireland, which resulted in fatalities. To determine the causes, the Department of Health issued new guidance for augmented care units on the best practice for prevention, risk management, risk reduction, and actions required if water systems are contaminated with *P. aeruginosa*. The recommendations appear in Guidelines 1–3.

EPIDEMIOLOGY

In the healthcare environment, surfaces, particularly those containing polymers, have been shown to leach carbon compounds, which can support and encourage the growth of the organism. Its key virulence factors are the production of alginate and quorum-sensing molecules that aid in its ability to exist within biofilms. Biofilms adhere to surfaces within water storage, delivery, and distribution systems particularly in areas of scaling, such as tap outlets. Although most bacteria will remain fixed within the biofilm, some will become detached resulting in planktonic forms that can contaminate the water layer above the biofilm and can thus be found within the last two meters before the point of discharge of water. In addition, devices fitted to, or close to, the tap outlet (for example, flow straighteners) may exacerbate the problem by providing the nutrients that support microbial growth, providing a surface area for oxygenation of water, and leaching nutrients. Outlets that are underused and have low water throughput form a greater risk of contamination by *P. aeruginosa* than those that are frequently used. Large concentrations of bacteria can also be found in washbasin bowl outlets, as well as within containers and nozzles of vessels such as cleaning agent bottles. The bacterium is inherently resistant to several disinfectants such as biguanides and quaternary ammonium compounds through the action of multidrug efflux pumps. In addition, the ability to form biofilm on a range of inanimate surfaces contributes to disinfectant resistance as well as impeding physical removal.

Medical equipment or products that come into contact with contaminated water serve as reservoirs for potential nosocomial transmission. This problem was confirmed in studies conducted in the 1950s and 1960s following reports of contamination by *P. aeruginosa* of eye drops, humidifiers, and disinfectants in neonatal and burns units, as well as neurosurgery. Since then, these sources of contamination have become widely acknowledged.

Although *P. aeruginosa* infections in the hospital setting are usually attributed to acquisitions from the environment, up to 7% of healthy humans carry the organism in the throat and nasal mucosa, and carriage rates as high as 24% have been reported in stool. Other common areas of colonization include moist skin surfaces, including the axilla, ear, and perineum.

There are several routes of transmission and infection that may exist in healthcare settings. In most cases, colonization will precede infection. Some colonized patients will remain well but can act as sources for transmission to other patients. As a microorganism that is often found in water, the more frequent the direct or indirect contact between a susceptible patient and contaminated water, and the greater the microbial load, the higher the potential for patient colonization or infection.

Infection can occur from contact with any moist reservoir in the hospital. Patients who are mechanically ventilated are frequently colonized in the upper airways, but only a small proportion will develop pneumonia. In the outbreak setting, *P. aeruginosa* is usually associated with a point source, such as a faucet or contaminated equipment. However, unlike other waterborne pathogens such as *Legionella* spp., case-to-case transfer can occur from contaminated hands of healthcare workers. *P. aeruginosa* has the ability to survive in moist environments for months with minimal nutritional requirements. In addition, direct delivery of organisms directly into the bloodstream from thawing of frozen blood products in an environment contaminated with *P. aeruginosa* can result in disease, even in the immunocompetent.

P. aeruginosa is encountered mainly as an opportunistic nosocomial pathogen. Recent voluntary surveillance data from Public Health England (PHE) has demonstrated a steady decrease in both the numbers of *Pseudomonas* spp. bacteraemia in England, Wales, and Northern Ireland, as well as the overall rate, from 7.0 per 100,000 in 2008 to 6.3 per 100,000 in 2013. Although these figures are for all reported *Pseudomonas* species, most were identified as *P. aeruginosa* (84% in 2013). Although a proportion may be community acquired, most cases are associated with healthcare and the reasons for the decrease in incidence are not entirely clear, although the use of antibiotic regimens directed at *Pseudomonas*, improved infection control practices, and control of organisms in the water supply all may play a role.

BIOLOGY

Pseudomonas spp. are aerobic, nonfermentative, Gram-negative bacteria. *P. aeruginosa* is the most common species associated with human disease. It is a saprophyte commonly found in soil, plants, water, and moist environments. It has the ability to adapt to a wide range of conditions with minimal nutritional requirements, and it can metabolize a large array of carbon energy sources. The organism is tolerant to temperatures as high as 45–50°C and can even grow in distilled water using dissolved carbon dioxide and residual sulfur, phosphorus, iron, and divalent cations as carbon and essential nutritional substrates.

P. aeruginosa is oxidase-positive and produces two pigments, pyocyanin and pyoverdin, giving the colonies a blue-green appearance. Pyoverdin is also fluorescent under UV light. It grows on selective media containing cetrimide (cetyl trimethylammonium bromide) and can grow at 42°C.

During an outbreak, molecular typing for discriminating different bacterial isolates of *P. aeruginosa* is essential. Preferred typing techniques for *P. aeruginosa* target genomic sequences and the most widely employed scheme is VNTR. However, isolates grouped by VNTR may be representative of a geographical distribution to a region but may not be indicative of a single locality clone. Therefore, where isolates appear similar by VNTR typing, and no epidemiological link is known, these are checked by the more discriminatory method of PFGE, which may identify similar isolates in other healthcare providers. PFGE compares strains through DNA size analysis. Because outbreaks of *P. aeruginosa* may not be confined to one augmented care unit, because of frequent interhospital transfers in these patients, this is an important method of comparing isolates to identify problems on a wider scale. The introduction of whole genome sequencing may provide even greater discrimination and may also indicate antibiotic resistance patterns.

DISEASE

P. aeruginosa is an opportunistic pathogen with the potential to cause infections in almost any organ system in those with physical, phagocytic, or immunologic defects in host defence mechanisms. Groups particularly affected are the immunocompromised (for example, HIV positive individuals and patients undergoing chemotherapy), neonates, and those with severe burns, diabetes mellitus, or cystic fibrosis. It rarely causes disease in healthy individuals. It is responsible for about 10% of healthcare associated infections causing pneumonia, including ventilator-associated pneumonia, central venous catheter infections, catheter associated urinary tract infection, wound infections, neurosurgical infections, and, more rarely, ecthyma gangrenosum and bacteraemia in neutropenic patients. The widespread use of broad-spectrum antibiotics also aids in the proliferation of *P. aeruginosa* in the hospital and highlights the increasing importance of antimicrobial stewardship programs.

In the community, *P. aeruginosa* infection can also occur in certain settings including skin infections following the use of hot tubs, whirlpools, and swimming pools. Ulcerative keratitis has been described related to contaminated contact lens solutions. Other infections include infective endocarditis in intravenous drug users and otitis externa after swimming in freshwater. The last may manifest as severe, potentially life-threatening malignant otitis externa in patients with diabetes and in immunocompromised individuals.

PATHOLOGY

Prior to the development of invasive disease, colonization can occur in up to half of individuals, although it is difficult to identify which patients with colonization will go on to develop invasive disease.

P. aeruginosa possesses a number of virulence factors, including the production of toxins and enzymes such as proteases and elastases, and phenazine pigments, such as pyocyanin. It possesses several different export systems that are involved in the secretion of virulence factors. Of particular importance is a type III secretion system that enables injection of effector proteins directly into the cytoplasm of host cells. These proteins have significant effects on epithelial barrier function and wound healing.

In patients with cystic fibrosis, *P. aeruginosa* colonizes the lung for periods of up to 40 years, perpetuating a chronic inflammatory process. The gradual loss of lung function results from the host response to the presence of the bacteria. Expression of an alginate coat probably protects against phagocytosis and contributes to biofilm formation. Down-regulation of the production of toxins, loss of lipopolysaccharide (LPS) O chains, and frequent loss of the flagellum and pili contribute to its chronic colonization. The role of biofilm formation is also important in infections associated with indwelling devices and other prosthetic material.

In contrast to cystic fibrosis, *P. aeruginosa* appears to display other virulence factors in the setting of acute invasive infection. These include secreted exotoxins, elastase, and cytotoxins, as well as surface bound organelles such as pili, flagella, and LPS.

QUESTIONS

1. In an outbreak of *P. aeruginosa* infection in an Intensive Care unit, the most likely source of the outbreak will be:
   a. Flower vases
   b. Ice machine
   c. Bathtub
   d. Handwashing basin used to wash patient equipment
   e. Bottled drinking water

2. Which of the following tests are used in the microbiology laboratory to identify *P. aeruginosa*:
   a. Oxidase test
   b. Fluorescence
   c. Hydrolysis of casein
   d. Production of pyocyanin
   e. All of the above

3. The method most often used for typing strains of *P. aeruginosa* is:
   a. Bacteriophage
   b. Serotyping
   c. Variable number tandem repeats (VNTR)
   d. Pyocyanin
   e. Random amplified polymorphic DNA (RAPD)

4. Routine testing of water outlets where patients are cared for on an augmented care unit has demonstrated the following results:
   Outlet 1: 56 cfu/mL *P. aeruginosa*
   Outlet 2: No detectable *P. aeruginosa*
   Outlet 3: 5 cfu/mL *P. aeruginosa*
   Outlet 4: 8 cfu/mL *P. aeruginosa*
   What further action needs to take place?
   a. Remove outlet 1 from service and retest all outlets with preflush and postflush samples
   b. Declare an outbreak and close the unit to all new admissions
   c. Do nothing
   d. Remove outlet 1 from service and continue to use all others
   e. Refer to the WSG (Water Safety Group) for risk assessment of the use of water in the unit

5. Best-practice advice relating to all clinical handwashing basins involves:
   a. Cleaning of the basin first, followed by the taps
   b. Placing soap dispensers directly above taps for ease of access
   c. Disposing of unused environmental cleaning agents directly into the drain outlet
   d. Do not flush taps in infrequently used handwashing basins, to avoid increasing the contamination of the bowl
   e. Using single-use bottles for antimicrobial hand rub and soap

GUIDELINES

1. Department of Health (2016) Health Technical Memorandum 04-01: Safe water in healthcare premises. Part A: Design, installation and commissioning. (www.gov.uk/government/publications/hot-and-cold-water-supply-storage-and-distribution-systems-for-healthcare-premises).

2. Department of Health (2016) Health Technical Memorandum 04-01: Safe water in healthcare premises. Part B: Operational management. (www.gov.uk/government/publications/hot-and-cold-water-supply-storage-and-distribution-systems-for-healthcare-premises).

3. Department of Health (2016) Health Technical Memorandum 04-01: Safe water in healthcare premises. Part C: Pseudomonas aeruginosa—advice for augmented care units. (www.gov.uk/government/publications/hot-and-cold-water-supply-storage-and-distribution-systems-for-healthcare-premises).

4. Health and Social Care Act 2008: Code of Practice on the prevention and control of infections and related guidance. (2015, www.gov.uk/government/publications/the-health-and-social-care-act-2008-code-of-practice-on-the-prevention-and-control-of-infections-and-related-guidance).

5. The Water Supply Regulations (2010, www.legislation.gov.uk/uksi/2010/991/contents/made).

6. Public Health England (2013). *Pseudomonas* spp. and *Stenotrophomonas* spp. bacteraemia in England, Wales and Northern Ireland. (www.hpa.org.uk/webw/HPAweb&HPAwebStandard/HPAweb_C/1200471675429).

7. Kelsey MC (2013) Control of waterborne microorganisms and reducing the threat from Legionella and Pseudomonas. In Decontamination in Hospitals and Healthcare (Walker J ed.). Woodhead Publishing.

8. The Regulation and Quality Improvement Authority Independent Review of Incidents of *Pseudomonas aeruginosa* Infection in Neonatal Units in Northern Ireland. Final Report. (2012, www.rqia.org.uk/publications/rqia_reviews.cfm).

REFERENCES

1. Loveday HP, Wilson JA, Pratt RJ et al. (2014) Epic3: national evidence-based guidelines for preventing healthcare-associated infections in NHS hospitals in England. *J Hosp Infect* **86** (Suppl 1):S1–70.

2. Kelsey M (2013) Pseudomonas in augmented care: should we worry? *J Antimicrob Chemother* **68**:2697–2700.

3. Berthelot P, Grattard F, Mahul P et al. (2001) Prospective study of nosocomial colonization and infection due to *Pseudomonas aeruginosa* in mechanically ventilated patients. *Intensive Care Med* **27**:503–512.

4. Cardenosa Cendrero JA, Sole-Violan J, Bordes BA et al. (1999) Role of different routes of tracheal colonization in the development of pneumonia in patients receiving mechanical ventilation. *Chest* **116**:462–470.

5. Public Health England (2013) Voluntary surveillance of pseudomonas and stenotrophomonas bacteraemia in England, Wales and Northern Ireland: 2013. Health Protection Report 2014 8. (**28**): HCAI.

6. Le VT & Diep BA (2013) Selected insights from application of whole-genome sequencing for outbreak investigations. *Curr Opin Crit Care* **19**:432–439.

7. Sabat AJ, Budimir A, Nashev D et al. (2013) Overview of molecular typing methods for outbreak detection and epidemiological surveillance. *Euro Surveill* **18**:20380.

8. Pfaller MA, Jones RN, Doern GV & Kugler K (1998) Bacterial pathogens isolated from patients with bloodstream infection: frequencies of occurrence and antimicrobial susceptibility patterns from the SENTRY antimicrobial surveillance program (United States and Canada, 1997). *Antimicrob Agents Chemother* **42**:1762–1770.

9. Centers for Disease Control and Prevention (2001) Pseudomonas dermatitis/folliculitis associated with pools and hot tubs--Colorado and Maine, 1999-2000. *JAMA* **285**:157–158.

10. Robertson DM, Petroll WM, Jester JV & Cavanagh HD (2007) Current concepts: contact lens related Pseudomonas keratitis. *Cont Lens Anterior Eye* **30**:94–107.

11. Sousa C, Botelho C, Rodrigues D, Azeredo J & Oliveira R (2012) Infective endocarditis in intravenous drug abusers: an update. *Eur J Clin Microbiol Infect Dis* **31**:2905–2910.

12. van Asperen IA, de Rover CM, Schijven JF et al. (1995) Risk of otitis externa after swimming in recreational fresh water lakes containing *Pseudomonas aeruginosa*. *BMJ* **311**:1407–1410.

13. Engel J & Balachandran P (2009) Role of *Pseudomonas aeruginosa* type III effectors in disease. *Curr Opin Microbiol* **12**:61–66.

ANSWERS

MCQ	Feedback
1. In an outbreak of *P. aeruginosa* infection in an Intensive Care unit, the most likely source of the outbreak will be: a. Flower vases b. Ice machine c. Bathtub d. Handwashing basin used to wash patient equipment e. Bottled drinking water	Most intensive care units in England prohibit the use of flower vases and ice machines in augmented care units, because of high rates of colonization with *P. aeruginosa*. Similarly, most units do not have baths, mainly because most patients are intubated and ventilated or have severe medical conditions precluding the use of baths. Although the use of commercially available bottled water has led to outbreaks of *P. aeruginosa* infection in critical care units, it is less likely than a handwashing basin used to wash patient equipment. In the recommendations made by the Department of Health in their 2013 document, handwashing basins should only be used to wash hands.
2. Which of the following tests are used in the microbiology laboratory to identify *P. aeruginosa*: a. Oxidase test b. Fluorescence c. Hydrolysis of casein d. Production of pyocyanin e. All of the above	*P. aeruginosa* is oxidase-positive, hydrolyses casein, produces pyocyanin, and is fluorescent. Occasionally atypical nonpigmented variants of *P. aeruginosa* occur. A pyocyanin-negative, casein-hydrolysis-positive, fluorescence-positive culture should be regarded as *P. aeruginosa*. Additional tests may be necessary to differentiate nonpigmented *P. aeruginosa* from *P. fluorescens* (such as growth at 42°C or resistance to C-390, 9-chloro-9-(4-diethylaminophenyl)-10-pheylacridan or phenanthroline, or use of more extensive biochemical tests).
3. The method most often used for typing strains of *P. aeruginosa* is: a. Bacteriophage b. Serotyping c. Variable number tandem repeats (VNTR) d. Pyocyanin e. Random amplified polymorphic DNA (RAPD)	In the past, typing of pyocyanin was in widespread use, but it has been superseded by more rapid and reliable molecular methods. Although RAPD is a well-recognized method of typing for *P. aeruginosa*, it is infrequently used in England for clinical isolates, and the reference laboratory will typically use VNTR with or without PFGE (pulsed-field gel electrophoresis).
4. Routine testing of water outlets where patients are cared for on an augmented care unit has demonstrated the following results: Outlet 1: 56 cfu/mL *P. aeruginosa* Outlet 2: No detectable *P. aeruginosa* Outlet 3: 5 cfu/mL *P. aeruginosa* Outlet 4: 8 cfu/mL *P. aeruginosa* What further action needs to take place? a. Remove outlet 1 from service and retest all outlets with preflush and postflush samples b. Declare an outbreak and close the unit to all new admissions c. Do nothing d. Remove outlet 1 from service and continue to use all others e. Refer to the WSG (Water Safety Group) for risk assessment of the use of water in the unit	Outlet 1 has an unsatisfactory bacterial load, but the location of the source needs to be ascertained, and this is done by microbiological sampling preflushing and postflushing. Before remedial measures can be put into place, the location of the contamination needs to be ascertained, as well as the geographic relationship to outlets 3 and 4. If both preflush and postflush samples from a particular outlet are >100 cfu/100 mL and other nearly outlets have no or low counts, this shows that the single outlet is heavily contaminated despite the high postflush count. If the sampling indicates that the water services are the problem, then most outlets would possibly be positive and other points in the system could then be sampled to assess the extent of the problem.

MCQ	Feedback
5. Best-practice advice relating to all clinical handwashing basins involves: a. Cleaning of the basin first, followed by the taps b. Placing soap dispensers directly above taps for ease of access c. Disposing of unused environmental cleaning agents directly into the drain outlet d. Do not flush taps in infrequently used handwashing basins, to avoid increasing the contamination of the bowl e. Using single-use bottles for antimicrobial hand rub and soap	Clinical handwashing basins are particularly high risk. During cleaning of basins and taps, there is a risk of contaminating tap outlets with microorganisms if the same cloth is used to clean the bowl of the basin or surrounding area before the tap. Wastewater drain outlets are particularly risky parts of the basin and system and are almost always contaminated. Bacteria may be of patient origin, so it is possible that bacteria could seed the outlet, become resident in any biofilm, and have the potential to be transmitted to other patients. Taps must be cleaned first before the rest of the clinical handwashing basin. Because the compounds in soap and antimicrobial hand rubs can be a source of nutrients for *P. aeruginosa*, dispensers should be located away from the tap to prevent soiling of the tap by drips from the dispensers or during the movement of hands from the dispensers to the basin when beginning handwashing. Bottles of partially used environmental agents may be contaminated with *P. aeruginosa*, and they should never be disposed of at clinical handwashing basins. All outlets that are infrequently used should be flushed regularly (at least daily in the morning for one minute) to prevent water stagnation. As *P. aeruginosa* has the ability to grow even within bottles of soap and detergent, nonfillable single-use bottles for antimicrobial hand rub and soap should always be used.

RABIES

Martine Usdin, Hilary Kirkbride, and Kevin E Brown

National Infection Service, Public Health England, UK.

In April 2008, an animal charity rescued five street dogs in Sri Lanka and imported them to the UK by air. They arrived at Heathrow Airport. The dogs had been cared for by a local veterinarian before departure and all had received a single rabies vaccination. They were imported by an accredited animal-carrying company and were assessed as healthy on departure. On arrival at Heathrow airport, the animals were held for routine assessment and quarantine, as per UK and European Union guidelines. The animals were arriving from a country considered to be high risk for rabies, and an animal import official noted that at least one of the dogs was unwell on arrival. The following day all five dogs were transferred to a nearby quarantine kennel for longer-term isolation and follow-up.

Six days after arrival, one of the animals, a puppy of uncertain age, began to show clinical signs consistent with rabies and became aggressive; a number of kennel staff members were bitten. The puppy died eight days after arrival, and the carcass was sent for immediate testing at the Animal Health and Veterinary Laboratories Agency (AHVLA, now Animal and Plant Health Agency, APHA). Public Health officials were notified the same evening by the Department of Environment, Farming and Rural Affairs (DEFRA) that the animal had tested positive for rabies by the initial fluorescent antibody test (FAT) and that the finding had been confirmed by polymerase chain reaction (PCR).

The other four dogs that had been housed with the index animal were all euthanized and tested for rabies. None of the other animals were found to be infected.

INVESTIGATION OF THE INCIDENT

An incident control team (ICT) was established to manage the public health response to the case. The initial priorities for the ICT were to identify and interview human contacts to determine rabies exposure risk, initiate post-exposure treatment, manage human health communications issues, and advise on future needs for occupational rabies vaccination.

PREVENTION OF HUMAN CASES

The immediate human health priority was to identify all individuals who may have been exposed to the puppy while it was potentially infectious and to initiate postexposure treatment if necessary. This is the responsibility of Public Health England, the executive agency of the Department of Health in the UK. Contacts would include those in the UK and Sri Lanka, as well as individuals involved in the infected puppy's transit. Thus the following actions were initiated immediately:

- identification of all possible exposed contacts
- risk assessment for postexposure management

Identification of human contacts

For dogs and cats the infectious period is up to 14 days before onset of clinical signs, so this was used to determine the period of concern for human exposures. In collaboration with all the agencies involved in both UK and Sri Lanka, 42 people were identified as potentially having been exposed to the puppy during its infectious period.

Risk assessment for postexposure treatment (PET)

Postexposure treatment with rabies vaccine and human rabies immunoglobulin (PET) is highly effective if initiated promptly following exposure, and it should be initiated as soon as possible following an individual risk assessment for all those assessed as exposed. When assessing a potentially exposed patient for rabies risk, several factors should be taken into account:

- species of animal
- the country of origin
- current health status of the animal
- nature of exposure
- immunity of individual exposed

In this case, the animal was a dog and was known to be infected. In investigations where the status of the animal is unknown, a risk assessment should be undertaken based on the likelihood of the species being infected. Rabies can be carried by any mammal species, although it is most often associated with dogs. Wild animals, especially foxes in part of Europe, are often reservoirs in countries where endemic rabies has been eliminated because of rabies vaccination programs in domestic animals. Primates and rodents sporadically have been reported to carry rabies and bites from these animals occurring in rabies-endemic areas should also be risk-assessed.

In the UK, bats can carry the closely related European bat lyssavirus (EBLV-2), which can cause rabies in humans. Bats may be asymptomatic but still transmit rabies, and therefore all bat exposures in the UK are considered a potential rabies risk. Bat bites can be painless and often go unnoticed, so a history of contact with bats should be assessed for risk, even if no bite is reported. The prevalence of EBLV-2 in the UK bats is approximately 0.1%.

According to UK guidance countries are classified by their rabies risk as high risk, low risk, or no risk for terrestrial rabies, based on the presence or absence of endemic rabies in companion animals, wildlife, or both. These classifications can be found on the rabies pages of the gov.uk website (see Guideline 2). Sri Lanka is considered a high-risk country. The UK is classified as rabies-free in terrestrial animals and therefore exposures from indigenous animals other than bats do not present a rabies risk and PET is not required.

Dogs or cats that are asymptomatic are less likely to be shedding rabies virus in their saliva. Once they become infectious, however, cats or dogs will become symptomatic within 10 days. In the UK, to be extra prudent, an observation period of 14 days is recommended. Thus, if the dog or cat can be observed for 15 days, and is confirmed to be free of symptoms after that time, it cannot have been infectious for rabies at the time of exposure. This timeframe does not apply to other terrestrial mammals, where the period between the start of viral shedding in the saliva and obvious clinical illness becomes apparent is less clearly defined. Bats may be infectious in the absence of symptoms and therefore an exposure to an apparently healthy bat is still considered a potential rabies risk.

In this case, the dog had a clear onset of illness and was considered potentially infectious (shedding virus in its saliva) for the maximum canine incubation period of 14 days prior to the onset of this dog's clinical signs.

The nature of the exposure allows classification of the level of risk and helps inform the actions to be taken. For terrestrial animals, touching or stroking an animal is classified as a category I exposure and no further treatment is needed; licks to intact skin, or bites or scratches that do not break the skin are considered category II; and licks to the broken skin, contamination of mucous membranes with saliva, or bites that break the skin are considered category III bites. All bat bites are considered category III bites.

The course of treatment recommended depends not only on the exposure risk, but also on whether the exposed individual has previously received a course of rabies vaccination and can be considered to be immune. This history was elicited from all contacts of the puppy. In the UK, a fully immunized individual is considered to be someone who has had at least three doses of vaccine as part of preexposure prophylaxis (PreP) or previous post-exposure treatment, or documented levels of rabies antibody of >0.5 IU/mL. In such circumstances, while PET is still required, only two doses of vaccine should be given (d 0 and days 3–7) and human rabies immunoglobulin (HRIG) is not required.

Once symptoms develop, rabies infection is virtually always fatal, and therefore it is essential that postexposure treatment is started promptly. The risk assessment described is used to select the most appropriate postexposure treatment. Figure 20.1 shows the UK algorithm for terrestrial animals. A similar algorithm is provided for individuals exposed to bats.

When a category III exposure involving an animal from a high-risk country has taken place unimmunized individuals should be offered a full five-dose schedule of vaccination (days 0, 3, 7, 14, 28–30) and a single dose of rabies immunoglobulin (RIG). The RIG should be given at the same time as the first dose of vaccine, or, if not, no later than seven days after the first vaccine. It should be infiltrated where possible around the site of the wound, and only if this is not possible, should be given as an intramuscular (i.m.) injection. If there is a category II exposure in an animal from a high-risk country, or category II and III exposure in a low-risk country, the RIG is omitted and a full five-dose course of vaccines is indicated.

The human incubation period is usually between three weeks and one year, although shorter and longer periods have been reported. This means that PET should still be given if indicated by the risk assessment, even if over a year since exposure. Priority should be

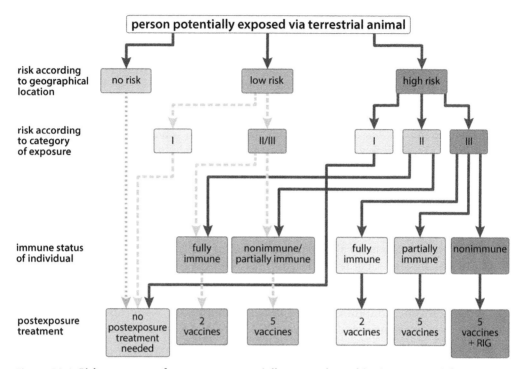

Figure 20.1. **Risk assessment for a person potentially exposed to rabies in a terrestrial animal.**
(From Rabies Risk Assessment: Treatment After Exposure to Terrestrial Animals (2015) PHE. www.gov.uk/
government/uploads/system/uploads/attachment_data/file/424262/Rabies_risk_assessment_for_exposure_
to_terrestrial_animals.pdf. Published under the Open Government License v3.0.)

given to those cases involving multiple bites and where the site of exposure is on the head
or neck. In these circumstances, PET should be started as soon as possible (even if out of
routine working hours). For other exposures, PET should be initiated the following working day.

In this incident of the 42 possibly exposed individuals, 12 had had direct physical contact
with the puppy's body fluids or had received a bite (that is, category II or category III
exposures): 11 individuals in the UK (four charity workers, one airport baggage handler,
and six staff working at the quarantine kennels) and the veterinarian in Sri Lanka. Six of
the contacts previously had a full preexposure course of three rabies vaccines, three had
received an incomplete course, and three were not previously immunized.

The two nonimmunized contacts with category III exposure required treatment with RIG
and a course of vaccine. The other individuals received vaccine only, either two doses or
five doses depending on the vaccination status at the time of the exposure (Figure 20.2).

Preexposure prophylaxis (PreP)

During an exposure incident, individuals at risk of exposure are reminded to maintain
up-to-date rabies immunization (preexposure prophylaxis) to protect against future
incidents. In the UK, this is recommended for individuals working with the virus in
laboratories (continuous risk group) and those who regularly handle potentially rabid
animals, for example, bat handlers or those working in quarantine kennels (frequent risk

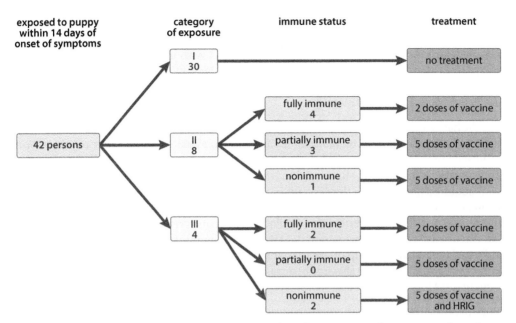

| exposed to puppy within 14 days of onset of symptoms | category of exposure | immune status | treatment |

Figure 20.2. **Outcome of treatment recommendations based on category of exposure and previous vaccination status.**

group). The UK recommendation for a complete course of PrEP is three doses of vaccine, given on days 0, 7, and 28, plus a booster at one year for those in the frequent risk group. Boosters are then recommended to maintain protective levels of rabies antibody, following the guidelines set out in the Green Book (Immunisation Against Infectious Diseases), together with the specific risk groups for whom PrEP is currently recommended or advised.

EPIDEMIOLOGY

Rabies is a zoonosis affecting a variety of mammalian species around the world. Terrestrial rabies is transmitted by a bite or scratch from an animal infected with rabies virus (RABV), a member of the *lyssavirus* genus. A similar disease caused by a related virus, European bat lyssavirus (EBLV), can be transmitted from bats to humans. Internationally, rabies remains a significant source of mortality and is a serious public health problem. The World Health Organization (WHO) estimates ~60,000 human fatalities globally each year, with most occurring in rural Africa and Southeast Asia. However, this official WHO figure is almost certainly an underestimate, given the poor surveillance data and infrastructure in many areas where rabies is endemic.

Most human rabies cases are reported in countries where canine rabies is endemic. Infections in these areas are reported as resulting from the bite of an infected dog, although other animals including cats, bats, foxes, and racoons have been implicated.

Virus is most efficiently transmitted via saliva of an infected animal, especially if inoculation occurs through deep penetrating bites involving nerve damage. The risk of developing rabies depends on the anatomical site, severity of the bite, and the species inflicting the wound; nontraumatic injury from infected terrestrial animals or slight scratches are of relatively low risk. The only well documented cases of human-to-human transmission have been through organ transplantation. However, there is a theoretical risk that transdermal bites or direct mucosal contact with infectious secretions (such as saliva) from symptomatic patients might carry an infection risk. There is no risk of transmission from patients who have been exposed to the virus and have received appropriate and timely post-exposure treatment.

The UK has been free of rabies in terrestrial animals since 1922, although EBLV-2 is found in some bat species. Strict regulations are in place to avoid reintroduction of rabies, with imported animals requiring proof of vaccination, and in some cases a period of quarantine. Since terrestrial rabies was eliminated in the UK, all subsequent human rabies cases have been acquired abroad, with the exception of one case of EBLV-2 infection acquired from a bat in 2002. The infrequent presentation of clinical disease in the UK often delays diagnosis in humans, but the effectiveness and availability of postexposure prophylaxis (PEP) means that the disease remains very rare in the UK and other countries with widely available PEP and a low incidence of companion animal rabies. Several other island countries and some parts of Western Europe are also free of terrestrial rabies, but in most other parts of the world, particularly Africa and Asia, animal rabies, particularly canine rabies, remains endemic.

Globally, the mainstay of rabies prevention in humans is through vaccination of wildlife and domestic animals. A one-health approach has been proposed, namely, an integrated strategy that includes wildlife, urban, and rural animal management, along with community engagement and a coordinated approach integrated with response to other health issues. Rabies can be successfully controlled in countries that have implemented active control programs and surveillance, and control of rabies in animals is considered the most effective way to prevent disease in humans. One promising control strategy being investigated is a combination of oral vaccine combined with a contraceptive agent to both control and vaccinate free-ranging dog populations.

Movement of animals across borders is highly controlled on a national level. Harmonization of standards across the EU has resulted in the development of the "Pet Passport" scheme (Guideline 5), within which animals moving between countries within the EU must be chipped with an RF device and show proof of vaccination given more than 21 days before travel. The animals must be older than six weeks at the time of vaccination. Animals that are not covered by the PET passport scheme still need to go into quarantine.

Bat lyssaviruses

Related lyssaviruses have been found in other species including bats, and, if humans are infected, can lead to human rabies. Lyssaviruses are found in bats in most countries, including Europe and the UK. In 2002, a Scottish bat handler developed rabies and died as a result of the European bat lyssavirus-2 (EBLV-2) contracted during routine work with bats. In the UK, EBLV-2 is found in Daubenton's bats and infects about 2% of the Daubenton bat population in the UK.

BIOLOGY

RABV is a member of the Rhabdoviridae family and *Lyssavirus* genus. Within the *Lyssavirus* genus, there are several closely related viruses, including viruses infecting fruit bats (Lagos bat virus) and insectivorous bats (Duvenhage virus, EBLV and Australian bat lyssavirus) in different parts of the world. All the lyssaviruses have a negative (anti-sense) stranded RNA genome of about 12 kb encoding just 5 proteins. Mature virus particles contain an RNA core, a polymerase complex made up of the polymerase, and L proteins responsible for transcription, as well as a phosphoprotein and the matrix and glycoproteins.

PATHOGENESIS

After a bite from an infected animal, the virus replicates first in muscle. Within muscles it binds to the nicotinic acetylcholine receptor on postsynaptic membranes, but it is not known how the virus infects the presynaptic nerve, from which it travels toward the central nervous system (CNS) by retrograde transport. One possible mechanism of internalization is through concentration of the virus in the synaptic cleft by the nicotinic receptors, which then facilitates uptake by the presynaptic terminal though a different, as yet unidentified receptor.

Once within the axon, the rabies virus travels along the nerve to the dorsal ganglion at a rate of approximately 250 mm/day. There is no evidence for haematogenous spread of the virus. Once the virus enters the CNS, there is rapid dissemination to other regions of the CNS, followed by centrifugal spread along sensory and autonomic nerves to other tissues, including sensory nerves to the skin and salivary glands.

DISEASE

Clinical presentation

Human rabies can present in two different forms: paralytic and furious. Furious rabies (which occurs in approximately two-thirds of cases) usually presents with typical signs of agitation, hypersalivation, and phobic or spasmic responses to drafts of air or presentation of water. Paralytic rabies is associated with muscle weakness and paralysis.

Clinical progression of rabies usually involves a prodromal phase followed by development of acute signs often beginning with pain and muscle weakness at the site of inoculation, which progresses to generalized meningoencephalitis, coma, and death. Rabies infection is almost always fatal once symptoms have begun; however, the long incubation period (usually 30–90 days with some reports of >1 year) and the effectiveness of postexposure prophylaxis make this disease largely preventable if appropriate postexposure measures are taken.

Diagnostic techniques

The gold standard test for rabies diagnosis in humans remains the postmortem examination of multiple parts of the brain via direct fluorescence antibody test. Antemortem tests are available that can be performed on saliva, cerebrospinal fluid, skin biopsies from the posterior region of the neck, and at the site of the bite. Antemortem testing may include reverse transcription polymerase chain reaction (RT-PCR), cell culture, indirect immunofluorescence, and virus neutralization depending on the sample submitted. Sequencing of the viral genome can be useful to confirm the likely source or country of infection, as strains with slightly different sequences are found in different parts of the world.

Where possible samples for diagnostic testing should include a nuchal biopsy and samples from the site of the injury, if it can still be identified. Often cerebrospinal fluid and blood samples are also tested, but usually only become positive later in the illness. Because the virus is not widespread throughout the body until after the onset of symptoms, rabies diagnostics should not be pursued unless a patient is exhibiting symptoms compatible with rabies and other, perhaps more common, causes of encephalitis have been eliminated as the cause of illness. It is also important to remember that a diagnosis of rabies is not always easy to confirm with antemortem samples, and so a patient with rabies may test negative via these methods, depending on the samples used for testing and the point during the course of illness when the testing is performed.

PATHOLOGY

Postmortem studies show relatively mild inflammatory changes in the CNS with mononuclear infiltration. Degenerative neuronal changes are often not prominent. Negri bodies (eosinophilic inclusions in brain neurons composed of rabies virus proteins and RNA) are likely the main site of replication within the brain and are pathognomonic for rabies infection.

QUESTIONS

1. True or false: Rabies:
   a. Is a common worldwide zoonotic infection
   b. Is caused by canine parvovirus
   c. Only affects dogs and cats
   d. Can be successfully treated once symptoms develop

2. True or false: Individuals at increased risk of rabies infection are:
   a. Travellers licked by a dog overseas
   b. Travellers bitten by bats overseas
   c. People bitten by a dog in the UK
   d. People bitten by bat in the UK
   e. Healthcare workers

3. True or false: Postexposure treatment may consist of:
   a. Two doses of rabies vaccine
   b. Five doses of rabies vaccine
   c. Human rabies immunoglobulin alone
   d. Five doses of vaccine and human rabies immunoglobulin

4. True or false: Following a dog bite in India (a rabies-endemic country), individuals should be advised to:
   a. Wash the wound and seek local medical advice
   b. Go to A&E (emergency room) immediately on return to the UK
   c. Be asked to prepare a list of their contacts
   d. Be closely monitored and hospitalized if necessary

5. True or false: Rabies preexposure prophylaxis is:
   a. Ineffective
   b. Offered to all those working with the virus
   c. Should be considered if visiting endemic countries
   d. Recommended for those regularly handling bats

GUIDELINES

1. UK guidelines for the management of rabies. www.gov.uk/government/collections/rabies-risk-assessment-post-exposure-treatment-management.

2. UK classification of countries by rabies risk. www.gov.uk/government/publications/rabies-risks-by-country

3. The guidelines on the use of preexposure and postexposure vaccination and human immunoglobulin is available in the Green Book, Chapter 27: www.gov.uk/government/publications/rabies-the-green-book-chapter-27

4. WHO international guidelines forming the basis of rabies response in most countries: www.who.int/rabies/PEP_prophylaxis_guidelines_June10.pdf
www.who.int/rabies/human/postexp/en

5. UK guidelines on pet travel and quarantine. www.gov.uk/pet-travel-quarantine#transporting-your-animal-to-the-uk

The incident has been based on a case reported in *Eurosurveillance*, Vol 13, April 2008, although some details have been modified for teaching purposes. The clinical advice and risk assessment is based on procedures in the UK. Other countries may have different risk assessments.

REFERENCES

1. Anderson LJ, Williams LP, Jr., Layde JB et al. (1984) Nosocomial rabies: investigation of contacts of human rabies cases associated with a corneal transplant. *Am J Public Health* **74**:370–372.

2. Baer GM, Shaddock JH, Houff SA et al. (1982) Human rabies transmitted by corneal transplant. *Arch Neurol* **39**:103–107.

3. Boland TA, McGuone D, Jindal J et al. (2014) Phylogenetic and epidemiologic evidence of multiyear incubation in human rabies. *Ann Neurol* **75**:155–160.

4. Dimaano EM, Scholand SJ, Alera MT & Belandres DB (2011) Clinical and epidemiological features of human rabies cases in the Philippines: a review from 1987 to 2006. *Int J Infect Dis* **15**:e495–e499.

5. Fekadu M, Shaddock JH & Baer GM (1982) Excretion of rabies virus in the saliva of dogs. *J Infect Dis* **145**:715–719.

6. Fooks AR, Banyard AC, Horton DL et al. (2014) Current status of rabies and prospects for elimination. *Lancet* **384**:1389–1399.

7. Hunter M, Johnson N, Hedderwick S et al. (2010) Immunovirological correlates in human rabies treated with therapeutic coma. *J Med Virol* **82**:1255–1265.

8. Jackson AC (ed) (2013) Rabies: Scientific Basis of the Disease and Its Management, 3rd ed. Academic Press.

9. McElhinney LM, Marston DA, Leech S et al. (2013) Molecular epidemiology of bat lyssaviruses in Europe. *Zoonoses Public Health* **60**:35–45.

10. Nathwani D, McIntyre PG, White K et al. (2003) Fatal human rabies caused by European bat Lyssavirus type 2a infection in Scotland. *Clin Infect Dis* **37**:598–601.

11. Pathak S, Horton DL, Lucas S et al. (2014) Diagnosis, management and post-mortem findings of a human case of rabies imported into the United Kingdom from India: a case report. *Virol J* **11**:63.

12. Smith JS, Fishbein DB, Rupprecht CE & Clark K (1991) Unexplained rabies in three immigrants in the United States. A virologic investigation. *N Engl J Med* **324**:205–211.

13. Tepsumethanon V, Lumlertdacha B, Mitmoonpitak C et al. (2004) Survival of naturally infected rabid dogs and cats. *Clin Infect Dis* **39**:278–280.

14. Un H, Eskiizmirliler S, Unal N et al. (2012) Oral vaccination of foxes against rabies in Turkey between 2008 and 2010. *Berl Munch Tierarztl Wochenschr* **125**:203–208.

15. Wallace RM, Stanek D, Griese S et al. (2014) A large-scale, rapid public health response to rabies in an organ recipient and the previously undiagnosed organ donor. *Zoonoses Public Health* **61**:560–570.

16. Yakobson B, Goga I, Freuling CM et al. (2014) Implementation and monitoring of oral rabies vaccination of foxes in Kosovo between 2010 and 2013—an international and intersectorial effort. *Int J Med Microbiol* **304**:902–910.

ANSWERS

MCQ	Feedback

1. True or false: Rabies:

   a. Is a common worldwide zoonotic infection

   b   Is caused by canine parvovirus

   c   Only affects dogs and cats

   d   Can be successfully treated once symptoms develop

   a. True. Rabies infection is one of the most common zoonotic infections in the world with ~60,000 fatalities each year.

   b. False. It is caused by rabies virus, and member of the *Rhabdoviridae* family, *Lyssavirus* genus.

   c. False. it can infect any warm-blooded animal.

   d. False. Rabies is almost universally fatal once symptoms develop.

2. True or false: Individuals at increased risk of rabies infection are:

   a. Travellers licked by a dog overseas

   b. Travelers bitten by bats overseas

   c. People bitten by a dog in the UK

   d. People bitten by bat in the UK

   e. Healthcare workers

   a. True. Especially if licked on an open wound or mucous membrane.

   b. True. Lyssaviruses in bats are found in most parts of the world.

   c. False. The UK is free of canine rabies.

   d. True. The UK has EBLV-2 in Daubenton's bats.

   e. False. Although human-to-human transmission is theoretically possible, this is almost unknown even in countries with endemic rabies.

3. True or false: Postexposure treatment may consist of:

   a. Two doses of rabies vaccine

   b. Five doses of rabies vaccine

   c. Human rabies immunoglobulin alone

   d. Five doses of vaccine and human rabies immunoglobulin

   a. True. For someone who has previously had PreP, PET, or has antibody levels >0.5 IU/mL.

   b. True. Category II exposures in a high-risk country or category II or III in a low-risk country.

   c. False. HRIG is never used as single-component therapy.

   d. True. For category III exposure in a high-risk country.

4. True or false: Following a dog bite in India (a rabies-endemic country), individuals should be advised to:

   a. Wash the wound and seek local medical advice

   b. Go to the emergency room immediately on return to the UK

   c. Be asked to prepare a list of their contacts

   d. Be closely monitored and hospitalized if necessary

   a. True. This should be done for all bites in any country.

   b. False. They should seek advice in India and start treatment there without delaying until their return to the UK. Completion of their course of treatment should then be through their GP.

   c. False. A list of contacts is only needed if they become symptomatic and rabies is confirmed.

   d. False. If promptly started on PET, they do not need close monitoring, but as with all potential exposures, they should be asked to report back to their GP if they become symptomatic.

5. True or false: Rabies preexposure prophylaxis is:

   a. Ineffective

   b. Offered to all those working with the virus

   c. Should be considered if visiting endemic countries

   d. Recommended for those regularly handling bats

   a. False. It is very effective.

   b. True. If working with the virus, they need PreP and regular monitoring of rabies antibody levels as described in the Green Book.

   c. True. Although it depends on what they are doing, where and for how long.

   d. True. PreP is recommended for all who are regularly handling bats.

AN OUTBREAK OF NONTYPHOID SALMONELLOSIS IN THE WORKPLACE

Stephen Woolley[1] and Emma Hutley[2]

[1]Specialist Registrar in Infectious Diseases and Medical Microbiology, Royal Liverpool Hospital and Royal Navy, UK.
[2]Consultant Medical Microbiologist, Frimley Park Hospital and Royal Army Medical Corps, UK.

On November 23 2014, public health authorities were informed of the absence from work of 16 staff with symptoms of gastroenteritis. There were unconfirmed reports that additional workers not scheduled to be on shift or scheduled to be in later were also affected. The following day a further 10 cases had been reported, of whom 3 had required attendance at the hospital for rehydration. A further 3 cases reported sick on November 25.

INVESTIGATION OF THE OUTBREAK

Stool samples were obtained from affected individuals and a preliminary questionnaire completed. Symptoms included diarrhoea, abdominal cramps, fever, nausea, and vomiting. Onset of symptoms ranged from 1800 hours on November 22 to 2300 hours on November 24. All affected personnel had eaten lunch in the staff canteen at work and 26 of the 29 had consumed the cottage pie for lunch.

A working case definition for the outbreak was made as follows:

- onset of 2 or more episodes of diarrhoea, abdominal cramps, fever, nausea, and vomiting
- after 1200 hours on November 22

A detailed questionnaire was produced with the intent of conducting a cohort study, interviewing all personnel who had eaten in the staff canteen in the two days prior to the outbreak. However, this amounted to almost 1100 staff members, so a case control study was considered a more reasonable option. A random sample of 85 personnel who had eaten lunch on November 22 was interviewed, with the intent of interviewing 2 controls for every case found.

In total, 114 staff members were interviewed as well as the initial 29 cases, and an additional 12 cases were identified. Symptoms were predominantly abdominal cramps and diarrhoea with the mean onset time being 23 hours after lunch on November 22. The epidemic curve is consistent with a point source and the implicated source was a lunch meal served in the staff canteen on November 22 (Figure 21.1). A food-specific attack rate was calculated for each foodstuff.

A total of 52 faecal samples were cultured and *Salmonella* Enteritidis PT4 was isolated from 17 of them.

An odds ratio of 110 [Confidence Interval (CI) 20–810] for illness associated with consumption of cottage pie served for lunch on November 22 strongly suggested this was the vehicle of infection.

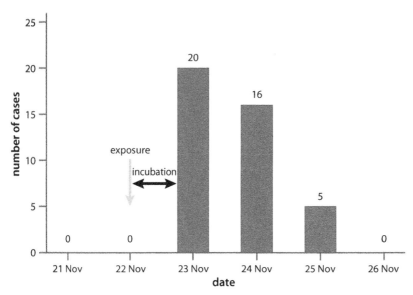

Figure 21.1. **Epidemic curve for the outbreak.** Incubation to pie refers to when the pie was cooked. Exposure to pie refers to when the pie was served to the consumers.

The catering staff involved in the preparation, cooking, and serving of food were identified and interviewed. The meal was cooked on the evening of November 21 with the pie mixture being poured into 8 serving tins (of approximately 130 servings) and covered with a layer of mashed potatoes. Raw eggs had been added to the mashed potatoes to give it a creamy texture and golden colour. The pies were then put onto trolleys and left in a side room for 13 hours before being reheated on November 23. They were not all refrigerated because the refrigerators lacked capacity during a period of refurbishment of the catering facilities.

The eggs used came from a batch laid and packaged on October 16 and had been stored at ambient temperature until use. There was poor stock control with the eggs with no mechanism to ensure that oldest eggs were used before newer eggs.

Of the random sample of 85 who ate in the canteen, 12 fulfilled the case definition, extrapolating that to all those who ate the meal (approximately 1100 people) suggests up to 150 employees may have been affected.

This outbreak of *Salmonella* Enteriditis PT4 was caused by a breakdown in standards and basic food hygiene. It is probable the salmonella, an organism associated with eggs, was added to the warm meat mixture with the mashed potatoes and then left at room temperature for 13 hours, sufficient time to allow a pathogenic quantity of bacteria to build up. The reheating for lunch may not have been at high enough temperature or for long enough to ensure the bacterial load could be killed.

CLINICAL MANAGEMENT

Gastroenteritis is usually a self-limiting condition not requiring treatment with anything other than oral rehydration therapy. Antibiotics are not usually indicated. However, in some cases, particularly at the extremes of age and in the immunosuppressed, patients become very dehydrated and may require intravenous therapy in the hospital.

Antibiotics that are used in the treatment of both typhoid and paratyphoid fever and those nontyphoidal salmonella (NTS) infections requiring antimicrobial treatment include ampicillin, ciprofloxacin, trimethoprim, or chloramphenicol depending on sensitivities. Third generation cephalosporins are being used increasingly to treat invasive NTS disease, in particular, cefotaxime or ceftriaxone in central nervous system (CNS) NTS disease because of their ability to penetrate the cerebrospinal fluid (CSF). As a result of the emergence of antimicrobial resistant NTS, azithromycin has been used increasingly as empirical treatment for NTS gastroenteritis, although there is documented evidence that NTS is becoming resistant to this antibiotic. A commonly reported reason for the emergence of antimicrobial resistant NTS is the extensive use of antimicrobials in the food production industry in the 1980s and early 1990s.

PREVENTION OF FURTHER CASES

Nontyphoid salmonella is usually acquired through the ingestion of contaminated food or water products. Uncooked shell eggs, pork, and beef are the most common sources of the infection. There have been major public health outbreaks of NTS infection in the UK and US through the consumption of infected chicken eggs and egg-based products. As a consequence of this, it is routine practice to use pasteurized eggs and avoid raw egg recipes. Restaurants and other food handling businesses have been advised to avoid pooling eggs and to use refrigeration when transporting and storing hens' eggs.

The advice given by public health agencies on minimizing contamination or risk of NTS transmission include the following:

- Cook meat thoroughly to a temperature of 75°C or until juices run clear.
- Do not purchase dry or cracked eggs.
- Use strict food handling procedures when preparing dishes containing raw or incompletely cooked egg products, such as homemade ice cream and mayonnaise.
- Store raw meat below ready to use food in the refrigerator.
- Keep cold food below 5°C and hot food above 60°C.
- Do not consume unpasteurized milk.
- Follow good handwashing and procedures for keeping food-preparation areas clean.
- Recognize the risk of *Salmonella* infections in pets: high-risk animals include chickens, ducks, tropical freshwater fish, and reptiles.
- Wash hands after handling raw meat.
- Always wash fruit and vegetables before eating. If home grown, wash them before bringing into the house.
- People infected with NTS should stay away from childcare, school, or work until diarrhoea-free for 24 h. If working as a food handler, then this should be 48 hours.

In the UK and US, it is a statutory requirement to report a suspected case of NTS to the regional public health authority. This allows a thorough investigation by the public health team to identify and manage any possible outbreaks.

EPIDEMIOLOGY

NTS infection are proved to be from a foodborne source in approximately 95% of cases. After *Campylobacter*, NTS is the most commonly isolated bacterial pathogen when a laboratory diagnosis of diarrhoea is investigated. Although, in 2002 the incidence of salmonellosis was reported as 17.7 per 100,000 population, which was documented as the highest incidence rate among 10 other potential food-borne pathogens under active surveillance, and varied little according to geography. The most commonly reported serotypes of NTS are *S.* Typhimurium and *S.* Enteritidis, accounting for 42% of all laboratory confirmed cases of NTS in 2000.

The bacteria are transmitted through a variety of recognized sources, including food sources (chicken eggs, pork, beef, and unpasteurized milk-based products), waterborne sources, pets (reptiles and birds), and nosocomial transmission. In addition, contaminated drugs or solutions have been reported in the literature. Infections are more common in the summer months and require ingestion of a large number of organisms in healthy adults.

NTS is commonly associated with chicken eggs through a series of NTS outbreaks in the UK and USA in the 1980s and 1990s. The serotype commonly associated with these outbreaks, especially in the UK was *S.* Enteritidis. It is estimated that 2.3 million of the 69 billion chicken eggs annually produced in the US are contaminated with *S.* Enteritidis. In the US between 1985 and 1995, *S.* Enteritidis accounted for 796 outbreaks, which resulted in 28,689 cases of infection; with 2839 being hospitalized and 79 deaths. Phage types 8 and 13a of *S.* Enteritidis are the most common phage types in the US, whereas phage type 4 is the most common in Europe. Since these highly publicized NTS outbreaks, improvements in food standards and practices have resulted in sharp declines in the incidences of NTS associated with chicken eggs in the US and UK.

Pets are now a well-documented reservoir for NTS, and between 3 and 5% of NTS originate from an animal reservoir. The literature comments upon as many as 90% of reptiles being carriers of NTS, with *S.* Typhimurium a common serotype associated with pet carriage of NTS.

Asymptomatic carriage is also another feature of NTS. In such cases, carriers secrete the NTS bacteria in their stool. This carrier state represents a public health issue especially if the carrier works in the food industry. The carrier states usually have a short duration; however, some patients can become chronic carriers (stool culture positive after 1 year). The group most likely to develop into chronic carriers is children under 5 years of age. An individual may be declared free from carriage of the NTS bacteria when three stool samples are negative.

In the UK the most common serotype of NTS infection is *S.* Enteritidis. In January 2015, there were 507 reported cases of NTS to Public Health England (PHE). Of those 507, 120 were infected with *S.* Enteritidis and 101 were infected with *S.* Typhimurium. Overall the incidence of NTS in the UK declined from nearly 18,000 infections in 2002 to 8355 infections in 2012.

BIOLOGY

Salmonella is a Gram-negative rod that is transmitted primarily through the improper handling and digestion of uncooked food. A large number of animal reservoirs have been identified including chicken, pigs, and cattle. *Salmonella* are facultative anaerobic, nonspore-forming mobile enterobacteria with peritrichous flagella.

There are two main species of *Salmonella*, which are *S. enterica* and *S. bongori*. *S. enterica* can be further split into six main subspecies *enterica, salamae, arizonae, diarizonae, houtenae,* and *indica*. There are reported to be over 2500 serovars of *Salmonella enterica* with many being reported to cause human infection. The most important serovars to comment upon are Typhi, Enteritidis, Paratyphi, Typhimurium, and Choleraesuis. Typhi and Paratyphi cause typhoid and paratyphoid fevers. Enteritidis and Typhimurium are the most common causes of human nontyphoid salmonellosis, although some serovars are more prevalent in specific geographical locations.

Salmonella is serotyped according to the O (somatic) antigens, Vi (capsular) antigens and H (flagellar) antigens (Figure 21.2). The Vi antigen is used to identify S. Typhi strains. Antigenic variation occurs on the O antigen. 99% of *Salmonella* species can be genus identified, using agglutination reactions of the O antigen, into groups AB, B, C, D and E. These groupings are not able to distinguish whether the organism is able to cause enteric fever. An example of this is that S. Enteritidis and S. Typhi are both group D; however S. Enteritidis causes NTS and S. Typhi causes enteric Typhoid fever.

The gold standard for the diagnosis of NTS infection is bacteriological culture, which can be from blood, urine, CSF, or more commonly stool. The samples are cultured onto a selective agar media at 37°C for between 18 and 24 hours. Stool samples are usually inoculated into a selective enrichment broth and incubated at 37°C for a similar time frame before being plated onto a selective agar (Figure 21.3).

Other more rapid methods of detection are available and are able to provide a result within two days. These tests include novel culture techniques, enzyme-linked immunosorbent assays (ELISA), as well as molecular techniques such as polymerase chain reaction (PCR) assays and DNA hybridization. Serological diagnostic tests for NTS infections have classically been via the Widal tube agglutination test. This test is very specific, although it is limited by poor sensitivity, an inability to distinguish between different antibody classes, and cross-reactivity with other *Salmonella* species. ELISA is not routinely used in Europe to diagnose NTS because of the lack of a standardized test.

Serotyping is used by the Centers for Disease Control and Prevention (CDC) and Public Health England (PHE) to track and monitor NTS outbreaks. Some serotypes are only

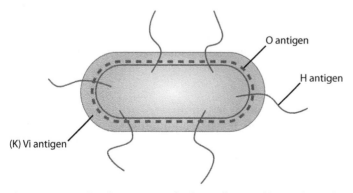

Figure 21.2. **Antigenic structure of salmonellae used in serological typing.**

Figure 21.3. **Culture of *Salmonella* demonstrating characteristic black colonies on a xylose lysine deoxycholate (XLD) selective agar plate.**

linked to one specific food item or one single geographical location, whereas others are found in multiple animals and some serotypes can cause more severe disease than others.

In addition to serotyping the CDC is using pulsed-field gel electrophoresis (PFGE) to create a DNA fingerprint of infections in regional reference labs. The reference labs are then able to cross-reference these fingerprints to help identify any potential outbreaks. In the UK, all positive cultures are sent to the national reference laboratory, which recently moved to the use of whole genome sequencing as routine for *Salmonella* isolates, because it is able to obtain enhanced discrimination of isolates that may initially appear to be the same.

Fluoroquinolone resistance in NTS species arises predominantly from topoisomerase mutations, although there is increasing evidence of three plasmid-mediated mechanisms that are conveying the bacteria with increased resistance to this important antimicrobial class. The three mechanisms are Qnr (quinolone resistance proteins), Aac(6')-Ib-cr, and QepA efflux. The Qnr proteins protect the DNA-gyrase from the quinolone, the Aac(6')-Ib-cr protein modifies the quinolone with a piperazinyl group, and the QepA protein is involved in active efflux.

DISEASE

Salmonella was first discovered in 1885 by an American medical researcher, Theobald Smith, at the US Department of Agriculture, who was working under a veterinary pathologist, Daniel Salmon. Like many pathogens, *Salmonella* causes most of its symptoms at the site of invasion, in this case, the small bowel, giving the patient an enterocolitis. However, because it is able to invade and replicate within macrophages, NTS can develop an invasive extraintestinal disease.

The enterocolitis associated with NTS commonly manifests as diarrhoea and abdominal pain with occasional fever. These symptoms are usually self-limiting and occur between 6 to 72 h after ingestion of the pathogen, although symptoms have been reported up to one week following exposure. The diarrhoea can be severe or chronic, causing both profound fluid and electrolyte disturbances.

Immunocompromised hosts are more susceptible in developing a more severe acute infection or chronic diarrhoea, or developing the symptoms of an invasive NTS salmonellosis.

Recurrent NTS bacteraemia is an acquired immunodeficiency syndrome (AIDS) defining illness with a mortality of up to 47%. Other immunocompromised hosts who are reported as being susceptible to NTS include children with sickle-cell anaemia, who are at risk of developing an aggressive NTS osteomyelitis following bacteraemia.

Extraintestinal manifestations of NTS are the result of an associated bacteraemia. These extra manifestations have clinical focal syndromes including osteomyelitis, meningitis, myocarditis, pneumonitis, pyelonephritis, and endovascular infections. The risk factors for developing a bacteraemia and invasive NTS salmonellosis are immunosuppression, HIV, malaria, and malnutrition. The development of invasive salmonellosis is associated with higher mortality rates, further septic manifestations, longer hospital stays, and prolonged periods of antimicrobial therapy.

Reported long-term complications of NTS salmonella are reactive arthritis and ulcerative colitis. There are also documented reports that patients with atherosclerotic disease or preexisting arterial aneurysms are more likely, compared with the general population, to develop vascular infections as a result of NTS salmonella. These infections can develop into aneurysms or enlarge existing aneurysms.

In the UK, NTS infections are conservatively managed because of the self-limiting nature of the disease. It is documented that antimicrobial regimens of 1 to 14 days do not decrease the positive rates of intestinal NTS after 2–3 weeks, but prolong NTS excretion. NICE (National Institute for Health and Clinical Excellence) comments on other issues with two key antimicrobial treatments—drug interactions such as skin rash with ampicillin, leucopaenia with cotrimoxazole and urticaria, severe headache, epigastric pain and dizziness with fluroquinolones. Emerging antimicrobial resistance to commonly used antibiotics is another major concern with the use of antibiotics in the treatment of NTS. The current advice for prescribing antimicrobials is for the following:

- Suspected or confirmed cases of septicaemia
- Patients with extraintestinal spread of NTS
- Children under the age of six months with NTS gastroenteritis
- Children who are malnourished or immunocompromised with NTS
- Immunocompromised adults

Salmonella Typhi and Paratyphi cause a systemic disease (typhoid or paratyphoid fever) that is always treated with antimicrobials. The disease may present with an apparent chest infection and constipation. There is a classical daily temperature rise followed by a fall with progressively increasing peak temperatures over subsequent days. Abdominal distention, colic, and diarrhoea may develop. Maculopapular lesions on the skin (rose spots) develop as does splenomegaly. There is a relative bradycardia with dicrotic pulse, tachypnea, confusion, and delirium. Bowel perforation may occur and without treatment the mortality is about 10%.

PATHOLOGY

Salmonella infections are contracted by oral ingestion of the pathogen, with the bacteria penetrating the intestinal epithelia in order to induce disease. A large number of organisms is required to cause disease in healthy adults because the organism is susceptible to stomach acid, which prevents most organisms reaching the site of action. The bacterium is able to exploit the host cells, and NTS species predominately invade the

M cells of the intestine. When in contact with the epithelium, the *Salmonella* bacterium induces degeneration of enterocyte microvilli. The subsequent loss of microvilli causes membrane ruffling localized to the area in direct contact with the bacterium. As a consequence, the epithelial cell undergoes macropinocytosis that leads to internalization of the *Salmonella* bacterium.

Salmonella invasion depends on several chromosomal genes clustered in a pathogenicity island. This is termed *Salmonella* pathogenicity island 1 (SPI1). SPI1 codes for several virulence factors and a type-III secretion system, which enables the bacterium to disrupt the host cell membrane, thus allowing subsequent penetration into the M cell. The bacterium also disrupts the tight junctions between the epithelial cells that alters the flow of ions, water, and immune cells in and out of the intestinal wall. The consequence of these two mechanisms is that the bacterium causes intestinal wall inflammation leading to diarrhoea.

NTS has adapted another pathogenicity island, *Salmonella* pathogenicity island 2 (SPI2), which allows this subspecies to invade macrophages. SPI2 codes for another type-III secretion system that provides up to 30 effector proteins, allowing NTS to replicate within a *Salmonella*-containing vacuole (SCV). The effector proteins work within the macrophage's endomembrane system and cytoplasm. This mechanism allows for the extraintestinal spectrum of disease for NTS.

QUESTIONS

1. Which species of NTS is commonly associated with chicken food products?
   a. *Salmonella* Enteritidis
   b. *Salmonella* Typhimurium
   c. *Salomonella bongori*
   d. *Salomonella* Indica
   e. *Salmonella Typhi*

2. What cells do NTS bacteria prefer to invade?
   a. Columnar epithelia of the trachea
   b. M-cells of the intestine
   c. Melanocytes
   d. Neutrophils
   e. MALT in the stomach

3. What trio of symptoms is commonly reported in NTS infection?
   a. Abdominal pain, rash, and haemoptysis
   b. Diarrhoea, rash, and cough
   c. Abdominal pain, fever, and diarrhoea
   d. Diarrhoea, vomiting, and rash
   e. Rash, anaemia, and headache

4. What is gold standard for the diagnosis of NTS?
   a. Serology
   b. PCR
   c. Culture
   d. ELISA
   e. Rapid diagnostic tests

5. For how long should an infected food worker be symptom free before returning to work?
   a. 12 h
   b. 24 h
   c. 36 h
   d. 48 h
   e. 72 h

GUIDELINES

1. National Institute for Health and Clinical Excellence (NICE). Gastroenteritis Management
cks.nice.org.uk/gastroenteritis

REFERENCES

1. Bennett JE, Dolin R, and Blaser MJ, Mandell, Douglas & Bennett's (2005) Principles and Practice of Infectious Diseases, 6th ed, Ch. 220. Elsevier.

2. Helms M, Ethelberg S, Molbak K (2005) International Salmonella Typhimurium DT104 infections, 1992-2001. *Emerg Infect Dis* **11**(6):859-867.

3. Jackson BR, Griffin PM, Cole, Walsh KA & Chai SJ (2013). Outbreak-associated *Salmonella enterica* serotypes and food commodities, United States, 1998-2008. *Emerg Infect Dis* **19**(8): 1239-1244

4. Bangtrakulnonth A, Pornreongwong S, Pulsrikarn C et al. (2004) Salmonella serovars from humans and other sources in Thailand, 1993-2002. *Emerg Infect Dis* **10**(1):131-136.

5. Goosney DL, Knoechel DG and Finlay BB. Enteropathogenic E.coli, Salmonella and Shigella: Masters of Host Cell Cytoskeletal Exploitation. Emerging Infectious Diseases. 1999. **5**(2)216-223

6. Varma JK, Greene KD, Ovitt J, Barrett TJ, Medalla F and Angulo FJ. Hospitization and Antimicrobial resistance in Salmonella outbreaks, 1984-2002. *Emerg Infect Dis* 2005. **11**(6);943-946

7. Onwuezobe A, Oshun PO & Odigwe CC (2012). Antimicrobials for Treating Symptomatic Non-Typhoidal Salmonella Infection. Cochrane Review, Wiley.

8. Hsu RB and Lin FY (2005) Risk factors for bacteraemia and endovascular infection due to non-typhoidal salmonella: a reappraisal. *Q J Med* **98**:821-827.

9. Ternhag A, Torner A, Svensson A, Ekdahl K & Giesecke J (2008) Short and long term effects of bacterial gastrointestinal infections. *Emerg Infect Dis* **14**(1).

10. Hassing R-J, Goessens WHF, Pelt W et al. (2014) Salmonella subtypes with increased MICs for azithromycin in travellers returned to the Netherlands. *Emerg Infect Dis* **20**(4):705-708.

ANSWERS

MCQ	Feedback

1. Which species of NTS is commonly associated with chicken food products?
  a. *Salmonella* Enteritidis
  b. *Salmonella* Typhimurium
  c. *Salmonella bongori*
  d. *Salmonella* Indica
  e. *Salmonella* Typhi
	Salmonella Enteritidis is most commonly associated with chicken-based food products. *S.* Typhimurium is most associated with pet carriage. *S. bongori* is a different species of *Salmonella* not associated with chicken-based products. The other answers are serotypes of *S. enterica*, which are not associated with chicken-based food products.

2. What cells do NTS bacteria prefer to invade?
  a. Columnar epithelia of the trachea
  b. M cells of the intestine
  c. Melanocytes
  d. Neutrophils
  e. MALT in the stomach
	NTS prefer to invade the M cells of the intestine.

3. What trio of symptoms ia commonly reported in NTS infection?
  a. Abdominal pain, rash, and hemoptysis
  b. Diarrhea, rash, and cough
  c. Abdominal pain, fever, and diarrhea
  d. Diarrhea, vomiting, and rash
  e. Rash, anemia, and headache
	NTS infection commonly presents with the trio of diarrhea, abdominal pain, and occasional fever. Tyhoid and paratyphoid are associated with rose spot rash.

4. What is gold standard for the diagnosis of NTS?
  a. Serology
  b. PCR
  c. Culture
  d. ELISA
  e. Rapid diagnostic tests
	The gold standard for the diagnosis of NTS infection is still culture of blood, urine, CSF, or stool. The drawback is that the results can take between three and five days. Other tests are in development but do not provide the same degree of certainty as bacterial culture does.

5. For how long should an infected food worker be symptom free before returning to work?
  a. 12 h
  b. 24 h
  c. 36 h
  d. 48 h
  e. 72 h
	Patients infected with NTS whose occupation is a food worker, should be symptom free for 48 h.

A CASE OF TOXOPLASMOSIS IN PREGNANCY

CASE 22

Alicia Yeap[1] and Nandini Shetty[2]

[1]Specialist Registrar, Public Health England, UK.
[2]Consultant Microbiologist, Public Health England, UK.

A 36-year-old woman in the first trimester of pregnancy was tested for serum antibodies to *Toxoplasma gondii* after developing a mild nonspecific illness. She had reported having ingested some undercooked meat at nine weeks' gestation.

INVESTIGATION OF THE CASE

Serum *T. gondii* IgG was negative from a test performed two years earlier. At 14 weeks' gestation, serum *T. gondii* IgG and IgM were strongly positive, with a positive Sabin–Feldman dye test result. Her antenatal clinic booking blood sample, which had been taken at 10 weeks' gestation, was subsequently tested in response to these results and showed a strongly positive IgM and a low-positive IgG. *T. gondii* IgG avidity testing was also performed. Table 22.1 shows additional details of her serological test results.

These results are strongly suggestive of acute infection with *T. gondii* around the time of or soon after conception. An amniocentesis was performed at 16 weeks' gestation and *T. gondii* DNA was not detected in the amniotic fluid by polymerase chain reaction (PCR).

Diagnostic tests

Detection of antibodies to *T. gondii* remains the mainstay of diagnosis. Serum IgM is generally detectable in the week after infection and plateaus at about 4 weeks after infection. IgM may persist for 6 months to several years but, in some, may become undetectable within 3 months. Serum IgG is generally detectable within a month of primary infection, plateaus within 8 to 12 weeks, and persists for life. As with serological tests for other infections, obtaining separate samples over time is crucial for accurate interpretation. A previous positive serum IgG result from up to 3 months before conception would make primary infection in pregnancy very unlikely.

Table 22.1. **Serological profile in a pregnant woman with suspected toxoplasmosis**

Gestational age (weeks)	Conceptional age (weeks)	IgG (1.6–3.0 IU/ml)	IgM	IgG avidity	Sabin–Feldman dye test (IU/mL)
10	8	5.0	Strongly positive	Uninterpretable (IgG level too low)	64
14	12	184	Strongly positive	<30%	1000

Several commercial assays are available, most of which are enzyme-linked immuno-sorbent assays (ELISA) and have varying performance characteristics. Assays based on indirect fluorescence and agglutination methods are also available. Tests that use only membrane surface antigens or whole parasites allow for earlier detection of IgG than those that use a mixture of cytosolic and surface antigens.

False-positive IgM results with commercial assays are common and can be problematic. The IgM-immunosorbent agglutination assay (ISAGA) is a highly sensitive and specific test for IgM and utilizes intact killed tachyzoites to detect antibodies bound to a solid phase. It is a good confirmatory test and useful for infant diagnoses, but it can persist for many months to years in adults.

IgG avidity testing can aid timing of infection. Avidity is a measure of the strength of anti-body binding to target antigens. Low-avidity antibodies are produced early in infection and as the infection progresses, avidity matures, that is, the antibodies produced become more specific to the target antigens and bind more strongly to them. The test uses mole-cules such as urea to displace weakly bound antibodies and requires a significant level of IgG to be detectable for interpretation. The presence of high-avidity IgG excludes recent infection in the 3- to 4-month period before the sample was taken. However, IgG avidity may remain low for longer periods in some circumstances, such as early treatment, and the absence of high-avidity IgG cannot be used as a reliable indicator of recent infection.

The Sabin–Feldman dye test is the gold standard reference method for antibody detection and mainly detects IgG. It is based on complement-mediated lysis of live organisms in the presence of antibody.

PCR for detection of DNA in amniotic fluid has improved the diagnosis and management of congenital infection. Assays based on the REP-529 sequence, which is repeated 200- to 300-fold throughout the organism's genome, are more sensitive than those that are based on the 35-fold-repeated B1 gene. Negative predictive values for these tests for foetal infec-tion are excellent. The risk of miscarriage with amniocentesis is approximately 1% and is greater early in pregnancy. Amniocentesis should be performed at ≥16 weeks' gestation and ≥4 weeks after infection is estimated to have taken place.

For neonatal diagnosis of congenital infection, IgA tests are more sensitive than IgM, although serological tests are generally less sensitive than for adults. Antibody responses may be attenuated by maternal treatment during pregnancy. Maternal IgG is transferred across the placenta and persists for 6 to 12 months after birth. Western blot analysis of specific banding patterns of IgG and IgM from paired mother and child serum samples may help to discriminate between passively transferred and actively produced antibodies in the neonate and can also be useful for looking for neonatal IgG production if neonatal IgM or IgA are absent.

CLINICAL MANAGEMENT

The couple was counselled about the implications of maternal primary toxoplasmosis infection in early pregnancy. The negative amniotic fluid PCR result indicated that it was likely that the foetus had not yet been infected. The risk of foetal infection at this early stage of pregnancy is low (below 10%), but the consequences are potentially severe, with a high risk of developing congenital defects such as epilepsy, learning difficulties, deafness, and blindness. Spiramycin, an oral macrolide antibiotic with few side effects, could be taken for the duration of the pregnancy to attempt to prevent foetal infection, but the efficacy of this treatment is unknown. Spiramycin does not cross the placenta and therefore will have no action on infection already transmitted to the foetus. As it acts mainly on the placenta, it prevents further transmission to the foetus for the duration of treatment.

If *T. gondii* DNA had been detected in amniotic fluid, it would have indicated that foetal infection was likely to have taken place. As spiramycin does not cross the placenta, the treatment of choice in this situation would be an oral combination regimen of pyrimethamine (an antagonist of dihydrofolate reductase) and sulfadiazine (a sulfonamide drug), along with folinic acid supplementation to prevent bone marrow suppression. This treatment should be continued in the mother through pregnancy. Trimethoprim-sulfamethoxazole has also been used. Clindamycin is an alternative agent if there has been a history of sulfonamide allergy. Other options are azithromycin, dapsone, and atovaquone, but experience of their use in this context is limited. Pyrimethamine is generally felt to be the pivotal drug in the treatment regimen but should not be given in the first trimester as it is teratogenic.

Combination therapy is sometimes used in late second and third trimester without confirmation of foetal infection because of the high rate of vertical transmission at these stages of pregnancy. The efficacy of combination therapy in preventing vertical transmission has not been clearly demonstrated. Combination therapy is also indicated for infants in whom infection has been proven or is highly likely for a minimum of 12 months after birth, as there is some evidence to suggest a reduction in sequelae with this approach.

The pregnancy would be closely monitored, with repeated foetal ultrasound scanning conducted. Foetal scans are unlikely to detect some abnormalities, such as hearing and visual loss, that usually become apparent only after birth. Long-term follow-up for the child would be necessary as sequelae such as hearing loss and learning difficulties could manifest late and chorioretinitis could be recurrent. The couple chose to terminate the pregnancy and were supported through this decision.

PREVENTION OF INFECTION

Isolation precautions for people who have recently or previously been infected with *T. gondii* are not necessary since horizontal human-to-human transmission does not occur. Avoidance of pregnancy after primary infection is advisable. The duration for this has not been clearly defined—3 to 6 months has been proposed as a reasonable period.

Women should be counselled about precautions to avoid infection with *T. gondii* early in pregnancy or before conception. Handling of cat faeces, soil, and raw meat should be avoided where possible, or gloves worn during contact and thorough handwashing performed after. They should avoid eating undercooked or raw meat and should wash fruit and vegetables before consumption, measures that may also help to prevent other infections. The efficacy of such health education measures is unknown.

Routine antenatal screening is performed in some countries (for example, France and Austria), but modelling has not demonstrated screening to be cost-effective in the UK. The estimated incidence of infection in pregnancy, the sensitivity and specificity of available tests, the efficacy of current treatments, and the consequences and management of false-positive test results are some of the many factors that should be considered before implementing national antenatal screening programs. Where carried out, seronegative pregnant women are followed up with repeat serological testing, foetal imaging, and amniotic fluid sampling, and treatment offered to those with suspected infection as detailed in the prior discussion of clinical management. France has demonstrated reductions in foetal infection, serious neurological sequelae, and clinical signs at 3 years of age after the introduction of various antenatal-screening interventions.

Neonatal screening is carried out in some countries (for example, Poland and Brazil). This involves routine serological testing of newborns, followed by treatment and clinical monitoring using different drug regimens and schedules. Denmark has abandoned this strategy as it failed to demonstrate a benefit with treatment comprising 3 months of pyrimethamine and sulfadiazine.

Public health measures implemented at national level have largely focused on the detection of infection in pregnant women and congenital infection so far. There has been a renewed interest in the public health control of toxoplasmosis globally in recent years, with the World Health Organization (WHO) and several countries undertaking studies to estimate the burden of infection. Broader strategies addressing transmission sources may reduce congenital infection as well as infection in the general population, which would benefit immunosuppressed individuals and potentially reduce ocular toxoplasmosis and severe strain-related infections. The relative efficacies of interventions aimed at reducing tissue cysts in meat for human consumption and those targeting environmental oocyst contamination depend on the contributions of these routes of transmission in any given setting. A study exploring this question using a novel serological test has shown that in the US, the proportion of congenital infections via oocyst ingestion is high at 78%. This proportion is likely to vary in other populations, depending on food and farming practices, as well as behavioural factors surrounding hygiene and cat contact.

Reducing the bioburden of *T. gondii* in the form of tissue cysts requires targeting meat production. Indoor rearing of pigs is felt to be a significant factor in reducing seroprevalence in the human population. Keeping animals indoors, preventing contact with cats, rodent control, and avoiding feed with offal and raw goat whey may reduce infection of animals reared for human consumption. These measures are not achievable in the growing organic and free-range meat industries and for game meat. Meat handling measures aimed at destroying existing tissue cysts such as freezing and gamma-irradiation may be effective but can be expensive. These interventions could be targeted at farms with high seroprevalence. There is a currently a live attenuated vaccine for sheep that prevents placental infection and could potentially reduce tissue cysts in meat but has yet to be fully evaluated for this purpose.

Prevention of environmental contamination with oocysts from cats can also be targeted. Limiting stray cat populations, reducing prey animal contact, restricting cat access to high-human-contact soil areas, and proper disposal of cat litter are some of the measures that may be useful. A candidate vaccine for cats has been developed.

It is likely that a range of interventions spanning the various transmission routes will be needed in order to cause a significant impact at a public health level.

EPIDEMIOLOGY

Toxoplasmosis can be found in human and animal populations across the world and is more common in tropical climates, at higher humidities, and at lower altitudes. Human seroprevalence increases with increasing age, with no significant gender difference. Seroprevalence varies widely across geographical areas and population groups and is likely to be influenced by dietary habits, sanitation, agricultural practices, and socioeconomic status. Estimates of seroprevalence are as high as 85% in antenatal clinic attendees. A recent study carried out in an antenatal clinic in London serving an ethnically diverse population reported seroprevalence of 17% in attendees. A general downward trend in seroprevalence has been observed in recent years in developed countries.

Humans acquire infection by two main routes: ingestion of viable tissue cysts in the meat of intermediate animal hosts and ingestion of infective oocysts from contaminated food or water. Cats and other felid species are the definitive hosts for *T. gondii* and are responsible for contaminating the environment with oocysts, which they shed in their faeces. The seroprevalence in wild felid species may be up to 100% in some areas. Oocysts can survive in moist environments for up to 18 months. Direct contact with cats does not pose an infection risk because oocysts require a period of sporulation in the environment before becoming infectious.

The proportion of infections attributable to each of these routes of transmission varies depending on food practices and hygiene conditions. Up to half of cases may have no identifiable source. Vegetarians acquire infection from consuming raw vegetables and fruits that have been contaminated with soil or water containing oocysts. Heat kills oocysts and tissue cysts within minutes at temperatures above 60°C. Microwaving may be less effective because of uneven heat distribution with this method of cooking. Freezing reduces the viability of tissue cysts but lower temperatures and longer durations are necessary to render meat non-infective. Infective oocysts can survive freezing for months. Refrigeration is ineffective at killing both infectious forms and methods of curing, such as salting and smoking, are unreliable.

Toxoplasmosis has a large animal reservoir and all warm-blooded animals can potentially be infected. Estimates of infection in livestock animals range from less than 1% to 93% and are likely to reflect variations in sterility of feeds, indoor or outdoor rearing, and exposure to cats, rodents, and birds. Freezing of meat after slaughter reduces the viability of tissue cysts and hence infectivity. Pork and lamb are generally felt to be the main meat sources of toxoplasmosis in humans. In cattle, seroprevalence estimates of 2% to 92% have been reported, but tissue cysts have rarely been found in beef. Spontaneous clearance of infection and natural resistance are thought to occur in cattle. Poultry are also less common sources of *T. gondii* infection and possible explanations for this are that the conditions of intensive farming tend to prevent infection and slaughtered poultry are usually stored frozen for long periods of time. The trend of ethical or free-range meat production may increase the likelihood of infection of livestock animals, because environmental exposure to infection is likely to be higher than with intensive farming methods. Similarly, game animals are also potential meat sources of infection.

Shellfish, such as mussels and clams, are filter feeders and can concentrate oocysts from contaminated freshwater runoff. Oocysts can survive in seawater for long periods of time. Contaminated freshwater-related outbreaks have also been described. Unpasteurized goats' milk has caused several outbreaks, and the likely infective stage implicated in this mode of transmission is the tachyzoite. Breastfeeding in humans is felt to be safe, with no associated cases of transmission ever reported.

Congenital infection occurs by transplacental spread of tachyzoites from mother to foetus, if primary infection occurs during pregnancy or shortly before conception. Congenital infection is estimated to occur in 3 in every 100,000 live births in the UK. The risk of infection of the foetus increases with increasing gestational age at the time of maternal infection, rising from less than 15% in the first trimester to approximately 60% in the third trimester. However, foetuses infected early in pregnancy are much more likely to be affected by congenital abnormalities—these are apparent in approximately 80% of foetuses infected in the first trimester, about 25% of those infected in the second trimester, and about 10% of those infected in the third trimester. Foetal mortality is also higher with infection in the first trimester at about 5% compared to less than 1% in the third trimester.

Transfusion of blood products contaminated with *T. gondii* tachyzoites has been described to cause outbreaks of primary toxoplasmosis. Another rare source of infection is solid organ transplantation. Infection of susceptible individuals can occur if they receive an organ containing *T. gondii* tissue cysts. The highest risk of infection is with heart transplantation. In haemopoietic stem cell transplantation, disease most often occurs as a result of reactivation rather than primary infection. *T.gondii* serostatus is usually sought before transplantation in anticipation of such events. These are unlikely routes of transmission in pregnancy.

Three clonal lineages have been described for *T. gondii*. Strains causing congenital infections largely belong to types I and II. Reactivation disease in those infected with human immunodeficiency virus (HIV) is mainly caused by type II strains, and infection in animals mostly by genotype III strains. Type I strains are considered to be the most virulent and can cause severe ocular disease in immunocompetent people. Some atypical strains have been reported to cause disseminated and severe disease in immunocompetent people, and some strains may be more likely to cause ocular and congenital disease. In immunocompromised people, the spectra of clinical disease and outcomes do not seem to differ between strains.

BIOLOGY

T. gondii is an obligate intracellular parasite belonging to the subclass Coccidia and phylum Apicomplexa, as are *Plasmodium* species, the causative agents of malaria. There are three infective stages of the parasite: the sporulated oocyst, the tissue cyst, and the tachyzoite. Sporulated oocysts are ovoid structures measuring about 12 μm in length. Each mature oocyst contains two sporocysts, which each contain four sporozoites. Oocysts have a multilayered wall that allows them to survive in the external environment. Tissue cysts may be found in any organ but are most common in brain and muscle tissue. They measure between 10 and 100 μm in diameter and are circular. Each tissue cyst can contain up to hundreds of bradyzoites and are surrounded by an outer membrane. Bradyzoites are elliptical, replicate slowly, and have a low metabolism, which allows for long-term survival and lifelong persistence of infection. Tachyzoites have a crescent shape and are about 5 μm in length (Figure 22.1). They are the rapidly dividing, highly metabolically active dissemination forms and are more vulnerable to physical or chemical changes in their environment than the other stages. Antimicrobial agents are only effective against these forms.

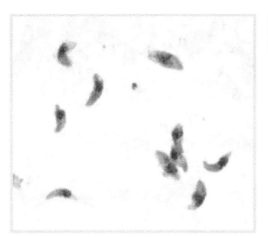

Figure 22.1. **Toxoplasma gondii tachyzoites (Giemsa stain).** (Courtesy of University College Hospital, London.)

The sexual reproductive cycle occurs only in cats and other felid species, beginning with ingestion of tissue cysts in infected intermediate hosts, such as rodents. Gastric enzymes break down the cyst walls to release bradyzoites. Bradyzoites penetrate enterocytes and form schizonts. Merozoites are produced within schizonts by asexual replication and these can infect other enterocytes when released. Schizonts can also undergo differentiation to become male or female gametes. Fertilization takes place within enterocytes, forming oocysts that pass into the bowel lumen with enterocyte rupture. An infected cat begins to excrete oocysts in its feces several days after infection and can shed millions of oocysts for two to three weeks. Once in the environment, oocysts take up to three weeks to develop sporozoites, and only sporulated oocysts are infective to intermediate animal hosts. Sporulated oocysts can also infect cats, but transmission via this route is less efficient.

When sporulated oocysts are ingested by intermediate hosts, sporozoites are released and these enter enterocytes and differentiate into tachyzoites. The tachyzoites divide rapidly and disseminate through the bloodstream within monocytes to the rest of the body. Ingestion of tissue cysts by intermediate hosts releases bradyzoites, which similarly penetrate enterocytes, differentiate into tachyzoites, multiply, and disseminate. Any cell type can be infected, but the most significantly infected organs are the eyes, brain, skeletal, and heart muscles. Tachyzoites survive intracellular killing by forming parasitophorous vacuoles. Within organs, tachyzoites can differentiate into bradyzoites and form tissue cysts (Figure 22.2). Bradyzoites are occasionally released when infected cells die and can infect adjacent cells or transform into tachyzoites in immunosuppressed hosts.

DISEASE

Primary infection with *T. gondii* in immunocompetent individuals is asymptomatic in up to 80% of patients. Symptoms are nonspecific: fever, general malaise, and lymphadenopathy, most commonly of the cervical region. Myalgia, sore throat, abdominal pain, a maculopapular rash, and hepatosplenomegaly may also manifest. Spontaneous resolution without treatment occurs in the majority, though symptoms may wax and wane for months and trigger investigations for other causes of illness. Myocarditis and polymyositis are severe manifestations of primary infection and, rarely, disseminated disease can

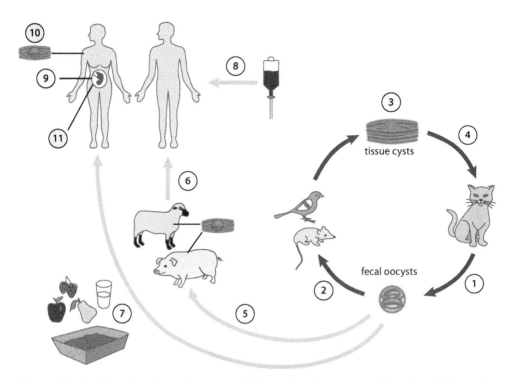

Figure 22.2. **The life cycle of Toxoplasma gondii.** Unsporulated oocysts are shed in the cat's faeces (1). Oocysts take 1–5 days to sporulate in the environment and become infective. Intermediate hosts (including birds and rodents) become infected after ingesting soil, water, or plant material contaminated with oocysts (2). After ingestion, oocysts transform into tachyzoites, where they localize in neural and muscle tissue and go on to develop into tissue cyst bradyzoites (3). Cats may become infected after consuming intermediate hosts harbouring tissue cysts (4); however, they may become infected directly by ingesting sporulated oocysts. Animals bred for consumption may become infected with tissue cysts after ingestion of sporulated oocysts in the environment (5). Humans can become infected by any of several routes: eating undercooked meat of animals harbouring tissue cysts (6), consuming food or water contaminated with sporulated oocysts (7), blood transfusion or organ transplantation (8), or transplacentally from mother to foetus (9). Diagnosis can be achieved in the following two stages. In the human host, the parasites form tissue cysts (10), which may remain throughout the life of the host. Diagnosis of congenital infections can also be achieved by detecting *T. gondii* DNA in amniotic fluid, using molecular methods such as PCR (11). (Courtesy of CDC-DPDx.)

develop. Acute chorioretinitis may be an indication of primary infection, though recurrence in people with ocular toxoplasmosis is common, occurring in up to 80% of cases.

The risk of foetal infection is unrelated to the presence or severity of maternal symptoms. Past maternal infection is generally felt to be protective for the foetus, though infection within three months of conception may confer a risk to the foetus. A single case of maternal re-infection with a second strain leading to subsequent foetal infection has been reported in the literature. There have been several reports in the literature of congenital infections resulting from toxoplasmosis reactivation in pregnancy, in the context of HIV infection, and in corticosteroid treatment with underlying systemic lupus erythematosus.

Before birth, foetal ultrasound scans demonstrating hydrocephalus, intracranial calcification, or hyperechoic mesentery should prompt testing for congenital toxoplasmosis. Foetal magnetic resonance imaging (MRI) has been used to confirm congenital infection

(Figure 22.3). Features of infection that may be evident at birth are myriad and include microcephaly, encephalitis, seizures, cerebellar signs, strabismus, rash, diarrhoea, hypothermia, jaundice, lymphadenopathy, pneumonitis, hepatosplenomegaly, anaemia, and thrombocytopenia. None of the symptoms and signs of congenital toxoplasmosis are pathognomonic and differential diagnoses such as cytomegalovirus (CMV) and rubella should be considered. Miscarriage and stillbirth may result from foetal infection. Epilepsy, psychomotor retardation, developmental delay, sensorineural deafness, and blindness may only present later in life, sometimes years after birth. Prospective studies suggest that up to 70% of untreated infected children have retinal lesions detected in the first 10 years of life and about 40% of those treated may experience recurrences of chorioretinitis after this period of time.

Apart from gestational age and treatment as mentioned previously, other factors that are likely to influence the presence and severity of sequelae in the infected child are parasite load, parasite strain, and immunological and host factors. There are indications that sequelae may be more likely in children born to HIV-infected mothers who have had primary infection in pregnancy. HLA-DQ3 has been significantly associated with hydrocephalus in infected children. Premature infected infants may have more ocular and neurological disease.

The main diagnostic tests and treatments have been discussed in the section on the Investigation of the Case. Histology of affected lymph nodes may aid maternal diagnosis but is rarely undertaken in this clinical context, unless sampling has been performed to investigate other potential causes of persistent lymphadenopathy. Cerebrospinal fluid (CSF) examination of infected neonates typically reveals a markedly high-level protein concentration, in contrast to reactivation-related toxoplasma encephalitis in immunosuppressed individuals where CSF protein levels are usually mildly, moderately raised, or normal. Histological examination of postmortem foetal samples may be useful in cases of miscarriage or termination. Culture of the organism using mouse inoculation or tissue culture is possible but the utility of these methods in clinical settings is severely limited by expense, time, and expertise required.

PATHOLOGY

Various mechanisms for evading intracellular killing enable survival of the parasite. Both cell-mediated and humoral immunity play a role in controlling *T. gondii* infection.

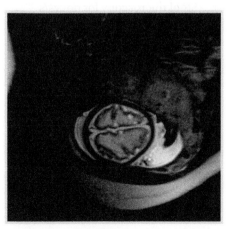

Figure 22.3. **Fetal MRI demonstrating marked intracranial calcification.**

Interferon γ (IFN), the interleukins (IL) IL-2, IL-10, IL-12, and tumour necrosis factor α (TNF) are the key cytokines implicated in pathogenesis. Generation of reactive nitrogen species and regulation of specific GTPases are important.

Histological examination of affected lymph nodes typically shows a combination of reactive follicular hyperplasia, epitheloid histiocytes clustering around the edges of germinal centres, and swollen sinuses filled with monocytes. Granulomata, microabscesses, necrosis, and organisms are seldom seen.

Patches of calcification affecting cerebral cortex, basal ganglia, or cerebellum may be seen on radiological imaging or postmortem samples in congenital infection. Obstructive hydrocephalus may be apparent. Periaqueductal and periventricular vasculitis with large areas of necrosis are typical histological findings in brain specimens. Focal necrosis of heart and skeletal muscle may be seen. Tachyzoites and tissue cysts may occasionally be demonstrated in tissue specimens and may be found in any organ.

QUESTIONS

1. The following precautions should be taken by pregnant women to avoid infection with *T. gondii*, except:
   a. Avoiding contact with cats
   b. Handwashing after handling raw meat
   c. Wearing gloves when gardening
   d. Thorough washing of raw vegetables
   e. Cooking meat thoroughly

2. The following are recognized routes of transmission of toxoplasmosis, except:
   a. Blood transfusion
   b. Ingestion of soil-contaminated vegetables
   c. Breastfeeding
   d. Ingestion of undercooked meat
   e. Ingestion of unpasteurized goats' milk

3. Which of the following statements is not correct?
   a. Spiramycin crosses the placenta effectively.
   b. Some strains may be more likely to cause severe disease.
   c. Bradyzoites are easily killed by pyrimethamine.
   d. Microwaving is a highly effective method for destroying tissue cysts.
   e. Screening for toxoplasmosis in pregnancy is universally practiced.

4. The following are tests that can be used for diagnosis of toxoplasmosis in humans except:
   a. Serum IgA
   b. Stool microscopy for oocysts
   c. PCR of amniotic fluid
   d. Serum IgM
   e. Mouse inoculation

5. The following feature is pathognomonic of congenital toxoplasmosis:
   a. Cerebral calcification
   b. Hepatosplenomegaly
   c. A combination of rash and microcephaly
   d. Psychomotor retardation
   e. None of the above

GUIDELINES

1. SOGC Clinical Practice Guidelines (2013) Toxoplasmosis in pregnancy: prevention, screening, and treatment. *J Obstet Gynaecol Can* **35**(1 eSuppl A):S1–S7.

2. Paquet C & Yudin MH (2013) Infectious Disease Committee. Toxoplasmosis in pregnancy: prevention, screening, and treatment. *J Obstet Gynaecol Can* **35**(1 eSuppl A):S1-7.

3. PHE toxoplasmosis: diagnosis, epidemiology and prevention

   www.gov.uk/guidance/toxoplasmosis

REFERENCES

1. Robert-Gangneux F & Dardé ML (2012) Epidemiology of and diagnostic strategies for toxoplasmosis. *Clin Microbiol Rev* 2012; **25**(2):264–296.

2. Montoya JG & Remington JS (2008) Management of *Toxoplasma gondii* infection during pregnancy. *Clin Infect Dis* **47**(4):554–566.

3. Montoya JG & Liesenfeld O (2004) Toxoplasmosis. *Lancet* **363**:1965–1976.

4. Opsteegh M, Kortbeek TM, Havelaar AH & van der Giessen JW (2015) Intervention strategies to reduce human *Toxoplasma gondii* disease burden. *Clin Infect Dis* **60**(1):101–107.

5. Robert-Gangneux F & Dardé ML (2012) Epidemiology of and diagnostic strategies for toxoplasmosis. *Clin Microbiol Rev* **25**(2):264–296.

6. Torgerson PR & Mastroiacovo P (2013) *Bull World Health Organ* **91**:501–508 (doi:http://dx.doi.org/10.2471/BLT.12.111732).

7. Feldman DM et al. (2016) Toxoplasmosis, parvovirus and cytomegalovirus in pregnancy. *Clin Lab Med* **36**(2):407–419 (doi: 10.1016/j.cll.2016.01.011).

8. McGovern OL & Carruthers VB (2016) Toxoplasma retromer is here to stay. *Trends Parasitol* pii: S1471-4922(16)30053-8 (doi: 10.1016/j.pt.2016.05.007).

9. Ferig RM & Nishikawa Y (2016) Towards a preventive strategy for toxoplasmosis: Current trends, challenges and future perspectives for vaccine development. *Methods Mol Biol* **1404**:153–164.

10. de Oliverira Azevedo CT et al. (2016) Performance of PCR analysis of the amniotic fluid of pregnant women for diagnosis of congenital toxoplasmosis: a systematic review and meta analysis. PLoS One **11**(4):e0149938 (doi: 10.1371/journal.pone.014993).

ANSWERS

MCQ	Feedback
1. The following precautions should be taken by pregnant women to avoid infection with *T. gondii*, except: a. Avoiding contact with cats b. Handwashing after handling raw meat c. Wearing gloves when gardening d. Thorough washing of raw vegetables e. Cooking meat thoroughly	Oocysts which have been freshly passed in cat feces are not infectious until they have sporulated in the external environment. This process may take up to three weeks. Sporulated oocysts are resistant to adverse environmental conditions and may survive in the environment for months to years. The other precautions stated are advisable during pregnancy. Further reading Robert-Gangneux F & Dardé ML (2012) Epidemiology of and diagnostic strategies for toxoplasmosis. *Clin Microbiol Rev* 25(2):264–296.
2. The following are recognized routes of transmission of toxoplasmosis, except: a. Blood transfusion b. Ingestion of soil-contaminated vegetables c. Breastfeeding d. Ingestion of undercooked meat e. Ingestion of unpasteurized goats' milk	There have not been any cases of proven transmission from breastfeeding. Mother-to-child transmission occurs *in utero*. Further reading Robert-Gangneux F & Dardé ML (2012) Epidemiology of and diagnostic strategies for toxoplasmosis. *Clin Microbiol Rev* 25(2):264–296.
3. Which of the following statements is not correct? a. Spiramycin crosses the placenta effectively. b. Some strains may be more likely to cause severe disease. c. Bradyzoites are easily killed by pyrimethamine. d. Microwaving is a highly effective method for destroying tissue cysts. e. Screening for toxoplasmosis in pregnancy is universally practiced.	Combination treatment with pyrimethamine and sulfadiazine is used in proven fetal infection because spiramycin is not transferred across the placenta. Bradyzoites have a low metabolism and divide very slowly—pyrimethamine acts on the folate pathway, which is important in DNA synthesis, and hence is ineffective at killing bradyzoites. Microwaving may produce uneven heating, making it an unreliable way of killing viable organisms present in food. Antenatal screening for toxoplasmosis has only been adopted by some, not all, countries. It is not currently practiced in the UK because of the lack of cost-effectiveness of this intervention. Further reading Robert-Gangneux F & Dardé ML (2012) Epidemiology of and diagnostic strategies for toxoplasmosis. *Clin Microbiol Rev* 25(2):264–296. Montoya JG & Remington JS (2008) Management of *Toxoplasma gondii* infection during pregnancy. *Clin Infect Dis* 47(4):554–566.
4. The following are tests that can be used for diagnosis of toxoplasmosis in humans except: a. Serum IgA b. Stool microscopy for oocysts c. PCR of amniotic fluid d. Serum IgM e. Mouse inoculation	Only cats and other felid species are definitive hosts, that is, sexual reproduction of *T. gondii*, and hence production of oocysts only occurs in these animals. Humans are incidental hosts. The other methods stated are all utilized to diagnose human toxoplasmosis. Further reading Robert-Gangneux F & Dardé ML (2012) Epidemiology of and diagnostic strategies for toxoplasmosis. *Clin Microbiol Rev* 2012; 25(2):264–296. Montoya JG & Remington JS (2008) Management of *Toxoplasma gondii* infection during pregnancy. *Clin Infect Dis* 47(4):554–566.
5. The following feature is pathognomonic of congenital toxoplasmosis: a. Cerebral calcification b. Hepatosplenomegaly c. A combination of rash and microcephaly d. Psychomotor retardation e. None of the above	No signs or symptoms are pathognomonic of congenital toxoplasmosis. Abnormalities found before or after birth should prompt investigation for all possible differential diagnoses. Further reading Montoya JG & Remington JS (2008) Management of *Toxoplasma gondii* infection during pregnancy. *Clin Infect Dis* 47(4):554–566. Montoya JG & Liesenfeld O (2004) Toxoplasmosis. *Lancet* 363:1965–1976.

VIRAL HAEMORRHAGIC FEVER

Colin Brown[1,2,3] and Emma Aarons[4]

[1]Department of Infection, Royal Free London NHS Foundation Trust, UK.
[2]King's Sierra Leone Partnership, King's Centre for Global Health, King's Health Partners and King's College London, UK.
[3]Reference Microbiology Services, National Infection Service, Public Health England, UK.
[4]Rare and Imported Pathogens Laboratory, National Infection Service, Public Health England, UK.

A 45-year-old Turkish man, a resident in the UK for the past 10 years, presents to his local district hospital on the London outskirts with a history of several days of fever, headache, myalgia, and gastrointestinal disturbance. The clinical history taken by the medical team in the accident and emergency department identifies that he has returned two days earlier from Sivas in northeastern Turkey, where he had taken his wife and children to visit relatives for a family celebration. No one else was unwell on their trip, and they report they ate no undercooked food, had no water exposure or insect bites, and were unaware of anyone locally with similar symptoms, although his wife has heard that typhoid fever was reported locally.

His background medical history is unremarkable except for asthma. Initial observations reveal a fever of 38.5°C and mild tachycardia (115 beats per minute), but he has normal blood pressure, is fully alert and orientated, and has no focal abnormal examination findings. His blood profile reveals mildly elevated liver function tests: alanine transaminase (ALT) of 122 IU/L, aspartate transaminase (AST) of 140 IU/L (normal limit <50), thrombocytopaenia (platelets of 140×10^9/L, normal limit >150×10^9/L), elevated C-reactive protein of 20 mg/L (normal limit <5 mb/L), and normal white blood cell parameters.

INVESTIGATION OF THE CASE

The patient is admitted under the general medical team with a presumed infection, which is possibly influenza, typhoid, or gastroenteritis. Because of the differential diagnosis, he is managed in a side room with enteric precautions. A nasopharyngeal aspirate is sent to virology with a request for routine respiratory virus testing, along with serum for hepatitis serology. Blood cultures taken in the Accident and Emergency department are processed for bacterial pathogens in the microbiology department, and stool is processed for standard gastrointestinal pathogens including shigella and salmonella. Serial blood samples for haematological and biochemical analyses are processed locally on autoanalysers.

Despite broad-spectrum antibiotics, his condition deteriorates over the next 72 hours, with persistent fever, a large rise in transaminases (ALT of 506 IU/L, AST of 1032 IU/L), progressive thrombocytopenia (23 platelets $\times 10^9$/L), and deranged clotting (prothrombin time of 24 sec). Disseminated intravascular coagulopathy is suspected secondary to overwhelming sepsis, but increasing amounts of bloody stool and significant bruising at venepuncture sites draws concern from the local microbiologist. The patient's principal infective investigations return negative and this prompts a phone call to the national Imported Fever Service (IFS) for expert advice. The duty consultant considers that all symptoms could be explained by a viral haemorrhagic fever (VHF), specifically Crimean-Congo haemorrhagic fever (CCHF) in view of the patient's recent return from Turkey. Urgent blood samples are couriered to the Rare and Imported Pathogens Laboratory (RIPL), and within 6 hours there is polymerase chain reaction (PCR) confirmation of CCHF virus infection. CCHF virus is a Hazard Group 4 pathogen.

CLINICAL MANAGEMENT

After careful discussion and evaluation of the risks of transfer, the patient is taken to the High Level Isolation Unit (HLIU) in central London by the Hazardous Area Response Team ambulance transfer, primarily to allow for isolation in a Trexler isolator because of ongoing bleeding including haematemesis, melena, and haemoptysis. Despite maximal supportive therapy and administration of ribavirin, the patient dies 48 hours after arrival at HLIU.

PREVENTION OF FURTHER CASES

VHFs are of particular infection control importance: they can spread readily within a hospital setting, have a high case-fatality rate, are difficult to recognize and detect rapidly, and there is little or no effective treatment. The following steps are taken to manage the infection risk.

Confirm the diagnosis

To establish the diagnosis, clinical material is dispatched to the national diagnostic laboratory for PCR testing. The samples are transported as Category A infectious substances with precautions to prevent leaks, including special packaging (in compliance with United Nations Packing Instruction 620) and use of a courier trained in the transport of Category A infectious substance.

Convene an incident management group

Once the diagnosis is established, a multidisciplinary team is convened to coordinate all the necessary infection control activity. Team members include local consultants in virology and in infectious diseases, senior nursing staff, the local health protection team and Trust communications, with advisory support from national VHF experts.

Control risk from the patient

Prior to the diagnosis being made, the patient had been isolated in a side room with en-suite toilet facilities in the local hospital, and the staff used standard personal protective equipment (PPE), namely plastic aprons and gloves, together with fluid-repellent surgical mask and eye protection for all potential splash-generating procedures. Following the diagnosis, the patient is transferred into a negative-pressure room and full VHF PPE is worn, as advised in the Advisory Committee on Dangerous Pathogens (ACDP) guidance, "Management of Hazard Group 4 viral haemorrhagic fevers and similar human infectious diseases of high consequence." Locally, this consists of double gloves, a fluid-repellent disposable gown, a full-length plastic apron, a surgical cap, an FFP3 respirator, a full face shield, and fluid-repellent boot covers.

All clinical waste is double bagged and incinerated. Small spills of blood and other body fluids are quickly addressed with freshly prepared sodium hypochlorite containing 5000 ppm available chlorine. The initial side room and subsequent negative-pressure room are first cleaned with detergent and then fumigated with hydrogen peroxide vapor. The patient's mattress, which is thought to have been contaminated with body fluids, is cut up and placed in hard plastic bins for disposal by an approved external company.

Control risk from clinical samples from the patient

Prior to CCHF being diagnosed, samples of blood and other body fluids have been processed at the referring hospital according to local laboratory health and safety procedures, with no reported exposure incidents. Over 50 samples are identified across five laboratories, and these are quarantined in Containment Level 3 facilities and then autoclaved. The autoanalysers used for processing of the patient's samples undergo their standard maintenance and decontamination procedures. Testing of all further samples is carried out with suitable containment. Tests needed promptly for clinical management, such as biochemical tests, are processed on-site in the HLIU laboratory. This is a negative-pressure, highly efficient particulate air (HEPA)-filtered room within the HLIU suite containing a flexible film isolator unit (comparable to a Class III Microbiological Safety Cabinet), the exhaust air from which is also HEPA-filtered. All of the analysers required to process Hazard Group 4 specimens are contained within the flexible film isolator. Specimens sent to the national laboratory for monitoring of viral load are managed in Containment Level 4.

Manage immediate contacts who may have been exposed

A list of family contacts and health care workers who have had close contact with the patient without appropriate PPE (two individuals who had examined the patient in Accident and Emergency department) is collated by the local health protection team, that contacts each individual and advises daily self-monitoring of temperature and symptoms compatible with CCHF. Laboratory staff that may have been exposed to infectious samples are also assessed for exposure risk and PPE use, and categorized accordingly. As the incubation period is up to 14 days, contacts are required to report their temperature to local health protection nurses for two weeks following the last unprotected contact with the patient or his body fluids. Any person who had direct skin or mucous membrane exposure to blood or body fluids when the patient was thought to be viraemic would be considered for ribavirin post-exposure prophylaxis (PEP).

Trace other contacts

National public health epidemiologists contact passengers who had been seated in close proximity to the patient on his return flight, but none of the contacts are deemed to have had high-risk exposure to the patient's body fluids. A National Health Service (NHS) 24-hour helpline is set up.

The main issues that were highlighted as a result of this incident were:

- Travel history is not included in pathology request
  This is a missed opportunity for the laboratory to identify possible VHF cases and may lead to laboratory staff unwittingly putting themselves at risk because some routine procedures are not deemed safe for use on samples that may contain VHF.

- Misdiagnosis of VHF is common and unrecognized infection is a significant risk

  It is important to seek early expert advice. Co-infection with malaria could be possible in travellers from VHF endemic areas, and therefore, the possibility of VHF should be reconsidered in patients failing to settle despite appropriate antimalarial treatment. There are considerable difficulties in balancing the treatment needs of the patient with the potential dangers and inherent risks in transporting to a specialist centre where both experience and viral containment are possible. Managing a patient with potential VHF in an open ward carries significant risk to other patients and staff.

- UK guidance indicates that routine samples can be processed locally

  Early in the care of a potential VHF case, hospitals should have protocols regarding which automated platforms will be used, based on expert guidance. They also must have the capacity to transport high-risk specimens safely to specialist laboratories.

- Adequate environmental decontamination is of particular concern

  All contaminated materials including infected bedclothes, swabs, and linens should be incinerated and affected rooms fumigated. If VHF diagnosis has been delayed, then the decontamination and disposal procedures used earlier may have been inadequate; a retrospective review may identify additional exposure to body fluids or devices that need full decontamination.

EPIDEMIOLOGY

Crimean haemorrhagic fever was first recognized in 1944 in the Crimean peninsula, with viral isolation in 1967; in 1969 it was found to be identical to Congo virus first seen in 1956, and thereafter the names have been combined. The early recognition that it encompassed two geographically disparate regions exemplifies the wide geographic range of the disease, and CCHF has among the widest reservoir of all known VHFs, being endemic in Africa, Eastern Europe including the Balkans, the Middle East, and Asian countries to the western edge of China—the geographical limit of the principal tick vector (Figure 23.1). The natural hosts of CCHF vary by location and include a wide range of wild and domestic animals, including ruminants (sheep, goats, cattle), cats, dogs, and even ostriches, although many birds are naturally resistant to infection. Ostriches, as well as other domestic ruminants, have been linked to outbreaks in several countries, both in abattoir workers and animal handlers. There is no evidence that any animal is clinically affected other than humans and baby mice.

Within Europe, CCHF has been endemic in Bulgaria since the 1950s, Greece had a 1% seroprevalence in the 1980s, cases have recently re-emerged within the southwest Russian Federation, and outbreaks in Kosovo and Albania have been recorded since 2000. The first CCHF cases were observed in Turkey in 2002 and by 2010 over 4400 recorded laboratory confirmed CCHF cases have been confirmed. Within CCHF endemic areas nearly one in five adults has serological evidence of exposure. By 2013, approximately 2000 further cases had been reported.

Nonendemic countries such as the UK may detect imported cases. CCHF is present in a number of countries from which international travellers to the UK may arrive. Prior to the West African Ebola outbreak in 2014, the last two VHFs diagnosed in the UK were CCHF, one imported from Afghanistan and one from Bulgaria. The severity of the symptoms at presentation of the first imported case resulted in quick diagnosis, but highlighted that, as with the scenario described here, patients may have time to return to the UK before becoming symptomatic and the increase in international travel has heightened the risk. Table 23.1 lists all diagnosed VHFs in the UK since 1970.

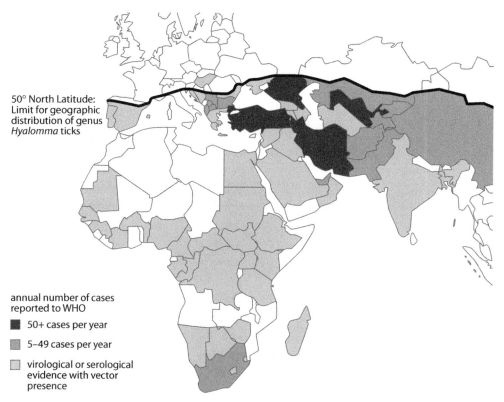

50° North Latitude:
Limit for geographic
distribution of genus
Hyalomma ticks

annual number of cases
reported to WHO

■ 50+ cases per year

▨ 5–49 cases per year

☐ virological or serological
evidence with vector
presence

Figure 23.1. **Geographic region where Crimea-Congo haemorrhagic fever is endemic, with density of notifications to the World Health Organization.** (Bente DA, Forrester N, Watts D et al. [2013] *Antiviral Res* **100**:159189. With permission from Elsevier.)

Infection can be transmitted by several routes:

- Bites from a tick that has recently feasted on a viraemic animal. In keeping with good clinical practice, patients should be examined from head to toe if returning unwell from an endemic country to look for bites or eschars.
- Direct contact with the fresh blood of animals. Animal slaughters and abattoir workers are particularly at risk. However, individuals who buy freshly butchered meat are at considerably lower risk of infection because of the rapid acidification of muscle tissue after death.
- Human to human contact via infected bodily fluids. Nosocomial transmission or between family members who care for ill relatives is common.

The incubation period depends on the route of transfer—for tick bites, 1–3 days is normal, with a maximum of 9 days; for bodily fluid contact, there is a prolonged incubation of 5–6 days, with a maximum of 13 days.

There is a significant risk of transmission to healthcare workers, both before the possibility of VHF is recognized and once diagnosis has been made because of a lack of available PPE. There are multiple examples of transmission to nurses, physicians, and other patients in countries including Sudan, Iran, India, and Pakistan. Early diagnosis is key to enable rapid appropriate isolation measures.

Table 23.1. **Viral haemorrhagic fevers that have been diagnosed in or repatriated to the UK since 1970**

Year of importation	Country	Case occupation	Disease
1971	Sierra Leone	Nurse	Lassa fever
1971	Sierra Leone	Physician	Lassa fever
1975	Nigeria	Physician	Lassa fever
1976	UK	Laboratory scientist	Ebola virus disease
1976	Nigeria	Engineer	Lassa fever
1981	Nigeria	Teacher	Lassa fever
1982	Nigeria	Diplomat	Lassa fever
1984	Sierra Leone	Geologist	Lassa fever
1985	Sierra Leone	Nurse	Lassa fever
2000	Sierra Leone	Peacekeeper	Lassa fever
2003	Sierra Leone	Peacekeeper	Lassa fever
2009	Nigeria	Visiting family	Lassa fever
2009	Mali	Rural worker	Lassa fever
2012	Afghanistan	Visiting family	CCHF
2014	Bulgaria	Tourist	CCHF
2014	Sierra Leone	Nurse	Ebola virus disease
2014	Sierra Leone	Nurse	Ebola virus disease
2015	Sierra Leone	Nurse	Ebola virus disease

A recent seroprevalence study in a university hospital in Turkey that has diagnosed nearly 1300 CCHF cases over the last decade demonstrated that the healthcare worker seropositivity was only 0.53%. This finding was attributed to educational campaigns and high staff usage of PPE. Available evidence indicates that the risk of CCHF transmission in laboratories routinely processing haematology and biochemistry samples is negligible.

BIOLOGY

CCHF is an arbovirus member of the *Nairovirus* genus (family Bunyaviridae), and is a single-stranded, enveloped, negative-sense RNA virus. There are over 300 species of Bunyaviridae within five genera: *Orthobunyavirus*, *Hantavirus*, *Phlebovirus*, *Nairovirus*, and *Tospovirus*. All Nairoviruses are thought to be transmitted by either the hard ixodid or soft argasid ticks, with *Hyalomma* ticks most responsible for CCHF transmission. Three species cause human illness, with CCHF by far the most important. Dugbe virus causes a mild febrile illness with thrombocytopenia, and Nairobi sheep disease virus causes fever, joint aches, and general malaise. Virions are 100 nm, spherical, with a lipid bilayer envelope derived from host cells through which virus-encoded glycoprotein spikes protrude. These spikes are responsible for virion binding to cellular receptors. The virus replicates in the host cell cytoplasm and virus particles are released from the infected cells by exocytosis.

DISEASE

CCHF shares many clinical similarities with other viral haemorrhagic fevers. After incubation, each has three phases: prehaemorrhagic, haemorrhagic, and convalescent. In the prehaemorrhagic phase the patient becomes suddenly unwell with predominant features of fever, severe headache, myalgia, and gastrointestinal symptoms, including nausea and diarrhoea. There is a wide range in severity of presentation. In severe cases there is rapid progression to the haemorrhagic phase with evidence of disseminated intravascular coagulation, significant bleeding and bruising, and haemodynamic shock. If the patient survives, convalescence is accompanied by a reduction in transaminases and leukocyte recovery. Duration and symptoms in these different phases vary significantly between individuals. High ALT and AST, as seen in this case, are commonly very elevated in fatal cases, along with high viral load on presentation. Overall the case fatality rate has been estimated to lie between 5 and 40%. Clearly there is a wide clinical spectrum of disease presentation, with one early model suggesting that only one-fifth of those infected have a clinical syndrome that would warrant seeking medical attention. Of the two cases imported to the UK, one was severely unwell and died, while the other was hospitalized and recovered with minimal supportive care.

Important differentials are malaria, rickettsial disease, leptospirosis, borreliosis, other geographically-overlapping haemorrhagic fevers such as Omsk, and other haemorrhagic disease manifestations such as meningococcal septicaemia and dengue haemorrhagic fever. Diagnosis can be achieved by detection of viral RNA (by real-time reverse transcriptase PCR), allowing for rapid detection or appearance of virus-specific IgM. In those who recover, detectable IgM is present within 7 days of infection and lasts for up to 4 months, whereas IgG remains detectable for at least 5 years. The key to timely diagnosis depends predominately on VHF being considered within the differential diagnosis and appropriate samples taken early and transported to the relevant laboratory.

Ribavirin is a guanosine analogue that acts as an inhibitor of the viral RNA, dependent RNA polymerase, and there is good evidence that ribavirin is effective *in vitro* against CCHF. However, evidence for clinical efficacy is lacking, although for some other viruses including VHFs, most notably Lassa fever and Old World hantavirus infection, there is good observational and randomized evidence of a beneficial effect. A retrospective case-control study assessing efficacy of oral ribavirin administration for CCHF demonstrated no treatment benefit. More recently randomized trial data have shown no clear benefit of administration, with potential delayed white cell recovery. The efficacy of convalescent plasma has not yet been proven, although it is theoretically plausible in keeping with other novel infections such as severe acute respiratory syndrome (SARS), and nonrandomized observation data have suggested improvement in clinical response. Vaccines have been developed for immunizing at-risk individuals, most notably in Bulgaria, using an inactivated vaccine that is "probably somewhat efficacious," although conclusive evidence of benefit is lacking.

Though there is conflicting evidence of the benefit of PEP, ribavirin administration has been recommended in some circumstances.

PATHOLOGY

As with many viral haemorrhagic fevers, one of the main pathologic effects of CCHF is vascular dysfunction, however whether this is a direct effect of the virus on the vasculature or the consequence of the resulting overwhelming inflammatory response is unclear, despite virus having been identified in vascular endothelial tissue on autopsy. Many believe the endothelial damage results from multiple host-induced mechanisms, including cytokines. Markers of endovascular activation and proinflammatory cytokines are increased in fatal cases, along with cytotoxic T cells. Prominent lymphoid depletion and apoptosis is seen in the few autopsy series reported, which may explain the lack of immunoglobulin response seen in fatal cases.

QUESTIONS

1. A febrile patient returns from Sudan with nonspecific symptoms including malaise, fever >38°C, muscle aches, and headache. You have been called at 2 am on a Saturday morning for diagnostic and treatment advice. The admitting junior doctor thinks the most likely diagnosis is either influenza or malaria. The patient has not had contact with any ill animals or individuals. He has been to Khartoum on a business trip. Would you be able to:
   a. Urgently process his samples as normal, including all routine pathology, malaria film, and blood cultures.
   b. Process his samples with great caution using the highest containment facilities available in your local laboratory (Containment Level 3), only if staff consent to proceed based on your risk assessment.
   c. Perform a malaria film, but if that is negative no other tests can be conducted until an urgent VHF test has been performed at the Defence Science and Technology Laboratory (Porton Down).
   d. Do nothing until an urgent VHF test has been ruled out and manage the patient symptomatically with appropriate fluid support and analgesia.

2. The patient in Question 1 has also visited rural areas and remembers being bitten by ticks. A malaria film has been processed and is negative, and bloods reveal a mild transaminitis, leucopenia, and thrombocytopenia. He has been admitted to a general bay on an acute medical ward and is still spiking a high fever. You are called for further advice and recommend the following:
   a. Repeat the malaria film—three negative films are required to exclude falciparum malaria, and that is the most likely diagnosis given his symptoms and travel.
   b. Isolate him in a side room. Process his samples as normal in your local laboratory on automated analysers but be aware of the risk of VHF. If the initial malaria test is negative, discuss with the imported fever service about the possibility of urgent VHF testing and notify the local health protection team.
   c. Place him in a negative-pressure isolation room, process further samples with caution according to local protocols, contact your local health protection team, and organize urgent VHF testing.
   d. Place him in a negative-pressure isolation room, identify and

discard all previous samples, contact your local health protection team, and do nothing further until you have the results of urgent VHF testing.

3. You receive a call from the duty consultant at the national specialist laboratory to say that the samples received from your patent have tested positive for CCHF by PCR. Though the patient is now isolated, you are worried about the pathology specimens that have already been processed. You determine that samples had been labelled as having a high possibility of VHF only after the history of tick bite was obtained. Do you now:

   a. Identify all stored samples, including any submitted before the patient was identified as high possibility of VHF and all disposable materials that may have been used in sample processing; double bag them and package them in secure containers; and then incinerate or autoclave them.

   b. Assume that standard procedures for safe disposal of routine pathology specimens are adequate, as any attempt to retrieve existing samples may be more likely to result in laboratory staff exposure, and discontinue any further sample processing from your patient.

   c. Halt all further pathology services at your hospital and implement contingency arrangements that allow for transfer of all specimens to a different site until all machines have been appropriately decontaminated.

   d. Discuss with your local BSL4 laboratory and HLIU about safe transfer of machines and stored samples to appropriately

resourced external facilities for decontamination and disposal.

4. A nurse who was caring for your patient before definitive diagnosis has reported that while cannulating the patient she received a superficial blood exposure. While removing the inner needle from the intravenous catheter, a drop of blood splashed on her arm despite wearing a plastic apron and gloves. She is very anxious and wants to know what to do. You advise, in addition to daily temperature monitoring:

   a. Remaining off work for the next 21 days

   b. Prophylactic ribavirin, and remaining off work for 14 days

   c. Prophylactic ribavirin, but can remain at work

   d. No special considerations at work, but active monitoring, including reporting daily temperatures to a health protection team

5. There has been a significant deterioration in your patient's condition. He has begun to haemorrhage actively, with melena and nose bleeds and extensive bruising at venepuncture sites. Do you:

   a. Decide that he is too unwell to transfer and ask for disease experts to come to your hospital to consult locally regarding his management.

   b. Decide that infection control concerns are now paramount and urgent transfer to HLIU is mandatory as the risk to your staff is too great.

   c. Discuss with the health and safety executive regarding what he or she wishes to happen.

   d. Transfer him to your intensive care unit for optimal monitoring and management.

GUIDELINES

The major UK guidelines are the ACDP guidance, "Management of Hazard Group 4 viral haemorrhagic fevers and similar human infectious diseases of high consequence." Jointly produced by the Health and Safety Executive and the Department of Health, these guidelines were updated in May 2012, November 2014, and again in November 2015 to allow for more testing to be performed locally, based on the risk assessment of the index case. The key initial decision is the patient risk assessment, which has several major components:

1. Has the patient had a fever (>37.5°C) in the past 24 hours and returned from a VHF-endemic area within the preceding 21 days.

2. In addition, has the patient had a significant epidemiological exposure. That is, has the patient:

   a. travelled to an area with a current VHF outbreak.

   b. lived or worked in a basic rural area where Lassa fever is endemic.

   c. visited caves or mines, or had contact with or eaten primates, antelopes, or bats in a Marburg or Ebola endemic region.

   d. travelled to a CCHF-endemic region and had a tick bite or crushed a tick with bare hands, or had close involvement with animal slaughter.

An algorithm dictating baseline investigations and management, specimen handling, infection control measures, including levels of personal protective equipment to be worn and public health actions, details the next steps based on the answers to these questions. This risk assessment is subject to regular review and the latest guidelines should be consulted for any possible case.

Other guidelines include the WHO and CDC management of viral haemorrhagic fevers in African settings, which detail how barrier precautions and disposal of waste can be conducted safely with limited resources.

UK Imported Fever Service: www.gov.uk/imported-fever-service-ifs

REFERENCES

1. UK Government. Imported Fever Service [cited 2016 June 5 2016]. (www.gov.uk/guidance/imported-fever-service-ifs).

2. Advisory Committee on Dangerous Pathogens (2015) Management of Hazard Group 4 viral haemorrhagic fevers and similar human infectious diseases of high consequence, London.

3. World Health Organization (2014) Guidance on regulations for the transport of infectious substances, 2013–2014.

4. Bell DJ. (2013) Experience of imported case of CCHF in Glasgow. In Viral Haemorrhagic Fevers (VHF) and Other Highly Infectious Diseases: A Practical Guide to the Management of an Imported Case. Highly Infectious Diseases Meeting, Dublin.

5. Swanepoel R & Burt F (2009) Bunyaviridae. In Principles and Practice of Clinical Virology 6e (Zuckerman AJ, Banatvala JE, Schoub BD, Griffiths PD & Mortimer P, eds), pp. 699–732. Wiley.

6. World Health Organization (2013) Crimean-Congo haemorrhagic fever. Factsheet No 208.

7. Mostafavi E, Chinikar S, Moradi M et al. (2013) A case report of Crimean Congo hemorrhagic fever in ostriches in Iran. *Open Virol J* 7:81–83.

8. Ergonul O & Whitehouse CA (2010) Introduction and historical perspective. In Crimean-Congo Hemorrhagic Fever: A Global Perspective (Ergonul A & Whitehouse C, eds), Springer.

9. Maltezou HC, Andonova L, Andraghetti R et al. (2010) Crimean-Congo hemorrhagic fever in Europe: current situation calls for preparedness. *Euro Surveill* 15(10):19504.

10. Bente DA, Forrester NL, Watts DM et al. (2013) Crimean-Congo hemorrhagic fever: history, epidemiology, pathogenesis, clinical syndrome and genetic diversity. *Antiviral Res* **100**(1):159–189.

11. Barr DA, Aitken C, Bell DJ et al. (2013) First confirmed case of Crimean-Congo haemorrhagic fever in the UK. *Lancet* **382**(9902):1458.

12. Lumley S, Atkinson B, Dowall S et al. (2014) Non-fatal case of Crimean-Congo haemorrhagic fever imported into the United Kingdom (ex Bulgaria). *Euro Surveill* **19**(30).

13. Johnston V, Stockley JM, Dockrell D et al. (2009) Fever in returned travellers presenting in the United Kingdom: recommendations for investigation and initial management. *J Infect* **59**(1):1–18.

14. Aradaib IE, Erickson BR, Mustafa ME et al. (2010) Nosocomial outbreak of Crimean-Congo hemorrhagic fever, Sudan. *Emerg Infect Dis* **16**(5):837–839.

15. Elata AT, Karsany MS, Elageb RM et al. (2011) A nosocomial transmission of Crimean-Congo hemorrhagic fever to an attending physician in North Kordufan, Sudan. *Virol J* **8**:303.

16. Gozel MG, Dokmetas I, Oztop AY et al. (2013) Recommended precaution procedures protect healthcare workers from Crimean-Congo hemorrhagic fever virus. *Int J Infect Dis* **17**(11):e1046–50.

17. Leblebicioglu H, Sunbul M, Guner R et al. (2014) Healthcare Acquired Crimean Congo Hemorrhagic Fever in Turkey 2002-2012—Low Risk of Transmission in Routine Diagnostic Laboratory Practice. In ID Week 2014, IDSA (idsa.confex.com/idsa/2014/webprogram/Paper47344.html).

18. Whitehouse CA (2004) Crimean-Congo hemorrhagic fever. *Antivir Res* 145–160.

19. Kraus AA & Mirazimi A (2010) Molecular biology and pathogenesis of Crimean–Congo hemorrhagic fever virus. *Future Virol* **5**(4):469–479.

20. Flick R (2007). Molecular Epidemiology, Genomics, and Phylogeny of Crimean-Congo Hemorrhagic Fever Virus. In (Whitehouse C, Ergonal O, eds) Crimean-Congo Hemorrhagic Fever: A Global Perspective, Springer.

21. Elevli M, Ozkul AA, Civilibal M et al. (2010) A newly identified Crimean-Congo hemorrhagic fever virus strain in Turkey. *Int J Infect Dis* **14**(Suppl 3):e213–216.

22. Ergonul O (2008) Treatment of Crimean-Congo hemorrhagic fever. *Antiviral Res* **78**:125–131.

23. Ergönül O (2006) Crimean-Congo haemorrhagic fever. *Lancet Infect Dis* **6**(4):203–214.

24. Cevik MA, Erbay A, Bodur H et al. (2008) Clinical and laboratory features of Crimean-Congo hemorrhagic fever: predictors of fatality. *Int J Infect Dis* **12**(4):374–379.

25. Saksida A, Duh D, Wraber B et al. (2010) Interacting roles of immune mechanisms and viral load in the pathogenesis of Crimean-Congo hemorrhagic fever. *Clin Vaccine Immunol* **17**(7):1086–1093.

26. Goldfarb LG, Chumakov MP, Myskin AA, Kondratenko VF & Reznikova OY (1980) An epidemiological model of Crimean hemorrhagic fever. *Am J Trop Med Hyg* **29**(2):260–264.

27. Atkinson B, Chamberlain J, Logue CH et al. (2012) Development of a real-time RT-PCR assay for the detection of Crimean-Congo hemorrhagic fever virus. *Vector Borne Zoonotic Dis* **12**(9):786–793.

28. McCormick JB, King IJ, Webb PA et al. (1986) Lassa fever. Effective therapy with ribavirin. *N Engl J Med* **314**:20–26.

29. Huggins JW, Hsiang CM, Cosgriff TM et al. (1991) Prospective, double-blind, concurrent, placebo-controlled clinical trial of intravenous ribavirin therapy of hemorrhagic fever with renal syndrome. *J Infect Dis* **164**(6):1119–1127.

30. Bodur H, Erbay A, Akıncı E et al. (2011) Effect of oral ribavirin treatment on the viral load and disease progression in Crimean-Congo hemorrhagic fever. *Int J Infect Dis* **15**(1):e44–47.

31. Keshtkar-Jahromi M, Kuhn JH, Christova I et al. (2011) Crimean-Congo hemorrhagic fever: current and future prospects of vaccines and therapies. *Antiviral Res* **90**(2):85–92.

32. Stockman LJ, Bellamy R & Garner P (2006) SARS: systematic review of treatment effects. *PLoS Med* **3**(9):e343.

33. Mardani M & Keshtkar-Jahromi M (2007) Crimean-Congo hemorrhagic fever. *Arch Iran Med* **10**(2):204–214.

34. World Health Organization (2014) Infection prevention and control guidance for care of patients in health-care settings, with focus on Ebola.

35. Leblebicioglu H, Sunbul M, Guner R et al. (2016) Healthcare-associated Crimean-Congo haemorrhagic fever in Turkey, 2002–2014: a multicentre retrospective cross-sectional study. *Clin Microbiol Infect* **22**(4):387.e1-4.

36. Leblebicioglu H, Ozaras R, Irmak H & Sencan I (2016). Crimean-Congo hemorrhagic fever in Turkey: Current status and future challenges. *Antiviral Res* **126**:21–34.

ANSWERS

MCQ	Feedback
1. A febrile patient returns from Sudan with nonspecific symptoms including malaise, fever > 38 °C, muscle aches, and headache. You have been called at 2 am on a Saturday morning for diagnostic and treatment advice. The admitting junior doctor thinks the most likely diagnosis is either influenza or malaria. The patient has not had contact with any ill animals or individuals. He has been to Khartoum on a business trip. Would you be able to:	In the UK, according to the latest ACDP VHF risk assessment algorithm, (a) is the correct answer based on his initial risk assessment. Using the available information (given there is no history of tick bite exposure), he would be classified as a low-possibility case of VHF and all routine pathology can be conducted locally in your hospital. In the absence of bleeding or significant bruising, there is no need to instruct laboratory staff to use special precautions in processing his samples.

a.   Urgently process his samples as normal, including all routine pathology, malaria film, and blood cultures.

b.   Process his samples with great caution using the highest containment facilities available in your local laboratory (Containment Level 3), only if staff consent to proceed based on your risk assessment.

c.   Perform a malaria film, but if that is negative no other tests can be conducted until an urgent VHF test has been performed at the Defence Science and Technology Laboratory (Porton Down).

d.   Do nothing until an urgent VHF test has been ruled out and manage the patient symptomatically with appropriate fluid support and analgesia.

| 2. The patient in Question 1 has also visited rural areas and remembers being bitten by ticks. A malaria film has been processed and is negative, and bloods reveal a mild transaminitis, leucopenia, and thrombocytopenia. He has been admitted to a general bay on an acute medical ward and is still spiking a high fever. You are called for further advice and recommend the following: | His risk assessment has subsequently changed with the added information that he has received a tick bite. His blood picture increasingly points toward the possibility of VHF (particularly CCHF). ACDP guidance therefore suggests that he be isolated in a side room while continuing with urgent local investigations. As he does not have bruising or bleeding at this stage, nor uncontrolled diarrhoea or vomiting, urgent discussion with HLIU is not yet warranted, though laboratory staff should be alerted to the high possibility of VHF for storage of samples and appropriate discard if a subsequent test is positive. Once a negative malaria test returns, the Imported fever service should be called and the local health protection team notified. Thus the appropriate answer is (b). |

a.   Repeat the malaria film—three negative films are required to exclude falciparum malaria, and that is the most likely diagnosis given his symptoms and travel.

b.   Isolate him in a side room. Process his samples as normal in your local laboratory on automated analysers but be aware of the risk of VHF. If the initial malaria test is negative, discuss with the imported fever service about the possibility of urgent VHF testing and notify the local health protection team.

c.   Place him in a negative-pressure isolation room, process further samples with caution according to local protocols, contact your local health protection team, and organize urgent VHF testing.

d.   Place him in a negative-pressure isolation room, identify and discard all previous samples, contact your local health protection team, and do nothing further until you have the results of urgent VHF testing.

MCQ	Feedback
3. You receive a call from the duty consultant at the national specialist laboratory to say that the samples received from your patent have tested positive for CCHF by PCR. Though the patient is now isolated, you are worried about the pathology specimens that have already been processed. You determine that samples had been labelled as having a high possibility of VHF only after the history of tick bite was obtained. Do you now:	In the UK, there is now clear guidance for disposal of existing samples following revision of the ACDP guidelines. Samples appropriately identified as having a risk for VHF will have been stored for safe disposal if a test returns positive. Answer (a) is correct. Note that where autoanalysers have been used for blood testing prior to diagnosis, there need be no particular concern about machine waste disposal or decontamination. Waste from these machines is not considered to pose a significant risk because of the small sample size and dilution step and requires no special waste disposal precautions.
a. Identify all stored samples, including any submitted before the patient was identified as high possibility of VHF and all disposable materials that may have been used in sample processing; double bag them and package them in secure containers; and then incinerate or autoclave them.	
b. Assume that standard procedures for safe disposal of routine pathology specimens are adequate, as any attempt to retrieve existing samples may be more likely to result in laboratory staff exposure, and discontinue any further sample processing from your patient.	
c. Halt all further pathology services at your hospital and implement contingency arrangements that allow for transfer of all specimens to a different site until all machines have been appropriately decontaminated.	
d. Discuss with your local BSL4 laboratory and HLIU about safe transfer of machines and stored samples to appropriately resourced external facilities for decontamination and disposal.	
4. A nurse who was caring for your patient before definitive diagnosis has reported that while cannulating the patient she received a superficial blood exposure. While removing the inner needle from the intravenous catheter, a drop of blood splashed on her arm despite wearing a plastic apron and gloves. She is very anxious and wants to know what to do. You advise, in addition to daily temperature monitoring:	Monitoring should be conducted in coordination with the local health protection team, local clinical virologist, clinical microbiologist or infectious disease physician, and occupational health provider. They would be classified as Category 3 contacts, but exposed individuals need not do anything further than temperature and symptom monitoring and reporting, particularly given that the exposure in this scenario carries minimal risk. Ribavirin administration has been suggested in febrile contacts, though there is no evidence that this has any protective effect. Therefore answer (d) is correct. For the later part of the 2013–2016 epidemic virus disease (EVD) outbreak in West Africa, 21-day work restrictions were enacted for Category 3 exposures.
a. Remaining off work for the next 21 days	
b. Prophylactic ribavirin, and remaining off work for 14 days	
c. Prophylactic ribavirin, but can remain at work	
d. No special considerations at work, but active monitoring, including reporting daily temperatures to a health protection team	
5. There has been a significant deterioration in your patient's condition. He has begun to haemorrhage actively, with melena and nose bleeds and extensive bruising at venepuncture sites. Do you:	Here there is no clear-cut answer to his scenario, which depends on many factors including whether there is local expertise in dealing with VHFs, local isolation facilities, and the local provision of adequate waste disposal by incineration. Where many staff members are likely to have ongoing exposure to infectious bodily fluids, it is highly recommended that urgent discussions with HLIU are undertaken to arrange transfer, assuming the patient is stable enough for the journey to be undertaken safely. Where transfer is not possible, samples may be processed in Containment Level 2 laboratories using routine autoanalysers with several additional precautions. The routine provision of intensive care has been discouraged outside of an HLIU setting.
a. Decide that he is too unwell to transfer and ask for disease experts to come to your hospital to consult locally regarding his management.	
b. Decide that infection control concerns are now paramount and urgent transfer to HLIU is mandatory as the risk to your staff is too great.	
c. Discuss with the health and safety executive regarding what he or she wishes to happen.	
d. Transfer him to your intensive care unit for optimal monitoring and management.	

A CASE OF VERO CYTOTOXIN-PRODUCING *ESCHERICHIA COLI* (VTEC)

CASE 24

Claire Jenkins and Gauri Godbole

Bacteriology Reference Department, National Infection Service, Public Health England, London, UK.

A previously fit and well five-year-old boy (Case A) presented to the general practitioner (GP) with a four-day history of watery diarrhoea, associated with abdominal cramps. On clinical examination he was mildly dehydrated; the GP advised oral rehydration fluid therapy and a stool specimen was taken for microbiological examination. Over the next three days, although his diarrhoea reduced in frequency, the boy continued to get severe cramps and abdominal pain and started vomiting. The parents were concerned and took the boy to the local hospital. On clinical examination at the hospital, he looked pale, his temperature was 37.6°C and his respiratory rate was 28 per min. He had tachycardia and his blood pressure was normal. There was no rash, oedema, or meningism. His abdomen was flat and soft with generalized tenderness in the lower abdomen but no rectal prolapse. Other systems were unremarkable.

The laboratory results were as follows: haemoglobin 78 g/L, haematocrit 0.246 l/L, platelet count 32 × 10⁹/L, white blood cell count was 17 × 10⁹/L, with neutrophilia; peripheral smear showed schistocytes, helmet cells, and polychromasia. The C-reactive protein (CRP) was 30 mg/L, sodium 133 mmol/L, potassium 5.9 mmol/L, urea 30.7 mmol/L, creatinine 669 μmol/L, lactate dehydrogenase (LDH) 840 IU/L, serum albumin was 40 g/L, serum bilirubin was 17 μmol/L, aspartate transaminase (AST) was 130 IU/L, alanine transaminase (ALT) was 136 IU/L, alkaline phosphatase was 37 IU/L, and coagulation studies were normal. While he was in Accident and Emergency, he had a single generalized tonic-clonic seizure.

He was diagnosed as a case of infection-induced haemolytic uraemic syndrome (HUS) and started on renal replacement therapy. Meanwhile, the stool culture was reported as negative by the local laboratory for *Campylobacter*, *Salmonella*, *Shigella* spp., and *Escherichia coli* serogroup O157 at 48 hours.

The following day, a six-month-old boy (Case B) was admitted to the same hospital with a history of bloody diarrhoea and symptoms of HUS. Of note was the family history: the three-year-old sibling of Case B (Case C) had diarrhoea but was relatively well. Faecal specimens from both cases (B and C) grew nonsorbitol fermenting (NSF) colonies on selective CT-SMAC (cefixime tellurite sorbitol MacConkey) agar that were biochemically identified as *E. coli* and agglutinated with sera raised to the lipopolysaccharide (LPS) of *E. coli* O157.

INVESTIGATION OF THE CASE

A serum and faecal specimen from Case A were submitted to the gastrointestinal bacteria reference unit (GBRU) for further testing, as recommended by the national Public Health England guidelines. Polymerase chain reaction (PCR) tests at GBRU showed the faecal specimen was positive for the following genes:

*vtx*2: encoding the Vero cytotoxin (VT) 2

eae (*E. coli* attaching and effacing): encoding the intimin protein involved in the intimate attachment of the bacterium to the host gut mucosa

*rfb*E O157: encoding part of the O157 LPS genes

By using an immunomagnetic separation technique incorporating magnetic beads coated with antibodies to the LPS of *E. coli* O157, a strain of *E. coli* O157 was isolated and cultured from the faecal specimen. The serum sample tested positive for antibodies to the LPS of *E. coli* O157 (combined IgG and IgM assay).

The culture isolated at GBRU from Case A and the cultures from Cases B and C submitted to GBRU from the local hospital laboratory were confirmed as Vero cytotoxin-producing *E. coli* (VTEC) O157 (also known as Shiga toxin-producing *E. coli* or STEC) and typed using phage typing (PT) and multilocus variable number tandem repeat (VNTR) analysis (MLVA). All three strains of phage typed as PT21/28. The MLVA profiles of each strain are shown in Table 24.1.

The local health protection team (HPT) was informed about the three cases and the parents of each case were interviewed using an enhanced surveillance questionnaire. The questionnaire focuses on date of onset of the various symptoms, the detailed food history, travel history, environmental exposure, animal exposure, water exposure, and consumption in the week prior to illness.

Their food histories were unremarkable, but both families had visited a petting farm in the week preceding the onset of symptoms. Cases A and C had fed the goats and sheep on the farm. The local authorities launched a full investigation of the petting farm and inspected the premises to look at handwashing facilities, interviewed the staff, and arranged a veterinary inspection to take faecal specimens from the animals.

Table 24.1 **Multilocus variable number tandem repeat (VNTR) analysis (MLVA) results from the molecular typing of isolates of VTEC O157 from cases and animals linked to the petting farm**

Source	MLVA profile
Case A	8-7-13-6-9-2-8-9
Case B	8-7-13-6-9-2-8-9
Case C	8-7-13-6-9-2-8-9
Sheep	8-7-13-6-9-2-8-9
Lamb	8-7-13-6-9-2-8-9
Goat kid	8-7-13-6-9-2-8-9

The following day, veterinary investigation officers visited the petting farm to collect samples from animals and environmental health officers collected environmental samples from around the farm for testing for VTEC O157. VTEC O157 PT21/28 was isolated from the sheep and goats.

CLINICAL MANAGEMENT

Clinical management of gastroenteritis caused by VTEC consists of adequate hydration therapy, correction of electrolyte balance, and supportive care. Nonsteroidal anti-inflammatory agents can worsen the renal function and delay recovery by reduction of renal blood flow, and therefore they should be avoided; antimotility agents and opioid narcotics should also be avoided in cases of VTEC, as these have been associated with a risk of HUS, neurological complications, or both. Antibiotics should be avoided as they do not improve the outcome of the illness and may increase the risk of HUS. Haemorrhagic colitis is treated with fluid resuscitation, blood products in compromised cases, pain control, and surgical management of complications, such as perforation and peritonitis.

Cases of HUS associated with diarrhoea (D+ HUS) are managed as follows: fluid resuscitation and hydration, treatment of acute renal failure with renal replacement therapy, management of haematological complications (packed red cells for anaemia and platelet transfusion for active bleeding), prevention of hypertension, nutritional support, and pain management. In cases that present with HUS in the absence of diarrhoea (D− HUS), atypical HUS (aHUS) should be ruled out as soon as possible.

PREVENTION OF FURTHER CASES

The strategies for prevention depend on suspected source of infection and vehicle of transmission.

Livestock and farms

When farm livestock are the suspected source of infection, on either an open or a private farm, the health protection team should work with local authorities to investigate and put measures in place to prevent further cases. The following measures should be considered:

- The farm owner should be alerted and warned about the risks posed by VTEC; residents, staff, and visitors should be made aware that the premises is a suspected source of infection.
- There should be no direct contact with animals in an open farm setting (especially ruminants) early in any potential outbreak and contact with animal manure should be avoided.
- Infection control measures should be reviewed, especially those associated with handwashing after animal or environmental contact, eating or preparing food, removal of work clothes and footwear before entering the home or food-preparation area.
- Managers of animal amenity premises (open farms, petting zoos, animal sanctuaries) should be directed to the health and safety executive and given other detailed official guidance on avoiding infection.

The Animal Health Veterinary Laboratories Agency (AHVLA) should be alerted as soon as a potential outbreak linked to a farm is identified. Consider sampling animal faeces, animal environments, manure and water, particularly if there is a potential for livestock contamination of private water supplies. If contaminated dairy products are implicated, sample primary filters or washings, farm pasteurizers if in use, and raw milk from bulk milk tanks on dairy farms.

- There should be signage to indicate risk of infection and the need for parents and caregivers to supervise children at all times. Special care must be taken with very young children who may suck pacifiers or thumbs or put objects in their mouths.
- Eating facilities should be clearly defined and segregation of animal and human areas should be maintained.
- Farm staff must be appropriately trained in the management of the premises in accordance with Health and Safety Executive (HSE), UK requirements.

Food

Introduction of VTEC into food items occurs during production, handling, or preparation. Transmission can result from incomplete cooking or pasteurization of food with viable organisms present in sufficient quantities to cause infection. Large outbreaks are often food-borne and are followed by an epidemiological investigation into the source of contamination. The investigation consists of generating a hypothesis as to the vehicle, source, and cause of infection and testing the hypothesis epidemiologically, microbiologically, and environmentally. A risk assessment is performed and control measures may be put in place, for example, the withdrawal of the suspect food from sale.

Water

Water supplied by water mains is chlorinated and should not be a source of VTEC infection unless there is damage or treatment failure. However, contamination of taps and standpipes by animal faeces in rural settings has been linked to infections. Private water supply can cause small family or community outbreaks. Temporary interventions include advising users to boil water prior to consumption. Water testing is performed to detect contamination. Surface water such as streams and lakes may be a source of VTEC from pollution from grazing animals' faeces and from seepage or runoff of agricultural slurries and sewage. If surface water is suspected as a source of infection environmental sampling may be appropriate.

Childcare settings

If a probable or confirmed case under five years of age attends preschool, nursery school, or daycare in the 7 days prior to onset, or while symptomatic, it is important to exclude the child from school until he or she is asymptomatic and has two negative stool cultures taken 24 hours apart. The institution should be contacted to find out if any others attending the nursery or class have had diarrhoea. Appropriate advice should also be given to support hygiene measures in the school, for example, supervised handwashing of children is recommended. Secondary spread of VTEC is common and, in a nursery outbreak, screening of all children and staff with direct contact with children should be considered.

EPIDEMIOLOGY

Cattle are the main animal reservoir of VTEC strains implicated in human disease, and foods of bovine origin, especially undercooked beef and unpasteurized milk, are major sources of human infection. Salad and raw vegetables and fruit cross-contaminated during production or preparation have also been linked to sporadic infection and outbreaks of illness in humans.

The disease is often nicknamed the "burger bug," but studies suggest that environmental risk factors involving direct or indirect contact with ruminants may be a more common source of infection. Analysis of VTEC O157 outbreaks in Scotland from 1994 to 2003 associated with either meat or dairy foods, or with environmental transmission, show that approximately 40% of these outbreaks were food-borne, 54% were environmental, and 6% involved both transmission routes. However, the largest outbreaks are food-borne. VTEC O157 is common in cattle; prevalence estimates in cattle range from 10 to 40%, and most farms have animals shedding VTEC at some time. Visits to dairy farms, petting farms, and open zoos are a common source of infection in the UK, as well as attendance at agricultural fairs and recreational use of pastures, such as camping. Outbreaks have also been associated with consumption of water, most commonly in private water supplies contaminated with bovine faeces, or with recreational use.

The incubation period for diarrheal illness caused by infection with VTEC is usually 3 to 4 days, but has been occasionally recorded as long as 14 days. Occasional reports of incubation periods of longer than 8 days may reflect secondary transmission rather than a prolonged incubation period. Cases under five years of age can excrete VTEC up to a month after onset of illness and prolonged excretion has been described up to 60 days. Transmission from person-to-person is by the faecal–oral route. The infective dose is low (10–100 organisms).

VTEC can survive in the environment in faeces and soil for several months, posing a risk of acquisition by exposed animals. VTEC are unusual among $E.\ coli$ because they are relatively acid tolerant, able to survive acid conditions down to pH 3.6. VTEC are also resistant to desiccation and are able to survive many drying and fermentation processes.

VTEC O157 is the most common serogroup of VTEC causing infections in the UK and the most likely $E.coli$ serogroup to cause bloody diarrhoea and HUS. The incidence rate of VTEC O157 is 1.80 per 100,000 in England and Wales, and 4.4 cases per 100,000 in Scotland. However this varies by age and is highest in those aged 1–4 years (7.63 per 100,000 in England and Wales). Over 40% of reported VTEC cases in England are in children aged below 15 years compared with 7.0% for cases of campylobacteriosis and 17.8% for cases of salmonellosis. Two-thirds of STEC cases are reported in the summer months of May–September, with the highest frequency of cases in August of each year. In England, the incidence of STEC is over four times higher in people living in rural areas than those in urban areas.

Current standard protocols for local hospital microbiology laboratories in England focus on the detection of VTEC O157, so the true incidence of non-O157 VTEC in the UK is not known.

VTEC has a worldwide distribution, although the incidence of VTEC, particularly VTEC O157 is highest in North and South America and Western Europe.

BIOLOGY

VTEC produces two different Vero cytotoxins, VT1 and VT2, both encoded by bacterio-phage. VT1 is very similar to the Shiga toxin produced by *Shigella dysenteriae* serotype 1. VT2 is genetically and immunologically distinct, and the presence of VT2 is significantly associated with symptoms of bloody diarrhoea and HUS. There are at least seven different subtypes of VT2 and the VT2a subtype is most commonly associated with HUS.

In addition to the production of VT, strains causing severe disease are able to attach to the host gut mucosa via one of two known attachment mechanisms. The most common mechanism involves a system of type III secreted proteins, encoded by genes on a pathogenicity island called the locus of enterocyte effacement, that mediate intimate attachment to the gut enterocytes and destroy the microvilli surrounding the point of attachment. A second attachment mechanism has recently been described in strains of VTEC harbouring plasmid-encoded genes that facilitate the aggregative attachment of the bacteria to the gut, including the aggregative adherence regulator (*aggR*), aggregative adherence transporter (*aat*), and aggregative fimbriae.

Strains of VTEC O157 are, generally, resistant to cefixime and tellurite, are unable to ferment sorbitol, and are identified as colourless colonies on selective agar (CT-SMAC). Strains of sorbitol-fermenting VTEC O157 do occur but are rare in the UK. Most strains of VTEC other than serogroup O157 are sorbitol fermenting and may be sensitive to both tellurite and cefixime, and so they are not detected in England using the current diagnostic algorithm. A molecular diagnostic approach to detection of VTEC directly from faecal specimens using a PCR targeting the VT genes (*vtx1* and *vtx2*) has been implemented in a number of local hospital laboratories.

DISEASE

VTEC infection can present in a number of ways.

Gastroenteritis is the most common manifestation of VTEC, and symptoms include watery diarrhoea with or without vomiting, abdominal cramps, or fever. Symptoms may last a few days and then disappear within a week or so.

Haemorrhagic colitis presents with bloody diarrhoea and severe abdominal pain, but fever may be absent. Symptoms can be severe and may last up to two weeks.

Haemolytic uremic syndrome (HUS) is a disease characterized by the triad of acute renal failure, thrombocytopenia, and microangiopathic haemolytic anaemia (Coombs' test negative). Up to 15% of cases infected with VTEC develop HUS after an initial period of gastroenteritis or haemorrhagic colitis (prodrome). Raised white blood cell count (WBC), neutrophilia, thrombocytopenia, raised LDH, and presence of vomiting are early indicators of development of HUS. Significant risk factors for HUS associated with VTEC O157 include children under 15 years of age (more common are children under 5 years of age) or over 65 years of age, hypochlorhydria, and use of coincidental antibiotics. In most cases of HUS, renal function recovers although hypertension and renal insufficiency can develop. Mortality in children with HUS ranges from 1.8 to 5.6%, however mortality can be as high as 86% in the elderly.

PATHOLOGY

The toxins released by VTEC strains in haemorrhagic colitis and their ability to adhere to and damage the intestinal epithelium can cause patients to develop an inflamed colon that bleeds, resulting in very bloody diarrhoea. Mesenteric ischaemia from circulating toxin may also contribute to bloody diarrhoea. The receptors for Vero cytotoxins (that is, glycosphingolipid globotriaosylceramide) are found in high concentration in the kidneys (specifically associated with renal glomerular endothelial, tubular epithelium, and mesangial cells) and in the central nervous system. The toxins also damage endothelial cells generating thrombin and fibrin deposits in the microvasculature. This goes on to cause leakage and tissue oedema. Erythrocytes are damaged as they pass through small vessels partially occluded by thrombus and haemolysis subsequently occurs. Bleeding is associated with thrombocytopenia, circulating fibrinogen concentrations are normal or high, and the prothrombin time is only slightly prolonged, unlike classic disseminated intravascular coagulation.

QUESTIONS

1. With respect to the features of HUS in children, which one of the following statements is true?
   a. Most cases are associated with *Shigella* infectious diarrhoea.
   b. Children under 4 years of age presenting with bloody diarrhoea should be prescribed a beta-lactam antimicrobial agent.
   c. Stools are usually watery; bloody stools are reported in a minority of cases.
   d. Central nervous system irritability and seizures occur in about one-third of patients.

2. Regarding the clinical symptoms of HUS in children, which one of the following statements is true?
   a. HUS is associated with fever in most cases.
   b. Infection-induced HUS can occur in patients in absence of diarrhoea.
   c. HUS is an uncommon cause of ARF in children.
   d. Anaemia is typically mild and haemoglobin never drops <90 g/L.

3. Which of the following strains would be most likely to cause HUS?
   a. *E. coli* serogroup O157, *vtx1* gene negative, *vtx2* gene negative, and *eae* gene positive
   b. *E. coli* serogroup O157, *vtx1* gene positive, *vtx2c* gene positive, and *eae* gene positive
   c. *E. coli* serogroup O104, *vtx1* gene negative, *vtx2* gene negative, *aggR* gene positive
   d. *E. coli* O26, *vtx1* gene negative, *vtx2a* gene positive, and *eae* gene positive

4. Which of the following do not require exclusion from work or school?
   a. A case diagnosed with VTEC O157 who works on the delicatessen counter in a supermarket
   b. Symptomatic eight-year-old sibling of a known case of VTEC O157
   c. Asymptomatic mother of a four-year-old case of VTEC O157 working in a call centre
   d. Asymptomatic father of a two-year-old diagnosed with VTEC O157 working in a care home for the elderly

5. A local hospital laboratory report on a serum sample and faecal specimen from a two-year-old with bloody diarrhoea and HUS states that the serum is positive for antibodies to the LPS of *E. coli* O157, the faecal specimen is PCR positive for *vtx* genes, but VTEC was not isolated. What is the least likely explanation of the PCR positive/culture negative faecal specimen?

    a. VTEC is the cause of the symptoms, but the patient has been given antibiotics that have affected the ability of the bacterium to grow.

    b. PCR is more sensitive than culture and will detect the presence of VTEC in the stool at levels too low for culture.

    c. The case has atypical HUS.

    d. Local protocols focus on the detection of NSF VTEC O157 and the case has a sorbitol-fermenting VTEC O157.

GUIDELINES

1. UK Standards for Microbiology Investigations (SMI) B 30 (2014) Investigation of faecal specimens for enteric pathogens. Public Health England www.gov.uk/government/collections/standards-for-microbiology-investigations-smi

2. Vero cytotoxin-producing Escherichia coli (VTEC): guidance, data and analysis (2014) Public Health England www.gov.uk/government/publications/vero-cytotoxin-producing-escherichia-coli-advice-for-public-health-management-teams

3. Avoiding infection on farm visits: advice for the public (2014). Public Health England. www.gov.uk/government/uploads/system/uploads/attachment_data/file/322846/Farm_visits_avoiding_infection.pdf

4. The management of acute bloody diarrhoea potentially caused by vero cytotoxin-producing Escherichia coli in children. A guide for primary care, secondary care and public health practitioners (2011) Public Health England www.gov.uk/government/uploads/system/uploads/attachment_data/file/342344/management_of_acute_bloody_diarrhoea.pdf

5. Taylor CM, Machin S, Wigmore S & Goodship THJ (2009). Clinical Practice Guidelines for the Management of Atypical Haemolytic Uraemic Syndrome in the United Kingdom, pp. 1–22. The Renal Association www.renal.org/guidelines/joint-guidelines#sthash.29lIW8fO.dpbs

6. Avoiding infection on farm visits: advice for the public (2014) Public Health England p 12. www.gov.uk/government/publications/farm-visits-avoiding-infection

REFERENCES

1. Jenkins C, Lawson AJ, Cheasty T & Willshaw GA (2012) Assessment of a real-time PCR for the detection and characterization of verocytotoxigenic *Escherichia coli*. *J Med Microbiol* **61**(Pt 8):1082–5 (cited 2014 Jul 27; www.ncbi.nlm.nih.gov/pubmed/22516135).

2. Byrne L, Elson R, Dallman TJ et al. (2015) Evaluating the use of multilocus variable number tandem repeat analysis of Shiga toxin-producing *Escherichia coli* O157 as a routine public health tool in England. *PLoS One* **9**(1):e85901. (cited 2015 Jan 10; www.pubmedcentral.nih.gov/articlerender.fcgi?artid=3895024&tool=pmcentrez&rendertype=abstract).

3. Byrne L, Jenkins C, Elson R & Adak G (2015) The epidemiology, microbiology and clinical impact of Shiga toxin-producing *Escherichia coli* in England, 2009-2012. *Epidemiol Infect* 2015 Dec;**143**(16):3475–87.

4. Wong CS, Jelacic S, Habeeb RL & Watkins SL (2000) The risk of the hemolytic-uremic syndrome after antibiotic treatment of *Escherichia coli* O157:H7 infections. *N Engl J Med* **342**(26):1930–1936.

5. Michael M et al. Interventions for haemolytic uraemic syndrome and thrombotic thrombocytopenic purpura (Review). Cochrane Database Syst Rev. 2009;(1).

6. Byrne L, Vanstone GL, Perry NT et al. (2014) The epidemiology and microbiology of Shiga-toxin producing *Escherichia coli* other than serogroup O157 in England 2009-2013. *J Med Microbiol* (cited 2014 Jul 27; www.ncbi.nlm.nih.gov/pubmed/24928216).

7. Tarr PI, Gordon CA & Chandler WL (2005). Shiga-toxin-producing *Escherichia coli* and haemolytic uraemic syndrome. *Lancet* (**365**):1073–1086.

8. Dundas S, Todd WTA, Stewart AI et al. (2001) The Central Scotland *Escherichia coli* O157:H7 outbreak: risk factors for the hemolytic uremic syndrome and death among hospitalized patients. *Clin Infect Dis* **33**(7):923-931.

9. Lynn RM, Brien SJO, Taylor CM et al. (2005) Childhood hemolytic uremic syndrome, United Kingdom and Ireland. *Emerging Infectious Diseases* **11**(4):590-6.

10. Strachan NJC, Dunn GM, Locking ME, Reid TMS & Ogden ID (2006) *Escherichia* coli O157: burger bug or environmental pathogen? *Int J Food Microbiol* **112**(2):129–137 (cited 2015 Jan 11; www.ncbi.nlm.nih.gov/pubmed/16934897).

ANSWERS

MCQ	Feedback
1. With respect to the features of HUS in children, which one of the following statements is true? a. Most cases are associated with *Shigella* infectious diarrhoea. b. Children under 4 years of age presenting with bloody diarrhoea should be prescribed a beta-lactam antimicrobial agent. c. Stools are usually watery; bloody stools are reported in a minority of cases. d. Central nervous system irritability and seizures occur in about one-third of patients.	Strains of *Shigella dysenteriae* serotype 1 produce Shiga toxin (VT1-like toxin) but is a rare cause of HUS. Although children presenting with watery diarrhoea may develop HUS, most cases report symptoms of bloody diarrhoea. Treatment with antibiotics is contraindicated. There is no convincing evidence to show that antimicrobial agents alter the course of infection or the duration of faecal excretion. There is concern that the use of subinhibitory concentrations of antibiotics may cause sublethal damage to VTEC with the subsequent liberation of VT. Neurological symptoms, including seizures, are a frequently described feature of childhood HUS.
2. Regarding the clinical symptoms of HUS in children, which one of the following statements is true? a. HUS is associated with fever in most cases. b. Infection-induced HUS can occur in patients in absence of diarrhoea. c. HUS is an uncommon cause of ARF in children. d. Anaemia is typically mild and haemoglobin never drops <90 g/L.	Fever may occur, but high fever with chills is rarely associated with HUS. Severe anaemia and low haemoglobin are diagnostic features of the condition. HUS is the most common cause of ARF in children. Children usually have a prodrome of vomiting, abdominal pain, and diarrhoea (frequently bloody), but VTEC infection can cause HUS even when early symptoms of diarrhoea are absent. Some children present with symptoms of a urinary tract infection, including blood in their urine and low urine output.
3. Which of the following strains would be most likely to cause HUS? a. *E. coli* serogroup O157, *vtx1* gene negative, *vtx2* gene negative, and *eae* gene positive b. *E. coli* serogroup O157, *vtx1* gene positive, *vtx2c* gene positive, and *eae* gene positive c. *E. coli* serogroup O104, *vtx1* gene negative, *vtx2* gene negative, *aggR* gene positive d. *E. coli* O26, *vtx1* gene negative, *vtx2a* gene positive, and *eae* gene positive	Strains of *E. coli* O157 negative for both *vtx* genes may cause gastrointestinal symptoms but are not associated with causing HUS. Strains of *E. coli* O104 positive for *aggR* gene and positive for the *vtx2* gene caused a large outbreak of HUS in Germany in 2011. However, strains belonging to this serogroup that do not contain the *vtx* genes are not associated with causing HUS, although they may cause persistent diarrhoea and severe abdominal pain. VTEC O157 VT1 and VT2c can cause severe disease, but those harbouring the *vtx2a* gene are more pathogenic and are significantly associated with HUS.
4. Which of the following do *not* require exclusion from work or school? a. A case diagnosed with VTEC O157 who works on the delicatessen counter in a supermarket b. Symptomatic eight-year-old sibling of a known case of VTEC O157 c. Asymptomatic mother of a four-year-old case of VTEC O157 working in a call centre d. Asymptomatic father of a two-year-old diagnosed with VTEC O157 working in a care home for the elderly	Confirmed cases whose work involves preparing or serving unwrapped food to be served raw or not subjected to further heating belong to risk group C and must be excluded from work. Cases not in a risk group should be excluded until 48 hours after their first normal stool. No routine testing or exclusion is required for symptomatic contact if they are not in a risk group. Contacts belonging to a risk group (such as contacts whose work involves caring for vulnerable patients and are in risk group D) should be excluded until microbiologically clear (two clear stool samples at least 24 hours apart).
5. A local hospital laboratory report on a serum sample and faecal specimen from a two-year-old with bloody diarrhoea and HUS states that the serum is positive for antibodies to the LPS of *E. coli* O157, the faecal specimen is PCR positive for *vtx* genes, but VTEC was not isolated. What is the least likely explanation of the PCR positive/culture negative faecal specimen? a. VTEC is the cause of the symptoms, but the patient has been given antibiotics that have affected the ability of the bacterium to grow. b. PCR is more sensitive than culture and will detect the presence of VTEC in the stool at levels too low for culture. c. The case has atypical HUS. d. Local protocols focus on the detection of NSF VTEC O157 and the case has a sorbitol-fermenting VTEC O157.	Antibiotic treatment is significantly associated with progression to HUS and often makes it difficult to grow the causative organism. PCR can detect between 10 and 100 colony forming units per mL (cfu/mL) of sample, whereas culture detects between 10^3 and 10^4 cfu/mL. Sorbitol-fermenting VTEC O157 is rare in the UK but has been associated with HUS cases, particularly in children reporting recent travel to mainland Europe, and would not be identified at the local laboratory using current standard operating procedures. Symptoms of D+ HUS and positive serology, a faecal specimen that is positive by PCR for the *vtx* genes, or both, clearly indicate a diagnosis of typical HUS.

A CASE OF VARICELLA-ZOSTER VIRUS IN A MATERNITY UNIT

Michelle Griffin, Meera Chand and Kevin E Brown

National Infection Service, Colindale, Public Health England, UK.

A 32-year-old woman, Mrs. N, at 27 weeks' gestation, presented to the maternal and foetal assessment unit of a large hospital with vaginal bleeding. She was accompanied by her husband, Mr. N, and her three-year-old daughter, and left them in the antenatal waiting area while she was assessed. She had mild vaginal bleeding, which resolved, with no concerns raised on foetal monitoring. She was admitted to the maternity ward for 24 hours of observation. Her husband and daughter accompanied her to the ward and stayed there for an hour. They then returned home, as her daughter was not feeling well.

The next morning, Mr. N called the ward and reported that their daughter had a temperature and had developed an itchy rash. She had been feeling unwell for the past day or two. The daughter attended a nursery where chickenpox was known to be circulating. The obstetrician advised Mr. N to telephone the general practitioner (GP), who made a diagnosis of chickenpox over the phone.

INVESTIGATION OF THE CASE

This incident is one that is commonly experienced. The child has chickenpox and has exposed a number of people. In this case, it included her pregnant mother and other patients and staff at the hospital. Although chickenpox is usually a mild illness, immunosuppressed patients, pregnant women, and neonates have an increased risk of severe disease. If they are not immune from past infection or immunisation, they may develop severe varicella-zoster infection after this exposure. The purpose of investigating this incident is to identify any exposed people who are in high-risk groups for severe disease and who are not immune, and to reduce their risk by offering them post-exposure prophylaxis where appropriate (Figure 25.1).

The first step is to identify who has been exposed. A significant exposure to chickenpox is defined as 15 min in the same room or any face-to-face contact. A case of chickenpox is infectious from 48 hours before the rash appears until all skin lesions are crusted over. In this case, in the 24 hours before the rash appeared, the child had been in the antenatal waiting room with pregnant patients and in the maternity ward, where there were pregnant women and postnatal women with neonates. It is therefore possible that she may have exposed staff members in these areas. She has also exposed her own mother and father. The family should also be asked about social contacts in the family setting during the infectious period.

Mr. and Mrs. N were, of course, exposed, but there were no other family contacts and they have no other children at home. An infection control practitioner worked with the

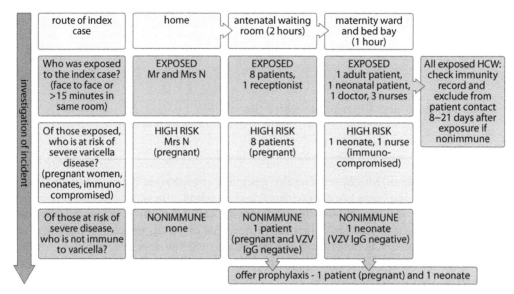

Figure 25.1. **Investigation of a varicella zoster incident in a healthcare setting.**

maternity unit to compile a list of the patients and staff who were exposed. Eight women had spent time in the antenatal waiting room at the same time as the chickenpox index case, all of whom were pregnant. A receptionist was also present in the room.

The child also came with her mother when she was admitted to the ward and spent an hour there. Mrs. N was in a four-bed bay. A postnatal woman with a two-day-old baby was in this bay, and there were two empty beds. The doors to the bay were kept closed and the child did not go outside the bay except when entering and leaving, at which point she did not go through any other patient areas in the ward.

No members of nursing staff interacted with Mr. N and the index case while they were in the waiting room. One receptionist spent approximately half an hour in the room. Three members of nursing staff and one obstetrician had been involved in looking after Mrs. N in the ward and had face-to-face contact with her daughter.

Mr. and Mrs. N, eight pregnant women, one neonate, and five members of staff therefore needed to be assessed. They were all contacted by the maternity unit nurse, working with an infection control practitioner.

The next step was to determine if any of those exposed to chickenpox were in high-risk groups that are more likely to develop severe varicella disease (pregnant, immunosuppressed, or a neonate under seven-days-old). Mrs. N is pregnant, and there are eight other pregnant women at risk and one neonate. The maternity nurse ascertained that none of the staff members exposed were pregnant. However, one of the nurses had an autoimmune condition and had been taking 40 mg of prednisolone daily for the past two weeks.

Ten adults and one under 7-day-old neonate were therefore required follow up to determine if they were immune to varicella. They were contacted by phone or interviewed in the ward by the maternity nurse, working with an infection control practitioner. They were first asked whether they recalled having chickenpox in the past. A previous history of chickenpox infection is 97–98% predictive of the presence of serum varicella antibodies, and a convincing history is considered adequate evidence of immunity by the Royal College of Obstetricians and Gynaecologists (RCOG). For those with no history of infection or two documented doses of varicella vaccine, or who are uncertain of their previous history, serum should be tested for the presence of varicella-zoster virus (VZV) IgG. If it is detected (>100 mIU/mLfor pregnant women), the patient is considered immune and does not require any further actions.

Mr. and Mrs. N and six of the eight other pregnant women gave good histories of childhood chickenpox. Two pregnant women were unsure if they had had chickenpox. They were tested for VZV IgG; one was positive and one was negative.

The immunosuppressed staff member was taking a dose of steroids, which qualifies her as high risk (Group A) for severe varicella infection (criteria are described in the Green Book). She had an uncertain history of childhood chickenpox and therefore it is recommended that the IgG is tested in this group of patients. This was done and she was found to have detectable VZV IgG (antibody level >150 IU/mL). The mother of the neonate was also uncertain of her history of chickenpox. The mother was tested for VZV IgG and it was not detected.

At the end of the investigation therefore, one pregnant woman and one neonate were considered to have been exposed, to be at high risk of severe disease, and to be nonimmune to varicella infection. They therefore qualified for post-exposure prophylaxis with varicella-zoster immunoglobulin.

Although all health care workers should know their varicella-zoster (VZ) immune status and have been offered vaccine if they were susceptible (see below), those in contact with the patient will also need a risk assessment and advice, regardless of their own personal risk of severe disease. Those who are nonimmune should be excluded from work involving patient contact during the time they may be developing varicella infection (8–21 days after exposure). All medics and nurses exposed were known to be immune to chickenpox in this case, but the receptionist had no history of chickenpox and tested VZV IgG negative. She was excluded from work in the antenatal clinic accordingly.

CLINICAL MANAGEMENT

A previously healthy child with uncomplicated chickenpox can be managed at home with supportive care including astringent soaks to reduce itch and good hygiene to prevent secondary infection.

It is possible that cases of varicella in pregnancy may arise after a child in the household contracts varicella, and it is important to be aware of the management of such cases. Varicella in pregnancy may be severe for the mother, as well as raising the possibility of transmission to the foetus. Any woman developing a vesicular rash should seek early medical advice and she is likely to require specialist management from an obstetrician regarding both treatment and the potential risks to the foetus. Pregnant patients with severe chickenpox should be admitted to the hospital for observation, for treatment, or both.

Aciclovir should be used for patients over 20 weeks of gestation, orally at first but intravenously if the patient is deteriorating. Aciclovir should be used cautiously in those under 20 weeks' gestation. After 36 weeks of gestation, intravenous (IV) aciclovir is the first choice and, if possible, delivery should be delayed to 7 days after the onset of the rash so that some protective maternal antibodies can transfer to the child. Mothers who have had chickenpox at any time during pregnancy, but especially within the first 28 weeks should be referred for foetal assessment and monitoring.

PREVENTION OF FURTHER CASES

Exposed patients belonging to the high-risk groups and with undetectable VZV IgG should be offered post-exposure prophylaxis, either a course of antivirals such as aciclovir or varicella-zoster immunoglobulin (VZIG), to attenuate infection and prevent severe complications. Although aciclovir can be given in pregnancy for treatment of varicella infection in women after 20 weeks' gestation, VZIG is recommended for post-exposure prophylaxis in women who are pregnant. VZIG does not prevent infection, and it is estimated that 50% of susceptible pregnant women given VZIG after a household exposure to chickenpox will develop clinical varicella (Green Book). Similarly, if given in early pregnancy, VZIG does not necessarily prevent congenital varicella syndrome. VZIG is a pooled plasma product derived from multiple (non-United Kingdom [UK]) donors with high varicella-zoster antibody titres. It is given by intramuscular injection. VZIG should be given within 10 days of the exposure for a pregnant woman (10 days of onset of rash if a household contact), and preferably within 7 days for an immunosuppressed patient (although can be given up to 14 days to attenuate infection if identified late).

Varicella is highly transmissible, primarily by the airborne route. The virus is present at high concentrations in vesicle fluid and the cutaneous lesions, vesicles, and crusts are also a source of infection.

All healthcare workers with direct patient contact should have their history of chickenpox recorded as part of their occupational health assessment. If in doubt, varicella-zoster IgG testing should be performed and those who are negative should be immunized with two doses of varicella vaccine.

Patients presenting to hospitals with suspected varicella should be admitted to an isolation side room (in the UK, a negative-pressure room or one with a positive-pressure lobby), with airborne infection precautions used during the infectious period (EN 149 FFP3 mask or equivalent, gown, gloves, eye protection). Ideally they should be cared for by staff who are known to be immune to varicella or who have received two doses of varicella vaccine. No special precautions are required for cleaning or waste. Once varicella-zoster infection is confirmed, isolation should be maintained until all skin lesions are crusted over.

Pregnant women and the immunosuppressed should be educated about the risks of chickenpox, including the need to seek early medical assistance if in contact with any case of rash illness.

EPIDEMIOLOGY

Chickenpox is highly contagious. It is transmitted by respiratory droplets and aerosolization of vesicle fluid or by direct contact with the skin lesions. Patients with shingles are not typically infectious from the respiratory tract but can still transmit from the skin lesions. Chickenpox cases are considered infectious from 48 hours before the onset of the rash until all skin lesions are crusted over.

Varicella-zoster is a typical herpesvirus with a large double-stranded DNA genome, an icosahedral capsid, and a lipid envelope. It is easily inactivated by lipid solvents, detergent heating at 60°C, and ultrasonic disruption. VZV is labile outside the host cell, and it survives in the external environment for a few hours and occasionally for a day or two.

Chickenpox is endemic in the UK and most of the rest of the world. The United States, Australia, and some countries in Europe, including Germany, Greece, and Luxembourg, have introduced routine childhood immunization with a consequently reduced incidence. Currently routine childhood varicella vaccination is not offered in the UK. However, it is recommended for healthy susceptible contacts of immunocompromised patients, where continuing close contact is unavoidable (for example, siblings of a leukemic child, or a child whose parent is undergoing chemotherapy), and susceptible healthcare workers (Green Book).

In temperate countries where there is no routine varicella vaccination program, 90% of cases occur in children under 13 years of age. More recent studies in the UK show that more than 75% of children are infected before the age of 5, and 40% of children attending daycare by the age of 3. Consequently, most adults raised in the UK are immune. The incidence of varicella is seasonal, with a UK peak between March and May. The epidemiology is less well understood in tropical areas where a higher proportion of adults are seronegative.

Live attenuated VZ vaccines are available and licensed in the Americas, Australasia, and most countries in Europe, including the UK. The vaccine is not suitable for pregnant women or the immunocompromised. A more potent VZ vaccine is also available for older adults to boost the immune response to VZ and to reduce the incidence and severity of shingles, and it was introduced into the UK routine vaccine program (for 70-year-olds) in 2013.

BIOLOGY

Varicella-zoster virus (VZV), also known as human herpes 3, is a member of the alpha herpesvirus family and is one of the eight herpesviruses that infect humans. The link between herpes zoster and chickenpox was first made in 1888 with the observation that children in contact with patients with shingles developed chickenpox, but it was only after the virus was cultured *in vivo* in 1958 that the link between the two infections was proven.

The virus is a double-stranded DNA virus, surrounded by a lipid-containing envelope with glycoprotein spikes. There are approximately 125,000 base pairs in the DNA, encoding approximately 75 proteins and 7 different glycoproteins (gB, gC, gE, gH, gI, gK, and gL). Neutralizing antibodies and the cell-mediated responses are predominantly targeted toward the glycoproteins.

DISEASE

Primary varicella-zoster infection is chickenpox. After an incubation period of 10–21 days, patients develop a prodrome of low-grade fever and malaise. These symptoms are followed by a rash that initially affects the head and trunk and then spreads distally. Lesions appear as macules and develop into vesicles, which may become necrotic or haemorrhagic. As the immune response develops, white blood cells and debris can cause the vesicle fluid to appear purulent. It will then crust over and eventually resolve.

Vesicles appear in crops over several days. A patient with varicella typically has lesions of several different ages at any point in the disease. (This feature can be used to differentiate it from smallpox.)

Chickenpox in children is typically mild and self-limiting. Adults are susceptible to more serious illness, which may include varicella pneumonia. The highest risk groups are the immunosuppressed, pregnant women, and neonates. Smokers and those with respiratory illness are also at increased risk of varicella pneumonia.

Laboratory confirmation of chickenpox is generally not required. If needed, vesicular fluid should be collected and tested for the presence of VZV DNA by polymerase chain reaction.

Complications of chickenpox

Pneumonia: About 1 in 400 immunocompetent adults with chickenpox develop varicella pneumonia. Smokers, pregnant women, and the immunosuppressed are more at risk. It can be severe and carries an associated mortality of 10–30%.

Neurological complications: Acute cerebellar ataxia is an uncommon complication of varicella in children. It is usually self-limiting with a complete recovery made. There is also a more rare form of diffuse varicella encephalitis. It is associated with poor outcomes with up to 10% mortality and 15% of patients left with permanent neurological deficits.

Secondary infection: Secondary bacterial infections of the skin lesions are common and can progress to severe systemic infections. Group A streptococcal infection (*Streptococcus pyogenes*) is particularly associated with varicella.

Maternal and foetal infections: Primary varicella infection in pregnancy can be transmitted across the placenta. Foetal infection in the first 20 weeks of gestation may result in congenital varicella syndrome, with skin scarring in a dermatomal distribution, eye pathology (microphthalmia, chorioretinitis, cataracts), limb hypoplasia, and neurological abnormalities. The risk of congenital varicella syndrome is <1% if infected in the first 12 weeks, and 2–3% at 12–20 weeks (Green Book). Congenital varicella syndrome does not occur with foetal infection after 20 weeks, though some less severe foetal abnormalities may still result. Varicella infection from one week before to one week after delivery is associated with severe disease in the mother and also in the neonate if infected.

Herpes zoster: Herpes zoster, also called shingles, is the reactivation of latent varicella-zoster infection. It is more common in older adults and in the immunosuppressed. It is typically a vesicular rash seen in a unilateral dermatomal distribution. It may be preceded or accompanied by very severe pain and changes in sensation. The pain may persist after the infection has subsided and can be difficult to manage.

PATHOLOGY

Humans are the only known natural host of VZV infection. VZV enters the body through the mucosal surfaces of respiratory tract, conjunctiva, or oropharynx. It replicates locally in regional lymph nodes, and then is disseminated to the rest of the body and to the reticuloendothelial system. It was originally though that VZV spread through sequential rounds of viremia, but is now thought to be more likely spread by infected peripheral blood mononuclear cells, particularly T lymphocytes. VZV infection then localizes to endothelial cells in the cutaneous capillaries and epithelial cells in the epidermis. It is the local replication in epithelial cells that leads to the characteristic vesicle formation, with vesicles containing a high concentration of infectious virus.

Sensory nerve axons in the dermis are infected from local spread, allowing viral transport to the sensory ganglia and the subsequent establishment of latent infection. Many years later, reactivation of the latent infection in sensory ganglia leads to replication in the ganglion and to the spread of virus back down the axon to the skin, with local replication in epithelial cells. The distribution of vesicles in the typical dermatomal pattern is characteristic of shingles or herpes zoster.

QUESTIONS

1. Which statement is false?
   a. Chickenpox is a common childhood infection.
   b. Chickenpox has increased prevalence in the tropics.
   c. Chickenpox can cause congenital infection.
   d. Chickenpox can be caused by contact with shingles.

2. True or false: Individuals at risk of severe infection are:
   a. Immune neonates
   b. Smokers
   c. Pregnant women in the second half of pregnancy
   d. Immunosuppressed children
   e. Healthcare workers

3. Immunosuppressed individuals in contact with a case of chickenpox should:
   a. Only be given treatment if they develop symptoms.
   b. Always have their VZV IgG measured even if they have a past history of chickenpox.
   c. Be given VZIG as soon as possible.
   d. Contact their physician for advice.

4. Pregnant women who develop chickenpox should not:
   a. Be given VZIG
   b. Be started on aciclovir if they are more than 20 weeks' gestation
   c. Be asked for a list of their contacts
   d. Be closely monitored and hospitalised if necessary

5. Shingles infection:
   a. Is non-infectious
   b. Should be treated with VZIG
   c. Only affects the elderly
   d. Its complications can be reduced by vaccination

GUIDELINES

1. Green Book, Ch. 34, Varicella www.gov.uk/government/uploads/system/uploads/attachment_data/file/148515/Green-Book-Chapter-34-v2_0.pdf

2. Green Book, Ch. 28a, Shingles www.gov.uk/government/uploads/system/uploads/attachment_data/file/357155/Green_Book_Chapter_28a_v0_5.pdf

3. VZIG chapter in Immunoglobulin Handbook https://www.gov.uk/government/uploads/system/uploads/attachment_data/file/638221/VZIG_Gudiance_Version_7_August_2017__.pdf

4. PHE guidance (2011) Rash in pregnancy www.gov.uk/government/publications/viral-rash-in-pregnancy

5. Royal College of Obstetricians and Gynaecologists (RCOG) (2015) Green-top guidance, Chickenpox in pregnancy 2007 www.rcog.org.uk/en/guidelines-research-services/guidelines/gtg13/

6. Royal College of Physicians (RCP) Guidance (2010) Varicella zoster management www.rcplondon.ac.uk/guidelines-policy/varicella-zoster-virus-occupational-aspects-management-2010

REFERENCES

1. Galetta KM & Gilden D (2015) Zeroing in on zoster: a tale of may disorders produced by one virus. *J Neurol Sci* **358**:38–45.

2. Cohen J & Breuer J (2015) Chickenpox: treatment. *BMJ Clin Evid* p. ii:0912.

3. Helmuth IG et al. (2015) Varicella in Europe—a review of the epidemiology and experience with vaccination. *Vaccine* **33**:2406–2413.

4. Kennedy PG et al. (2015) A comparison of herpes simplex virus type 1 and varicella zoster virus latency and reactivation. *J Gen Virol* **96**:1581–1602.

5. Bialas KM et al. (2015) Perinatal cytomegalovirus and varicella zoster virus infections: epidemiology, prevention and treatment. *Clin Perinatol* **42**:61–75.

6. Swamy GK & Heine RP (2015) Vaccinations for pregnant women. *Obstet Gynecol* **125**:212–226

7. Ogunjimi B et al. (2015) Integrating between host transmission and within host immunity to analyse the impact of varicella vaccination on zoster. *eLife* 4 (doi 10.7554/eLife. 07116).

8. Weinert LA et al. (2015) Rates of vaccine evolution show strong effects of latency: implications for varicella zoster virus epidemiology. *Mol Biol Evol.* **32**:1020–1028 (doi 10.1093/molbev/msu406).

9. Heaton PR et al. (2015) Evaluation of two multiplex real-time PCR assays for the detection of HSV1/2 and varicella zoster virus directly from clinical samples. *Diag Microbiol Infect Dis* **81**:169–170.

10. Carpenter JE et al. (2015) Defensive perimeter in the central nervous system: predominance of infected astrocytes and astrogliosis during recovery from varicella zoster virus encephalitis. *J. Virol* p.ii:JVI02389–15.

ANSWERS

MCQ	Feedback
1. Which statement is false? a. Chickenpox is a common childhood infection. b. Chickenpox has increased prevalence in the tropics. c. Chickenpox can cause congenital infection. d. Chickenpox can be caused by contact with shingles.	Chickenpox is a very common childhood infection. In the absence of immunization, 90% of children from temperate countries have seroconverted by the age of 13. However, there is a significantly lower seroprevalence in tropical countries. Infection is caused by the varicella-zoster virus, and maternal infection in the first trimester can lead to congenital varicella syndrome. Chickenpox can be acquired from contact with either chickenpox or shingles.
2. True or false: Individuals at risk of severe infection are: a. Immune neonates b. Smokers c. Pregnant women in the second half of pregnancy d. Immunosuppressed children e. Healthcare workers	a. False b. True c. True d. True e. False Smokers and pregnant women are at risk of varicella pneumonitis, as are susceptible immunosuppressed children. Neonates should be protected from varicella if they are immune. Healthcare workers are not at increased risk of severe infection, but if they acquire chickenpox infection are at risk of transmitting to their patients.
3. Immunosuppressed individuals in contact with a case of chickenpox should: a. Only be given treatment if they develop symptoms. b. Always have their VZV IgG measured even if they have a past history of chickenpox. c. Be given VZIG as soon as possible. d. Contact their physician for advice.	The risk of immunosuppressed individuals will need to be assessed depending on the contact and the level of immunosuppression, and this assessment should not wait until they have developed symptoms. The assessment may include determining the VZV IgG level on a recent blood sample, but it depends on their degree of immunosuppression and their previous history of chickenpox or shingles. VZIG is of limited benefit in patients who have detectable antibodies and so the level of susceptibility should be determined.
4. Pregnant women who develop chickenpox should not: a. Be given VZIG b. Be started on aciclovir if they are more than 20 weeks' gestation c. Be asked for a list of their contacts d. Be closely monitored and hospitalised if necessary.	Treatment with VZIG is too late if a pregnant woman has developed chickenpox, and she should be started on aciclovir and closely monitored for signs of pneumonitis. If she has developed chickenpox her contacts will also need to be assessed for their risk.
5. Shingles infection: a. Is non-infectious b. Should be treated with VZIG c. Only affects the elderly d. Its complications can be reduced by vaccination	Shingles is a reactivation of VZV infection and is infectious. Treatment with VZIG is ineffectual. Although it is more prevalent in the elderly, it can affect individuals of any age. There is a shingles vaccine given to those aged between 70 and 80 to reduce the incidence of shingles. This vaccine has a higher antigen content than the varicella vaccine and acts by boosting antibody levels. It is particularly effective at reducing complications including postherpetic neuralgia (PHN).

Printed and bound by CPI Group (UK) Ltd, Croydon, CR0 4YY

17/10/2024

01775699-0001